Smart Materials in Additive Manufacturing, Volume 2: 4D Printing Mechanics, Modeling, and Advanced Engineering Applications

Additive Manufacturing Materials and Technologies Series

Smart Materials in Additive Manufacturing, Volume 2: 4D Printing Mechanics, Modeling, and Advanced Engineering Applications

Edited by

**Mahdi Bodaghi
Ali Zolfagharian**

ELSEVIER

Elsevier
Radarweg 29, PO Box 211, 1000 AE Amsterdam, Netherlands
The Boulevard, Langford Lane, Kidlington, Oxford OX5 1GB, United Kingdom
50 Hampshire Street, 5th Floor, Cambridge, MA 02139, United States

Copyright © 2022 Elsevier Inc. All rights reserved.

No part of this publication may be reproduced or transmitted in any form or by any means, electronic or mechanical, including photocopying, recording, or any information storage and retrieval system, without permission in writing from the publisher. Details on how to seek permission, further information about the Publisher's permissions policies and our arrangements with organizations such as the Copyright Clearance Center and the Copyright Licensing Agency, can be found at our website: www.elsevier.com/permissions.

This book and the individual contributions contained in it are protected under copyright by the Publisher (other than as may be noted herein).

Notices
Knowledge and best practice in this field are constantly changing. As new research and experience broaden our understanding, changes in research methods, professional practices, or medical treatment may become necessary.

Practitioners and researchers must always rely on their own experience and knowledge in evaluating and using any information, methods, compounds, or experiments described herein. In using such information or methods they should be mindful of their own safety and the safety of others, including parties for whom they have a professional responsibility.

To the fullest extent of the law, neither the Publisher nor the authors, contributors, or editors, assume any liability for any injury and/or damage to persons or property as a matter of products liability, negligence or otherwise, or from any use or operation of any methods, products, instructions, or ideas contained in the material herein.

ISBN: 978-0-323-95430-3

For information on all Elsevier publications
visit our website at https://www.elsevier.com/books-and-journals

Publisher: Matthew Deans
Acquisitions Editor: Dennis McGonagle
Editorial Project Manager: Mariana L. Kuhl
Production Project Manager: Surya Narayanan Jayachandran
Cover Designer: Christian Bilbow

  Working together to grow libraries in developing countries

www.elsevier.com • www.bookaid.org

Typeset by STRAIVE, India

Dedication

This book is dedicated to my great parents, Mohammad and Kokab, who raised me with love, compassion, and a sense of appreciation for knowledge and innovation; and to my lovely wife for her patience, support, and adventurous spirit.
— Mahdi Bodaghi

This book is dedicated to my parents and sisters for their consistent support, to my wife for her patience and company, and to my son, Alan, for motivating me to keep going.
— Ali Zolfagharian

Contents

Contributors	**xi**
Editors biography	**xv**
Preface	**xvii**
Acknowledgments	**xix**

1 4D printing mechanics, modeling, and advanced engineering applications **1**

Ali Zolfagharian and Mahdi Bodaghi

Introduction	1
4D printing electro-induced shape memory polymers	2
4D printing modeling using ABAQUS: A guide for beginners	3
4D printing modeling via machine learning	4
4D-printed pneumatic soft actuators modeling, fabrications, and control	4
4D-printed auxetic structures with tunable mechanical properties	5
4D-printed shape memory polymers: modeling and fabrication	6
4D textiles—Materials, processes, and future applications	7
Closed-loop control of 4D-printed hydrogel soft robots	8
Hierarchical motion of 4D-printed structures using the temperature memory effect	9
Manufacturing highly elastic skin integrated with twisted coiled polymer muscles: Toward 4D printing	10
Multimaterial 4D printing simulation using grasshopper plugin	11
Origami-inspired 4D RF and wireless structures and modules	12
Shape-reversible 4D printing aided by shape memory alloys	12
Variable stiffness 4D printing	13
References	14

2 4D printing electro-induced shape memory polymers **19**

Rytis Mitkus, Ferdinand Cerbe, and Michael Sinapius

Introduction	19
Materials, working principle, and similar structures in 4D printing	21
Integration of conductive PLA	26
Investigation of SMP structures	33
Conclusions	47
Acknowledgments	48
References	48

viii Contents

3 4D printing modeling using ABAQUS: A guide for beginners 53
Hamid Reza Jarrah, Ali Zolfagharian, Bernard Rolfe,
and Mahdi Bodaghi
Introduction 53
Methodology 54
Results and discussions 64
Conclusion 71
References 71

4 4D printing modeling via machine learning 73
Ali Zolfagharian, Lorena Durran, Khadijeh Sangtarash,
Seyed Ebrahim Ghasemi, Akif Kaynak, and Mahdi Bodaghi
Introduction 73
Methodology 74
FEM modeling using hyperplastic material constitutive laws 78
Results and discussions 83
Discussions 96
Conclusion 99
References 100

5 4D-printed pneumatic soft actuators modeling, fabrication,
and control 103
Charbel Tawk and Gursel Alici
Introduction 103
4D-printed pneumatic soft actuators 105
Discussion 125
Conclusion 127
References 128

6 4D-printed structures with tunable mechanical properties 141
Wael Abuzaid, Mohammad H. Yousuf, and Maen Alkhader
Introduction 141
Shape memory polymer material 146
Stability and functional properties of 4D-printed specimens 146
Summary and concluding remarks 187
References 189

7 4D-printed shape memory polymer: Modeling and
fabrication 195
Reza Noroozi, Ali Zolfagharian, Mohammad Fotouhi,
and Mahdi Bodaghi
Introduction 195
4D printing programming 197
Constitutive equations 199
Fabrication and modeling 4D-printed elements 204

	Case studies	209
	Conclusion	225
	References	225
8	**4D textiles: Materials, processes, and future applications**	**229**
	David Schmelzeisen, Hannah Kelbel, and Thomas Gries	
	Introduction	229
	State of the art	231
	Printing method	232
	Form giving through surface tessellation	242
	Applications	243
	Conclusion and outcomes	247
	References	248
9	**Closed-loop control of 4D-printed hydrogel soft robots**	**251**
	Ali Zolfagharian, Mahdi Bodaghi, Pejman Heidarian,	
	Abbas Z Kouzani, and Akif Kaynak	
	Introduction	251
	Motion mechanism of the soft actuator	252
	Materials and methods	254
	Results and discussions	257
	Conclusion	275
	References	276
10	**Hierarchical motion of 4D-printed structures using the temperature memory effect**	**279**
	Giulia Scalet, Stefano Pandini, Nicoletta Inverardi, and Ferdinando Auricchio	
	Introduction	279
	Temperature memory effect: Basics and literature review	280
	Experimental testing	289
	Constitutive modeling	298
	Case study	300
	Conclusions and perspectives	305
	Acknowledgments	306
	References	306
11	**Manufacturing highly elastic skin integrated with twisted and coiled polymer muscles: Toward 4D printing**	**311**
	Armita Hamidi and Yonas Tadesse	
	Introduction	311
	Materials	314
	Manufacturing	319
	Results and discussion	321
	Conclusion	323
	References	325

| x | Contents |

12 Multimaterial 4D printing simulation using a grasshopper plugin — 329
Germain Sossou, Hadrien Belkebir, and Frédéric Demoly
Introduction — 329
Computational design for 4D printing — 330
Case studies — 335
Conclusion and future work — 341
Appendix: The *distribution Computation* component — 342
References — 344

13 Origami-inspired 4D tunable RF and wireless structures and modules — 347
Yepu Cui, Syed A. Nauroze, and Manos M. Tentzeris
Introduction — 347
Origami-inspired inkjet-printed FSS structures — 353
Fabrication process of 4D-printed origami-inspired RF structures — 359
4D-printed origami-inspired frequency selective surfaces — 363
4D-printed chipless RFID pressure sensors for WSN applications — 368
4D-printed origami-inspired deployable and reconfigurable antennas — 373
Conclusion — 381
References — 383

14 Shape-reversible 4D printing aided by shape memory alloys — 387
Saeed Akbari, Amir Hosein Sakhaei, Sahil Panjwani, Kavin Kowsari, and Qi Ge
Introduction — 387
Materials and methods — 389
Simulation of actuation cycle — 393
Numerical and experimental results — 397
Conclusions — 403
References — 404

15 Variable stiffness 4D printing — 407
Yousif Saad Alshebly, Marwan Nafea, Khameel Bayo Mustapha, Mohamed Sultan Mohamed Ali, Ahmad Athif Mohd Faudzi, Michelle Tan Tien Tien, and Haider Abbas Almurib
Introduction — 407
Design and working principle — 410
Fabrication process and experimental setup — 415
Results and discussion — 416
Discussion — 425
Conclusion — 429
Acknowledgment — 430
References — 430

Index — 435

Contributors

Wael Abuzaid Department of Mechanical Engineering, American University of Sharjah, Sharjah, United Arab Emirates

Saeed Akbari RISE Research Institutes of Sweden, Mölndal, Sweden

Mohamed Sultan Mohamed Ali School of Electrical Engineering, Faculty of Engineering, Universiti Teknologi Malaysia, Johor Bahru, Johor, Malaysia

Gursel Alici School of Mechanical, Materials, Mechatronic and Biomedical Engineering, Applied Mechatronics and Biomedical Engineering Research (AMBER) Group, ARC Centre of Excellence for Electromaterials Science, University of Wollongong, Wollongong, NSW, Australia

Maen Alkhader Department of Mechanical Engineering, American University of Sharjah, Sharjah, United Arab Emirates

Haider Abbas Almurib Department of Electrical and Electronic Engineering, Faculty of Science and Engineering, University of Nottingham Malaysia, Semenyih, Selangor, Malaysia

Yousif Saad Alshebly Department of Electrical and Electronic Engineering, Faculty of Science and Engineering, University of Nottingham Malaysia, Semenyih, Selangor, Malaysia

Ferdinando Auricchio Department of Civil Engineering and Architecture, University of Pavia, Pavia, Italy

Hadrien Belkebir ICB UMR 6303 CNRS, University Bourgogne Franche-Comté, UTBM, Belfort, France

Mahdi Bodaghi Department of Engineering, School of Science and Technology, Nottingham Trent University, Nottingham, United Kingdom

Ferdinand Cerbe Institute of Mechanics and Adaptronics, TU Braunschweig, Braunschweig, Germany

Yepu Cui School of Electrical and Computer Engineering, Georgia Institute of Technology, Atlanta, GA, United States

Frédéric Demoly ICB UMR 6303 CNRS, University Bourgogne Franche-Comté, UTBM, Belfort, France

Lorena Durran School of Engineering, Deakin University, Geelong, VIC, Australia

Mohammad Fotouhi Department of Materials, Mechanics, Management & Design (3MD), Delft University of Technology, Delft, Netherlands

Qi Ge Digital Manufacturing and Design Center, Singapore University of Technology and Design, Singapore, Singapore; Department of Mechanical and Energy Engineering, Southern University of Science and Technology, Shenzhen, China

Seyed Ebrahim Ghasemi Department of Engineering Sciences, Hakim Sabzevari University, Sabzevar, Iran

Thomas Gries Institut für Textiltechnik of RWTH Aachen University, Aachen, Germany

Armita Hamidi David L. Hirschfeld Department of Engineering, Angelo State University, San Angelo, TX, United States

Pejman Heidarian School of Engineering, Deakin University, Geelong, VIC, Australia

Nicoletta Inverardi Department of Mechanical and Industrial Engineering, University of Brescia, Brescia, Italy

Hamid Reza Jarrah Department of Engineering, School of Science and Technology, Nottingham Trent University, Nottingham, United Kingdom

Akif Kaynak School of Engineering, Deakin University, Geelong, VIC, Australia

Hannah Kelbel Institut für Textiltechnik of RWTH Aachen University, Aachen, Germany

Abbas Z Kouzani School of Engineering, Deakin University, Geelong, VIC, Australia

Kavin Kowsari Digital Manufacturing and Design Center, Singapore University of Technology and Design, Singapore, Singapore

Rytis Mitkus Institute of Mechanics and Adaptronics, TU Braunschweig, Braunschweig, Germany

Ahmad Athif Mohd Faudzi School of Electrical Engineering, Faculty of Engineering, Universiti Teknologi Malaysia, Johor Bahru, Johor, Malaysia

Khameel Bayo Mustapha Department of Mechanical, Materials and Manufacturing Engineering, Faculty of Science and Engineering, University of Nottingham Malaysia, Semenyih, Selangor, Malaysia

Marwan Nafea Department of Electrical and Electronic Engineering, Faculty of Science and Engineering, University of Nottingham Malaysia, Semenyih, Selangor, Malaysia

Syed A. Nauroze School of Electrical and Computer Engineering, Georgia Institute of Technology, Atlanta, GA, United States

Reza Noroozi Department of Engineering, School of Science and Technology, Nottingham Trent University, Nottingham, United Kingdom; Faculty of Engineering, School of Mechanical Engineering, University of Tehran, Tehran, Iran

Stefano Pandini Department of Mechanical and Industrial Engineering, University of Brescia, Brescia, Italy

Sahil Panjwani Digital Manufacturing and Design Center, Singapore University of Technology and Design, Singapore, Singapore

Bernard Rolfe School of Engineering, Deakin University, Geelong, VIC, Australia

Amir Hosein Sakhaei School of Engineering and Digital Arts, University of Kent, Canterbury, United Kingdom

Khadijeh Sangtarash School of Engineering, Deakin University, Geelong, VIC, Australia

Giulia Scalet Department of Civil Engineering and Architecture, University of Pavia, Pavia, Italy

David Schmelzeisen Institut für Textiltechnik of RWTH Aachen University, Aachen, Germany

Michael Sinapius Institute of Mechanics and Adaptronics, TU Braunschweig, Braunschweig, Germany

Germain Sossou ICB UMR 6303 CNRS, University Bourgogne Franche-Comté, UTBM, Belfort, France

Yonas Tadesse Mechanical Engineering Department, The University of Texas at Dallas, Richardson, TX, United States

Charbel Tawk School of Mechanical, Materials, Mechatronic and Biomedical Engineering, Applied Mechatronics and Biomedical Engineering Research (AMBER) Group, ARC Centre of Excellence for Electromaterials Science, University of Wollongong, Wollongong, NSW, Australia; Faculty of Engineering and Information Sciences, University of Wollongong in Dubai, Dubai, United Arab Emirates

Manos M. Tentzeris School of Electrical and Computer Engineering, Georgia Institute of Technology, Atlanta, GA, United States

Michelle Tan Tien Tien Department of Electrical and Electronic Engineering, Faculty of Science and Engineering, University of Nottingham Malaysia, Semenyih, Selangor, Malaysia

Mohammad H. Yousuf Faculty of Engineering and Information Sciences, University of Wollongong Dubai, Dubai, United Arab Emirates

Ali Zolfagharian School of Engineering, Deakin University, Geelong, VIC, Australia

Editors biography

Dr. Mahdi Bodaghi, BSc, MSc, PhD, PGCAP, FHEA, CEng, MIMechE, is Senior Lecturer in the Department of Engineering, School of Science and Technology at Nottingham Trent University. Mahdi heads the 4D Materials and Printing Laboratory (4DMPL) that hosts a broad portfolio of projects focusing on the electro-thermo-mechanical multi-scale behaviors of smart materials, soft robots, and 3D/4D printing technologies. In the recent 12 years, he has been working toward the advancement of state-of-the-art smart materials and additive manufacturing, which has led him to cofound the 4D Printing Society and to coedit the Smart Materials in Additive Manufacturing book series. His research has led to the publication of more than 120 scientific papers in prestigious journals on mechanics, manufacturing, and materials science as well as the presentation of his work at major international conferences. Mahdi has also served as Chairman and Member of Scientific Committees for 10 International Conferences, as Guest Editor for 10 journals, as Editorial Board Member for 8 scientific journals, and as Reviewer for more than 130 journals. Mahdi's research awards include the Best Doctoral Thesis Award (2015), the CUHK Postdoctoral Fellowship (2016), the Annual Best Paper Award in Mechanics and Material Systems presented by the American Society of Mechanical Engineers (2017), Horizon Postdoctoral Fellowship Award (2018), and the IJPEM-GT Contribution Award (2021) recognized by the Korea Society for Precision Engineering.

Dr. Ali Zolfagharian, BSc, MSc, PhD, ADPRF, GCHE, is Senior Lecturer in the Faculty of Science, Engineering and Built Environment, School of Engineering at Deakin University, Australia. He has been among the 2% top-cited scientists listed by Stanford University and Elsevier (2020), the Alfred Deakin Medallist for Best Doctoral Thesis (2019), and the Alfred Deakin Postdoctoral Fellowship Awardee (2018). He has been directing the 4D Printing and Robotic Materials laboratory at Deakin University since 2018. Ali is the cofounder of the 4D Printing Society, the coeditor of the Smart Materials in Additive Manufacturing book series published by Elsevier, and a technical committee member of five international conferences. From 2020 to 2022, he has received more than AUD 200k research funds from academic and industrial firms. Ali's research outputs on flexible manipulators, soft grippers, robotic materials 3D/4D printing, and bioprinting include 71 articles, being editor of 2 journals, 15 special issues, and 5 books.

Preface

This is the second book in the series Smart Materials in Additive Manufacturing focusing on four-dimensional (4D) printing mechanics, modeling, and advanced engineering applications. This book presents recent practical advances and innovations in 4D printing from the perspectives of engineering design and modeling in emerging directions, with the goal of being useful for students and early career researchers in the field at the higher education level.

Significant progress has been made in 4D printing by researchers in leading universities and laboratories throughout the world, integrating developments in additive manufacturing of dynamic structures with stimuli-responsive materials, known as smart materials, to build smart structures and mechanisms. However, as 4D printing technology is still in its early phases of research and development, industries will most likely learn more about it in the coming years. With the Fourth Industrial Revolution well underway, industries that take advantage of its enormous opportunities will surely benefit commercially. With this information, institutions, researchers, and students who invest in cutting-edge 4D printing technology have enormous potential.

Currently, there are relatively limited tools and resources available for including 4D printing in university courses and curricula, as most studies use the bottom-up approach to find the target design. Therefore, the research in 4D printing applications toward a strategic top-down modeling and designing approaches are ongoing demand. This book is primarily aimed at students and educators studying multidisciplinary fields, either as a self-contained course or as a module within a larger engineering course. There is sufficient depth for an undergraduate- or graduate-level course, with many references to point the student further along the path.

The current book, focusing on 4D printing mechanics, modeling, and advanced engineering applications, has case studies, including the design, modeling, and experimental steps, to help readers improve their understanding of 4D printing by repeating and carrying out the exercises presented. Researchers in 4D printing may also find this text useful in helping them understand the state of the art and the opportunities for further research. Entrepreneurs and research and development experts will also find this book useful as it will guide them in developing new start-ups based on engineering applications of 4D printing.

In this book, recent techniques, structural design, modeling, and simulation tools and software to realize the applications based on 4D printing are presented. The chapters present case studies of research where the 4D printing has provided useful applications for smart structures and mechanisms in the textile industry, soft robotics, bioprinting, origami, auxetics and metamaterials, micromachines, sensors, and wireless devices. The readers will have access to the detailed design, modeling, simulation, and manufacturing steps for applications in different fields that will be suitable for the technical level of 4D printing teaching and learning required for diverse educational levels.

Putting this book together was an intense collaborative process for the authors and editors and they hope that readers will benefit from it. Working on this book has been an incredible experience, and it has allowed us to expand our network of specialists with whom we can collaborate to advance this area even further and continue the Smart Materials in Additive Manufacturing book series.

Mahdi Bodaghi
Ali Zolfagharian

Acknowledgments

The coeditors express their sincere gratitude and deep appreciation to all the authors for their significant and excellent contributions to this book. Their enthusiasm, commitment, and technical expertise have made this pioneering book possible. We are also grateful to Elsevier for supporting this book project, and we extend our special thanks to the Editorial Team, Dennis, Mariana, and Surya, for their proactive support and cooperative attitude, with both the publishing initiative in general and the editorial aspects.

4D printing mechanics, modeling, and advanced engineering applications

1

Ali Zolfagharian[a] and Mahdi Bodaghi[b]
[a]School of Engineering, Deakin University, Geelong, VIC, Australia
[b]Department of Engineering, School of Science and Technology, Nottingham Trent University, Nottingham, United Kingdom

Introduction

4D-printed structures may fold or unfold in predetermined forms if activated by an external stimulus, thereby opening the door to various exciting applications. To do so, a geometric code that offers instructions about how a structure moves or shifts until a stimulus activates it should be programmed during the printing stage. This facilitates the design of intelligent and sensitive structures that can be adapted to complex environmental factors. This book has collected and classified recent advances and innovations in modeling and programming of 4D-printed smart materials from an engineering point of view and explores the potential of 4D-printed structures in emerging directions.

In the application of 4D printing, given a selected material, a printing approach, and a programming method, the final step for establishing 4D printing is modeling the original structure and predicting its shape- and property-changing behavior (Shen, Erol, Fang, & Kang, 2020). New structural designs and advanced simulation and modeling tools can further advance 4D printing. For example, porous materials with multiscale pore size enable faster diffusion and mass transport for water and chemical-based actuation, or multimaterial, functionally graded, and bilayer structures are extensively inspiring researchers for 4D programming. Indeed, simulating the smart material dynamic process requires multiphysics simulations. The mechanical calculation, or finite-element analysis (FEA), needs to be combined with extra physical fields such as thermal, electrical, and magnetic, depending on the stimuli mechanism. Additionally, the inhomogeneity and anisotropy of printed structures based on different printing methods and parameters is a current research direction. For shape actuation involving multiphysics problems, finite-element modeling (FEM) could be used to accurately predict the shape transformation of the printed shape under external stimuli through an inverse design problem. Yet, this is a challenging problem even with the full capability of modeling of the material system. From optimization theory, the solution to such an inverse problem could be unstable or even nonexisting. To this end, artificial intelligence methods, machine learning and deep learning algorithms,

---

Smart Materials in Additive Manufacturing, Volume 2: 4D Printing Mechanics, Modeling, and Advanced Engineering Applications
https://doi.org/10.1016/B978-0-323-95430-3.00001-4
Copyright © 2022 Elsevier Inc. All rights reserved.

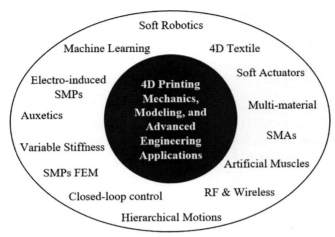

Fig. 1.1 The 4D-printed structures covered in this book.

offer a powerful approach to optimize material distribution for the rational design of smart materials for controlled shape shifting. Topology optimization is also used to optimize the design of 4D-printed smart structures based on the target architecture with reduced weight. Currently, there are limited tools available on this topic. Most studies are using numerical tools to find the target design, while analytical calculations could also be applied in some simple geometries. Therefore, the research efforts and progress in 4D-printed structures and systems toward strategic top-down modeling and designing approaches are ongoing.

4D printing finds its applications in textile industries (Koch, Schmelzeisen, & Gries, 2021), soft robotics (Zolfagharian, Kaynak, & Kouzani, 2019; Zolfagharian, Kaynak, Noshadi, & Kouzani, 2019), bioprinting (Askari et al., 2021), origami (Mehrpouya, Azizi, Janbaz, & Gisario, 2020), auxetic (Yousuf, Abuzaid, & Alkhader, 2020), micromachines (Spiegel et al., 2020), sensors (Faller, Krivec, Abram, & Zangl, 2018; Khosravani & Reinicke, 2020), and wireless devices (Cui, Nauroze, & Tentzeris, 2019).

This book discusses some areas of research where the application of 4D-printed materials has proved to be useful in engineering applications, so that the readers will be introduced to various fields of 4D-printed structures and applications (Fig. 1.1). The following introduce the chapters presented in this book, which present different mechanics, modeling techniques, and engineering applications of 4D-printed smart materials and structures.

4D printing electro-induced shape memory polymers

A novel approach to the internal heating of 4D-printed SMP structures is to integrate heating elements into the structures (Kačergis, Mitkus, & Sinapius, 2019). This approach uses 3D printable conductive materials to change the stiffness of the structure, taking advantage of the Joule heating effect. The use of this approach in soft robotics enables a change in the stiffness of the structure (Bodaghi, Damanpack, & Liao, 2017; Bodaghi & Liao, 2019; Hu, Damanpack, Bodaghi, & Liao, 2017). The use of conductive filaments allows the application of an electric current to generate Joule heating, thus changing the stiffness of the conductive material itself and of surrounding materials. Furthermore, the deformation of conductive traces changes their resistance, so they can be used for various sensing purposes as well. In such systems, there are several factors to be considered, including different methods to attach cables to printed electrodes to find a connection with the lowest resistance, the influence of printing speed and extrusion temperature on the resistance of the material, heating performance, and the influence of the maximum bending of 4D-printed structures.

In this chapter, we present an approach to combining plain polylactic acid (PLA) with conductive PLA (CPLA) in a single part to activate SMP structures with electric current, also known as Joule heating. The CPLA is encapsulated inside the structure and covered with nonconductive PLA. The SMP structures reported here with conductive traces can be activated with 60 V and offer full shape-morphing in a few minutes, where the speed of deformation depends on the applied electric current. These smart structures can be fully printed with a single machine and do not require any specific tools or devices. Embedding electronics opens new possibilities for controlled deformation and sensing for various applications, such as flexible grippers. The SMP structures presented here are one of the first of such kinds that use printable conductive materials for heating purposes to activate the morphing of the structure. The printing parameters and their influence on the performance of 4D-printed structures are presented. Further, different cable connection methods are of interest to apply high current to the structure. Since conductive PLA has quite a large resistance, the influence of printing parameters on the resistance of conductive PLA is investigated to achieve the highest performance possible. Also, the influence of placement of the conductive PLA layer within a 4D-printed shape-morphing structure is investigated. For this characterization, a special testing stand, capable of measuring maximum free deformation and blocking force, is designed.

4D printing modeling using ABAQUS: A guide for beginners

The advent of 4D printing in AM, which requires knowledge in multiphysics, chemistry, and engineering skills, is bringing many engineering applications. This serves as a significant multidisciplinary research and training platform for both academia and the professional world. Due to reduced prices, more design flexibility, and faster manufacturing techniques, this newly developed technology has become a viable commercial prospect. With the increasing use of AM and smart materials in industry, industries are looking to academics with interdisciplinary skills and expertise.

This chapter presents an easy-to-implement and follow approach for students and early career researchers new to the field of 4D printing modeling and simulation, providing them with a significant edge in an increasingly competitive job market and higher degree research. This chapter provides a straightforward step-by-step approach, including design, multiphysics modeling principles, and finite-element method (FEM) simulation of a shape memory polymer and hydrogel composite 4D-printed structure using ABAQUS software for new researchers and students to the field of 4D printing. This study may be used in a variety of cross-disciplinary engineering courses and classes to educate and practice theoretical modeling and simulation for 4D printing.

4D printing modeling via machine learning

One of the current challenges in 4D printing of soft robots and actuators is the modeling and prediction of their motion, particularly due to the nonlinearity of the materials. To achieve a target shape in such systems, an inverse nonlinear problem should be solved to estimate the shape change of 4D-printed systems in response to the amount and time of stimuli (Zolfagharian et al., 2021). However, using machine learning (ML) techniques trained via FEM numerical results could help with saving time and effort during design improvement and modification to handle the variable physical responses of incorporated materials.

This chapter presents a case study guiding the development of ML models for predicting the behavior of nonlinear 4D printing problems. This includes creating an FEM to accurately simulate experimental actuation and running a series of simulations to obtain training data for the ML modeling. First, the effectiveness of using a linear model to estimate the bending angle of a hyperelastic material is analyzed. Then, in the ANSYS software package, an FEM method using a hyperelastic material model is developed to accurately model the deflection of the actuator in response to the input pressure for variable geometrical features. More than a thousand training samples from the FEM simulations are generated to use as training data for the ML model. The ML model is developed to predict the bending angle of the 4D-printed system. The ML model accurately predicts FEM and experimental data and proves to be a viable solution for 4D printing modeling of soft robots and dynamic structures. This work helps the readers understand how to best develop models for nonlinear 4D printing problems using ML methods.

4D-printed pneumatic soft actuators modeling, fabrications, and control

Soft actuators are one of the primary and critical components of a soft robot. Soft robots demand dexterous, fast, and reversible soft actuators that must be able to contract, extend, bend, and/or twist with favorable performance while delivering sufficient output forces (Tawk & Alici, 2021). A soft actuator converts an input stimulus into a useful mechanical output. In general, soft actuators can be classified according to the input stimulus required for their activation (Zolfagharian, Kaynak, & Kouzani, 2020). For example, chemical species are required for electroactive polymer actuators, air for pneumatic actuators, liquid for hydraulic actuators, heat for shape memory polymers and alloys, light for light-responsive actuators, an electrical field for dielectric elastomers, and a magnetic field for magnetorheological fluids and polymers (Tawk & Alici, 2021). Pneumatic soft actuators are one of the most commonly used types of soft actuators for soft robotic systems since they can be designed with internal deformable hollow chambers that dictate a desired output motion and can be directly fabricated using various additive manufacturing techniques (Zolfagharian et al., 2020). Pneumatic soft actuators based on volumetric expansion and contraction are two main classes that include positive pressure soft actuators requiring compressed air to be activated or inflated upon their actuation and negative pressure soft actuators requiring a vacuum source for actuation (Tawk, Gillett, in het Panhuis, Spinks, & Alici, 2019). There is a type of soft actuator known as four-dimensional (4D)-printed soft actuators that are based on 3D-printed passive deformable geometries (e.g., pneumatic soft actuators) or active smart and functional materials (e.g., shape memory alloys, shape memory polymers, hydrogels, variable stiffness materials, and others) that can change their shape with time in response to an external stimulus such as moisture, electric field, magnetic field, light, temperature, pH, chemical species, or pneumatic pressure (Zolfagharian et al., 2016).

Pneumatic soft actuators are 4D-printed due to their unique characteristics such as safety, light weight, and compliance, while they do not require complex controllers (Tawk, in het Panhuis, Spinks, & Alici, 2018). This chapter presents 4D-printed pneumatic soft actuators that are fabricated directly using various additive manufacturing techniques to generate a useful mechanical output for the purpose and function they are designed and fabricated for. The actuators are presented in terms of their types, modeling, materials, fabrication, sensing, control, capabilities, and applications in diverse soft robotic systems. The presented 4D-printable soft actuators are discussed, along with a list of what is required to develop robust and functional soft pneumatic actuator models for soft robots, including various hyperelastic material models that can be developed, such as the Mooney-Rivlin model, Ogden model, Yeoh model, and Neo-Hookean model for use in FE simulations (Zolfagharian, Mahmud, et al., 2020). This chapter emphasizes the importance of implementing a top-down approach to stimulate more interaction between materials scientists and robotic researchers so that 3D- and 4D-printable soft materials that meet the function and application requirements of a robotic system are developed to fully exploit such synthesized materials by roboticists and use them in practical robotic systems and devices (Tawk & Alici, 2021).

4D-printed auxetic structures with tunable mechanical properties

3D printing offers significant simplifications in the design cycle and manufacturing of complex geometries (Khosravani & Zolfagharian, 2020; Khosravani, Zolfagharian, Jennings, & Reinicke, 2020). However, the resulting components have a fixed set of properties that depend on the geometric design and inherent material properties of the printing material. In 4D printing, the rigidity in properties is removed, thus enabling the synthesis of adaptive structures with tunable and functional properties in dynamic environments (Abuzaid, Alkhader, & Omari, 2018). Introducing this extra dimension has tremendous advantages for the development of applications for deployable and morphing structures, such as variable stiffness and Poisson's ratios (Yousuf et al., 2020). As a technology, 4D printing can potentially be used to fabricate a wide range of complex geometries that exhibit functionally triggered behaviors, particularly those that belong to the class of materials known as periodic cellular solids or lattice structures. The ability to alter the geometry of the 4D-printed periodic structure through programming could be employed to achieve tunability in the mechanical properties, for example, in the auxetic structures.

This chapter focuses on auxetic 4D printing structures with programmable properties using SMPs. A comprehensive assessment of the challenges associated with the use of this smart material, including dimensional stability issues and incomplete local recovery, is presented. The complex interactions between material deposition direction and dimensional changes and shape recovery are also being investigated. In addition, both simple structures and complex meta-structures (2D auxetic cellular solid) cases with tunable mechanical properties, structural stiffness, and Poisson's ratio are thoroughly discussed. This chapter also sheds insight into the functional fatigue properties of the 4D-printed structures following multiple programming (tuning) and recovery cycles, through the detailed experimental results, to provide a deep insight into the properties, challenges, and future research and optimization needs to further improve the mechanical properties of thermoplastic 4D-printed structures.

4D-printed shape memory polymers: modeling and fabrication

Active materials, such as shape memory polymers, should be used in the printing process to print structures with dynamic properties (Bodaghi, Damanpack, & Liao, 2018). Shape memory polymers (SMPs) have received more attention due to their lower cost, lower density, high recoverable strain, simple shape programming process, and excellent controllability over the recovery process (Bodaghi, Damanpack, & Liao, 2016). The fabrication and modeling of 4D printing components are extremely important, and many researchers have done various studies on 4D printing and its applications. Therefore, in this chapter, by considering the shape memory effect (SME) and fused deposition modeling (FDM) processes, the concept of 4D printing programming is described.

Due to the viscoelastic nature of polymers, thermo-viscoelastic modeling can be employed to describe and predict SMP behaviors. Consequently, constitutive equations are based on a rheological model in which parameters are time and temperature dependent (Baniasadi, Yarali, Bodaghi, Zolfagharian, & Baghani, 2021; Bodaghi & Liao, 2019). Dashpots, springs, and frictional elements are common components of thermo-viscoelastic models. Viscosity is one of the most important features of the SMPs that is due to the mobility between polymer chains, which in such modeling has been considered. Polymers are a combination of many chains with different lengths and angles, which each have different entropy and temperature transitions. During the cooling process, the phase transition first occurs in the chain with the lower entropy, so SMPs have a phase heterogeneity consisting of glassy and rubbery phases (Bodaghi, Noroozi, Zolfagharian, Fotouhi, & Norouzi, 2019). These phases could change into each other under thermal conditions, and during the thermal process constantly. Since SMPs are a combination of two phases, a new class of constitutive equations was proposed by researchers. In contrast to the thermo-viscoelastic model, or a rheological model, the phase transition model is a phonological model with a physical meaning. These SMP models not only cover the viscosity of the polymers but also predict the phase transformation.

In this chapter, initially based on the thermo-mechanical properties of smart materials, a constitutive model is presented for the shape memory feature. Due to the complexity and difficulties of employing shape memory constitutive equations in finite-element software to model 4D-printed elements, a simple method based on the 4D printing concept is proposed. Subsequently, the fabrication of 4D-printed elements based on the FDM process is described, and a simple finite-element method (FEM) is introduced to predict their self-morphing features. The implementation of constitutive equations in commercial finite-element software packages such as ABAQUS, COMSOL, and ANSYS has been gaining attention due to its extensive application. Due to their complexity, the constitutive equation of SMPs has a lot of difficulties in finite-element software. To settle this issue by considering the concept of SME and 4D printing, a simple method for modeling the thermo-mechanical behaviors of the 4D-printed self-folding element is proposed. The proposed model is calibrated based on experiments and, as case studies, some of the smart structures are designed and modeled in the finite-element software. The chapter then demonstrates some of the 4D printing applications in smart structures such as grippers, adaptive dynamic structures, and smart composites as case studies and their behavior under external stimuli is investigated. First, a soft actuator under thermal stimulus is fabricated and modeled. Second, 4D-printed composites with self-assembly and self-folding features are investigated. Finally, by using passive and active elements, adaptive meta-structures with variable bandgap regions are proposed.

4D textiles—Materials, processes, and future applications

4D printing could replace common electronic sensors, digital controllers, and motors acting as actuators, which are required for the operation of wearable electronics and smart textile systems responding to stimuli in robotics and medical technology, with less complexity (Koch et al., 2021). However, the current materials, manufacturing, and design techniques are still inadequate. Common multimaterial printing processes are time-consuming and costly. Limited print bed dimensions limit the overall dimensions of the structures. This makes the scalability of 4D-printed structures difficult in terms of geometry, weight, and cost-effectiveness (Narula et al., 2018). What is new is the purposeful change in the shape of the hybrid material over time as a result of the interaction of a prestressed, anisotropic textile carrier structure and printed heterogeneous beam structures as reinforcement (Stapleton et al., 2019). In hybrid manufacturing processes, the advantages of textile manufacturing processes can be combined with the advantages of generative manufacturing processes. A hybrid 4D structure consisting of prestressed textile surfaces with printed thermoplastic reinforcements enables the engineering of the hybrid material. Tension energy can be introduced into textiles by stretching an elastic textile. The tension energy can be transferred to the 4D-textile system via the form and material closure and can multiply the effect of previous 4D structures in the hybrid 4D textile system (Koch et al., 2021). Small changes in shape because of external stimuli can lead to large structural changes. In addition, the geometry, weight, and economy of the overall system can be scaled by the targeted use of 4D structures in textile carrier systems. It could be shown that 4D textiles contribute to the development of 4D-printed structures by broadening the material usage and therefore especially the energy that can be stored in textiles.

In this chapter, models for the calculation, selection, design, and characterization of possible solutions and test methods are discussed for the elements of 4D textiles and their relations. A design methodology based on "Design Thinking" for 4D structures is proposed. The hybrid material 4D textile is validated on four application examples (gripper, glove, orthosis, and sportswear). An open innovation concept in the form of a collaborative prototype workshop is proposed for the introduction of technology.

Closed-loop control of 4D-printed hydrogel soft robots

Polyelectrolyte hydrogels consist of highly polar polymer chains that form hydrophilic networks with desirable features such as biocompatibility, biodegradability, tunable mechanical and electrical properties; hence, they could be promising materials for 3D-printed biomedical applications (Zolfagharian, Kouzani, Khoo, Nasri-Nasrabadi, & Kaynak, 2017; Zolfagharian, Kouzani, Khoo, Noshadi, & Kaynak, 2018; Zolfagharian, Kouzani, Moghadam, et al., 2018; Zolfagharian, Kouzani, Maheepala, Khoo, & Kaynak, 2019). Progress in materials science and 3D printing can provide the platform to pursue innovative applications of polyelectrolytes in 4D printing and soft robotics where the shape-changing characteristics of such materials in response to external energy stimuli can be utilized in manufacturing specialized functional materials (Zolfagharian, Kaynak, & Kouzani, 2019; Zolfagharian, Kaynak, Khoo, & Kouzani, 2018; Zolfagharian, Kaynak, Noshadi, & Kouzani, 2019; Zolfagharian, Kouzani, Khoo, et al., 2018; Zolfagharian, Kouzani, Moghadam, et al., 2018).

Through access to 3D printing and smart materials, scientists can come up with creative designs by incorporating specific functions to develop biodegradable actuators and renewable biocompatible soft robots for accomplishing designated tasks (Al-Qatatsheh et al., 2020; Heidarian et al., 2021; Heidarian, Kouzani, Kaynak, Paulino, et al., 2020; Heidarian, Kouzani, Kaynak, Zolfagharian, & Yousefi, 2020; Nasri-Nasrabadi et al., 2018). Hydrogels, being natural polymers, have biocompatibility and biodegradability, which are desirable properties for soft actuators

(Heidarian, Kouzani, Kaynak, Paulino, et al., 2020; Heidarian, Kouzani, Kaynak, Zolfagharian, & Yousefi, 2020). These soft mechanisms are often referred to as "throw-away" robots due to their biodegradable nature (Rossiter, Winfield, & Ieropoulos, 2016). Biodegradable natural hydrogels are suitable materials for developing sustainable and renewable soft actuators. Conversely, the biocompatibility in compliance with some synthesized hydrogels limits their applications in the biomedical and pharmaceutical sectors.

The fabrication of soft actuators with intricate geometrical designs, particularly in miniature scales, faces traditional manufacturing constraints. However, this can be circumvented by 4D printing through the production of complex designs imported from a computer-aided design (CAD) model, which significantly reduces fabrication time, postprocessing effort, and material wastage (Zolfagharian, Kaynak, & Kouzani, 2020). Also, customized geometry, specific functions, and control properties can be directly incorporated into the 4D printing of the morphing structure without the need to print each part individually and subsequently assemble them in a process involving multiple steps.

This chapter presents a detailed instruction of design, fabrication, characterization, optimization, modeling, and control of a common polyelectrolyte hydrogel, chitosan, in a 4D printing study with a soft robotic direction. To illustrate the principle of polyelectrolyte hydrogel actuator manufacturing using 4D printing, chitosan was chosen as a biocompatible and readily accessible hydrogel. The methodology of optimization of the 3D printing parameters is described. The performance of the 4D-printed polyelectrolyte is compared to the conventional cast film replica. A systematic approach to modeling and control methods including other 4D-printed polyelectrolyte gel soft robots is also presented for reproducing the similar studies in additive manufacturing of smart materials and structures in tertiary courses.

Hierarchical motion of 4D-printed structures using the temperature memory effect

4D printing is possible through a thermo-mechanically history-based approach (Scalet et al., 2019). It is demonstrated that proper adoption of parameters such as the deformation temperature, the strain rate, and the time under load could be a useful strategy to control the triggering temperature and the recovery rate of SMP materials and structures (Pandini et al., 2020). Indeed, it is in the framework of thermo-mechanical tailoring that a particular effect, termed "temperature memory effect" (TME), becomes relevant (Inverardi et al., 2020). The temperature memory effect (TME) refers to the ability of a SMP to display recovery around the temperature at which its predeformation occurred so that the material expresses its shape memory response not only in terms of shape but also as it concerns the deformation temperature (Scalet, Pandini, Messori, Toselli, & Auricchio, 2018). This peculiar effect, displayed only by certain classes of polymers, allows control of the triggering temperature for the SME as well as multiple shape memory responses for specific, properly designed predeformation histories.

The TME is displayed by certain SMP systems possessing a broad transition region (Boatti, Scalet, & Auricchio, 2016). Interestingly, these materials are not just capable of actively undergoing a temporary-to-permanent shape transformation, but they also memorize the temperature at which they have been deformed in the temporary configuration. Therefore, for these systems, the TME may be defined as the possibility to activate the recovery process in the desired temperature region, imposed by carrying out the programming procedure at that specific temperature. Moreover, when combined with 3D printing, such an effect opens new and powerful perspectives for designing autonomous structures with customized architectures and programmable/controllable shape changes. However, the design of such structures and of their active responses is not trivial and requires careful attention at different levels, i.e., during printing, experimental characterization, modeling, and simulation.

This chapter focuses on 4D-printed structures exhibiting the TME, and it aims at providing the reader with both an analysis and a discussion, helpful in guiding toward the design of functional structures capable of controlled motions in a hierarchical manner. A methodological approach is proposed and includes the three main stages: evaluation of material properties, experimental characterization of printed structures, and modeling and simulation. A discussion about the steps of each stage is provided, together with an overview of the current state-of-the-art, and a case study is presented. Potential application fields and future perspectives are also explored and discussed.

Manufacturing highly elastic skin integrated with twisted coiled polymer muscles: Toward 4D printing

Developing soft robots integrated with active materials has been gaining popularity due to the ability of these smart structures to interact safely with humans and the ability to function in an inaccessible environment (Wu et al., 2017). Twisted and coiled polymer (TCP) muscles embedded in soft silicone skin showed great potential in robotics and bio-robotics by attaining morphed complex structures that resemble the flexible appendages of living creatures (Saharan & Tadesse, 2016). Advanced manufacturing methods such as 4D printing enable manufacturing adaptive structures with complex morphing capabilities. Fabrication of these structures is extremely dependent on the overall design and the materials used. Hence, a functionalized network of flexible, responsive fibers embedded in a single conformable substrate allows sensing, actuation, and stiffness control of certain systems (Hamidi & Tadesse, 2019). Therefore, the well-established 4D printing techniques, which could embed artificial muscles in between soft or functionally graded layers, have a high potential to make breakthroughs into commercial products. A combination of active and responsive artificial muscles in any number of arrangements, controlled by a robust 4D printing technique, yields an immense design space for this concept.

This chapter presents the use of TCPs and silicone elastomer, and proposes a method of embedding TCPs in silicone, in an effort toward 4D printing by adding the prefabricated artificial muscles during the 3D printing process of silicone elastomer. The printed structures acquired high flexibility and customization and can achieve morphed shapes by actuation and retain their original shape after actuation. The materials used, custom-made 3D printer setup, the manufacturing process parameters, and the resulting structure/products are reported and discussed.

Multimaterial 4D printing simulation using grasshopper plugin

Design for 4D printing is a research topic in manufacturing, materials, and stimuli control, which targets this new manufacturing trend from a design perspective (Sossou, Demoly, Montavon, & Gomes, 2018a, 2018b). In comparison with the "Design for Additive Manufacturing," the literature is rather scarce in 4D printing in this area. Looking at 4D printing, particularly heterogeneous shape-changing objects, from a design perspective is part of the road to its broader dissemination and particularly to its industrial adoption. Given the capability of some AM machines to print objects with nearly continuously varying material properties, the potential of heterogeneity-based 4D printing is largely unexploited. A reason for this could be the difficulty in both design (there is no provision in most of today's CAD software to model material variations in 3D objects) and simulation of such heterogeneous objects (Sossou et al., 2018a, 2018b). Furthermore, predicting how a given heterogeneous object made among others of shape-changing printed smart materials and structures would deform is way more tedious and less intuitive than predicting how a homogeneous part could deform under loading. Knowing how a given spatial arrangement of SMs would behave upon exposure to the right stimulus is not as intuitive as knowing how a homogenous part would deform when subjected to a given load. For this reason, and to ease the exploration of several materials' arrangements of the same object, an appropriate modeling technique should also allow for a rapid simulation of the object's mechanical behavior, namely, how it would deform upon exposure to stimulus should be quickly found.

In this chapter, a voxel-based computational design tool for 4D printing is presented. The authors propose a modeling scheme based on voxels. As essentially a volumetric pixel, a voxel can be used as a base element for representing three-dimensional objects in a discrete manner (Hiller & Lipson, 2010). In the proposed modeling scheme, voxels are cubic and aligned with the Cartesian coordinate system. Any 4D-printed structure could therefore be represented as a 3D stack of equally sized voxels. The tool encompasses modeling, simulation, automated design, and manufacturing of heterogeneous objects made of shape-changing materials. At the core of the proposed toolbox is the vision that material heterogeneity is the key to instilling a desired behavior or functionality in an object. A vision similar to how the arrangement of four chemical bases defines an organism's DNA and ultimately its traits (Sossou et al., 2019). The tool is based on the graphical algorithm editor Grasshopper, a plugin of the CAD software Rhino3D.

Origami-inspired 4D RF and wireless structures and modules

Modern mobile and wireless devices require multiple communication modules and sensors that allow them to communicate with other devices as well as collect information about the environment around them for applications such as smart cities, smart skins, 5G, internet-of-things, and quality of life (Jeong, Cui, Tentzeris, & Lim, 2020). Typically, these modules require packaging several multiband radio frequency (RF) components to reduce overall size and cost while improving system efficiency. The number of RF components in a module can be significantly reduced by realizing tunable multilayer RF structures (such as antennas, filters, matching networks) that can change their electromagnetic (EM) behavior in response to external stimuli (Bahr, Tehrani, & Tentzeris, 2017). 4D-printed origami-inspired RF structures are realized by first 3D-printing extremely complex origami structures using novel 3D-printed flexible materials, thereby making them durable, strong, and highly repeatable (Nauroze, Novelino, Tentzeris, & Paulino, 2017). Traditionally, these structures were mostly realized on paper-based substrates that are prone to absorb moisture, tear, and feature high dielectric losses. Moreover, these structures required manual cutting and folding of the paper substrate, making it laborious and harder to replicate. The RF structures are inkjet-printed on the 3D-printed origami structure to realize wideband, continuous-range tunability (Cui et al., 2019). These structures can be used to realize a wide range of tunable RF structures such as multilayer tunable metamaterials, reconfigurable antennas, wireless sensors, and dielectric reflectors.

This chapter presents state-of-the-art techniques to realize origami-inspired RF structures using 4D printing technologies that exhibit unprecedented continuous-range wideband tunability characteristics. First, the design process for mathematically modeling the origami-inspired structures for kinematic and full-wave electromagnetic simulations is described. Next, a detailed investigation into material characterization and fabrication processes is discussed. Lastly, various techniques to realize robust actuation mechanisms are presented that would facilitate a continuous range of tunability for the origami structure.

Shape-reversible 4D printing aided by shape memory alloys

4D-printed structures fabricated from shape memory polymers (SMPs) are typically one-way actuators, i.e., for each actuation cycle, they must be programmed to deform from the original (as-printed) shape to a secondary (programmed) shape (Akbari, Sakhaei, Panjwani, Kowsari, & Ge, 2021; Bodaghi & Liao, 2019). This is done by applying a combination of thermal and mechanical loads. Then, they restore the initial shape during the actuation process by applying a thermal load. Two-way 4D-printed actuators are fabricated by embedding shape memory alloy (SMA) wires into the printed SMP structures (Akbari et al., 2019; Bodaghi et al., 2018). If a certain thermo-mechanical load is applied to the shape memory alloys (SMAs), they can deform into a secondary elongated shape, and then restore their original length. At a microstructural level, this is a result of phase transformations, because at low temperatures, the SMA is in the martensite phase, while at high temperatures, it is in the austenite phase. This category of smart materials has been extensively exploited over the past decades to fabricate shape memory active structures with various applications in robotics, aerospace, and biomedical engineering (Sakhaei & Thamburaja, 2017).

A major drawback of the SMAs is that their maximum recoverable strain is only 4%–8%. This limits their use in applications that require large deformations (Rodrigue, Wang, Han, Kim, & Ahn, 2017). To address this issue, the SMA wires can be integrated into an elastomeric soft matrix to transform the SMA's small deformations into deformations of at least an order of magnitude larger. This principle can be used in fabricating bending, twisting, and coupled bending-twisting actuators. 4D printing is a more flexible fabrication process that does not require cumbersome molding and assembly processes. Hence, this is an alternative approach to fabricating multimaterial soft SMA actuators. The 4D-printed SMA-SMP composite actuators demonstrate three features simultaneously: variable stiffness, shape retention, and shape restoration.

In this chapter, the printing process of a two-way bending actuator whose bi-layer hinges consist of stiff SMPs as well as elastomers with low modulus is described in detail. Joule heating is employed to modulate the hinges' bending stiffness. An electrical current is applied to the resistive wires inserted into the hinge SMP layer to control their temperature. Indeed, thermo-mechanical programming of the SMA wires, which are integrated into the actuator, provided the bending actuation force. The fabricated actuator could bend, maintain the deformed shape, and recover the as-fabricated shape in a fully automated manner. Furthermore, the potential of this design methodology is assessed using a nonlinear finite-element model. The model incorporated user-defined subroutines to incorporate the complex material behaviors of SMAs and SMPs.

Variable stiffness 4D printing

Stiffness-tunable structures, also known as variable stiffness structures, are an interesting stillbirth of 4D printing (Baniasadi et al., 2021). Variable stiffness or compliance is a feature that is well known in many naturally occurring materials, ranging from wood and bone (Ghazali et al., 2020; Shirzad, Zolfagharian, Matbouei, & Bodaghi, 2021). Variable stiffness allows for: (i) nimble motion; (ii) safer human-robot interactions by reducing the effect of impact with human operators; (iii) energy-efficient actuation; (iv) embodied intelligence; and (v) flexible robots to exhibit different behaviors than traditional structures designed exclusively with rigid members. For a long time, replicating this characteristic in man-made materials without prohibitive cost was an arduous task due to limitations imposed by the manufacturing process (Kadir, Dewi, Jamaludin, Nafea, & Ali, 2019). However, the recent rise in additive manufacturing has facilitated the generation of functional materials that can be endowed with temporal and spatial stiffness. Four-dimensional (4D) printing offers the ability to develop variable stiffness structures by controlling their compliance using heterogeneous materials, such as SMPs (Bodaghi et al., 2016). The stiffness of these structures can be adjusted to introduce passive sensing to control the stiffness of the structures based on the variation in the stimulus. SMPs can be easily processed and printed while offering a high stiffness change of more than a hundred times when exposed to stimuli, such as temperature.

This chapter provides a brief historical account of research into variable stiffness in 4D printing using programmed SME of the structures. This chapter investigates the effect of FDM fabrication parameters on the response of polymeric parts designed with variable stiffness in both the axial and transverse directions. Three types of variable stiffness 4D-printed structures that are realized using single and multiple materials are analyzed. The single material type is achieved by changing the amount of infill percentage inside the materials, using different printing patterns of different profiles to change the directional stiffness of the materials, and using hinges that are made of different printed patterns. Then, the multimaterial actuators are accomplished by using multiple materials with different stiffness values and bending angles on the same structures.

References

Abuzaid, W., Alkhader, M., & Omari, M. (2018). Experimental analysis of heterogeneous shape recovery in 4d printed honeycomb structures. *Polymer Testing*, *68*, 100–109.

Akbari, S., Sakhaei, A. H., Panjwani, S., Kowsari, K., & Ge, Q. (2021). Shape memory alloy based 3D printed composite actuators with variable stiffness and large reversible deformation. *Sensors and Actuators A: Physical*, *321*, 112598.

Akbari, S., Sakhaei, A. H., Panjwani, S., Kowsari, K., Serjouei, A., & Ge, Q. (2019). Multimaterial 3D printed soft actuators powered by shape memory alloy wires. *Sensors and Actuators A: Physical*, *290*, 177–189.

Al-Qatatsheh, A., Morsi, Y., Zavabeti, A., Zolfagharian, A., Salim, N., Kouzani, A. Z., et al. (2020). Blood pressure sensors: Materials, fabrication methods, performance evaluations and future perspectives. *Sensors*, *20*(16), 4484.

Askari, M., Naniz, M. A., Kouhi, M., Saberi, A., Zolfagharian, A., & Bodaghi, M. (2021). Recent progress in extrusion 3D bioprinting of hydrogel biomaterials for tissue regeneration: A comprehensive review with focus on advanced fabrication techniques. *Biomaterials Science*, *9*(3), 535–573.

Bahr, R., Tehrani, B., & Tentzeris, M. M. (2017). Exploring 3-D printing for new applications: Novel inkjet-and 3-D-printed millimeter-wave components, interconnects, and systems. *IEEE Microwave Magazine*, *19*(1), 57–66.

Baniasadi, M., Yarali, E., Bodaghi, M., Zolfagharian, A., & Baghani, M. (2021). Constitutive modeling of multi-stimuli-responsive shape memory polymers with multi-functional capabilities. *International Journal of Mechanical Sciences*, *192*, 106082.

Boatti, E., Scalet, G., & Auricchio, F. (2016). A three-dimensional finite-strain phenomenological model for shape-memory polymers: Formulation, numerical simulations, and comparison with experimental data. *International Journal of Plasticity*, *83*, 153–177.

Bodaghi, M., Damanpack, A., & Liao, W. (2016). Self-expanding/shrinking structures by 4D printing. *Smart Materials and Structures*, *25*(10), 105034.

Bodaghi, M., Damanpack, A., & Liao, W. (2017). Adaptive metamaterials by functionally graded 4D printing. *Materials & Design, 135*, 26–36.

Bodaghi, M., Damanpack, A., & Liao, W. (2018). Triple shape memory polymers by 4D printing. *Smart Materials and Structures, 27*(6), 065010.

Bodaghi, M., & Liao, W. (2019). 4D printed tunable mechanical metamaterials with shape memory operations. *Smart Materials and Structures, 28*(4), 045019.

Bodaghi, M., Noroozi, R., Zolfagharian, A., Fotouhi, M., & Norouzi, S. (2019). 4D printing self-morphing structures. *Materials, 12*(8), 1353.

Cui, Y., Nauroze, S. A., & Tentzeris, M. M. (2019). Novel 3d-printed reconfigurable origami frequency selective surfaces with flexible inkjet-printed conductor traces. In *2019 IEEE MTT-S International Microwave Symposium (IMS), IEEE*.

Faller, L.-M., Krivec, M., Abram, A., & Zangl, H. (2018). AM metal substrates for inkjet-printing of smart devices. *Materials Characterization, 143*, 211–220.

Ghazali, F. A. M., Hasan, M. N., Rehman, T., Nafea, M., Ali, M. S. M., & Takahata, K. (2020). MEMS actuators for biomedical applications: A review. *Journal of Micromechanics and Microengineering, 30*(7), 073001.

Hamidi, A., & Tadesse, Y. (2019). Single step 3D printing of bioinspired structures via metal reinforced thermoplastic and highly stretchable elastomer. *Composite Structures, 210*, 250–261.

Heidarian, P., Kaynak, A., Paulino, M., Zolfagharian, A., Varley, R., & Kouzani, A. Z. (2021). Dynamic nanocellulose hydrogels: Recent advancements and future outlook. *Carbohydrate Polymers, 118357*.

Heidarian, P., Kouzani, A. Z., Kaynak, A., Paulino, M., Nasri-Nasrabadi, B., Zolfagharian, A., et al. (2020). Dynamic plant-derived polysaccharide-based hydrogels. *Carbohydrate Polymers, 231*, 115743.

Heidarian, P., Kouzani, A. Z., Kaynak, A., Zolfagharian, A., & Yousefi, H. (2020). Dynamic mussel-inspired chitin nanocomposite hydrogels for wearable strain sensors. *Polymers, 12*(6), 1416.

Hiller, J., & Lipson, H. (2010). Tunable digital material properties for 3D voxel printers. *Rapid Prototyping Journal, 16*(4), 241–247. https://doi.org/10.1108/13552541011049252.

Hu, G., Damanpack, A., Bodaghi, M., & Liao, W. (2017). Increasing dimension of structures by 4D printing shape memory polymers via fused deposition modeling. *Smart Materials and Structures, 26*(12), 125023.

Inverardi, N., Pandini, S., Bignotti, F., Scalet, G., Marconi, S., & Auricchio, F. (2020). Sequential motion of 4D printed photopolymers with broad glass transition. *Macromolecular Materials and Engineering, 305*(1), 1900370.

Jeong, H., Cui, Y., Tentzeris, M. M., & Lim, S. (2020). Hybrid (3D and inkjet) printed electromagnetic pressure sensor using metamaterial absorber. *Additive Manufacturing, 35*, 101405.

Kačergis, L., Mitkus, R., & Sinapius, M. (2019). Influence of fused deposition modeling process parameters on the transformation of 4D printed morphing structures. *Smart Materials and Structures, 28*(10), 105042.

Kadir, M. R. A., Dewi, D. E. O., Jamaludin, M. N., Nafea, M., & Ali, M. S. M. (2019). A multi-segmented shape memory alloy-based actuator system for endoscopic applications. *Sensors and Actuators A: Physical, 296*, 92–100.

Khosravani, M. R., & Reinicke, T. (2020). 3D-printed sensors: Current progress and future challenges. *Sensors and Actuators A: Physical, 305*, 111916.

Khosravani, M. R., & Zolfagharian, A. (2020). Fracture and load-carrying capacity of 3D-printed cracked components. *Extreme Mechanics Letters, 37*, 100692.

Khosravani, M. R., Zolfagharian, A., Jennings, M., & Reinicke, T. (2020). Structural performance of 3D-printed composites under various loads and environmental conditions. *Polymer Testing*, *91*, 106770.

Koch, H. C., Schmelzeisen, D., & Gries, T. (2021). 4D textiles made by additive manufacturing on pre-stressed textiles—An overview. *Actuators*, *10*(2), 31. (Multidisciplinary Digital Publishing Institute).

Mehrpouya, M., Azizi, A., Janbaz, S., & Gisario, A. (2020). Investigation on the functionality of thermoresponsive origami structures. *Advanced Engineering Materials*, *22*(8), 2000296.

Narula, A., Pastore, C. M., Schmelzeisen, D., El Basri, S., Schenk, J., & Shajoo, S. (2018). Effect of knit and print parameters on peel strength of hybrid 3-D printed textiles. *Journal of Textiles and Fibrous Materials*, *1*, 2515221117749251.

Nasri-Nasrabadi, B., Kaynak, A., Nia, Z. K., Li, J., Zolfagharian, A., Adams, S., et al. (2018). An electroactive polymer composite with reinforced bending strength, based on tubular micro carbonized-cellulose. *Chemical Engineering Journal*, *334*, 1775–1780.

Nauroze, S. A., Novelino, L., Tentzeris, M. M., & Paulino, G. H. (2017). Inkjet-printed "4D" tunable spatial filters using on-demand foldable surfaces. In *2017 IEEE MTT-S International Microwave Symposium (IMS), IEEE*.

Pandini, S., Inverardi, N., Scalet, G., Battini, D., Bignotti, F., Marconi, S., et al. (2020). Shape memory response and hierarchical motion capabilities of 4D printed auxetic structures. *Mechanics Research Communications*, *103*, 103463.

Rodrigue, H., Wang, W., Han, M.-W., Kim, T. J., & Ahn, S.-H. (2017). An overview of shape memory alloy-coupled actuators and robots. *Soft Robotics*, *4*(1), 3–15.

Rossiter, J., Winfield, J., & Ieropoulos, I. (2016). Here today, gone tomorrow: Biodegradable soft robots. In *Electroactive polymer actuators and devices (EAPAD) 2016, International Society for Optics and Photonics*.

Saharan, L., & Tadesse, Y. (2016). Robotic hand with locking mechanism using TCP muscles for applications in prosthetic hand and humanoids. In *Bioinspiration, biomimetics, and bioreplication 2016, international society for optics and photonics*.

Sakhaei, A. H., & Thamburaja, P. (2017). A finite-deformation-based constitutive model for high-temperature shape-memory alloys. *Mechanics of Materials*, *109*, 114–134.

Scalet, G., Niccoli, F., Garion, C., Chiggiato, P., Maletta, C., & Auricchio, F. (2019). A three-dimensional phenomenological model for shape memory alloys including two-way shape memory effect and plasticity. *Mechanics of Materials*, *136*, 103085.

Scalet, G., Pandini, S., Messori, M., Toselli, M., & Auricchio, F. (2018). A one-dimensional phenomenological model for the two-way shape-memory effect in semi-crystalline networks. *Polymer*, *158*, 130–148.

Shen, B., Erol, O., Fang, L., & Kang, S. H. (2020). Programming the time into 3D printing: Current advances and future directions in 4D printing. *Multifunctional Materials*, *3*(1), 012001.

Shirzad, M., Zolfagharian, A., Matbouei, A., & Bodaghi, M. (2021). Design, evaluation, and optimization of 3D printed truss scaffolds for bone tissue engineering. *Journal of the Mechanical Behavior of Biomedical Materials*, *120*, 104594.

Sossou, G., Demoly, F., Belkebir, H., Qi, H. J., Gomes, S., & Montavon, G. (2019). Design for 4D printing: A voxel-based modeling and simulation of smart materials. *Materials & Design*, *175*, 107798.

Sossou, G., Demoly, F., Montavon, G., & Gomes, S. (2018a). An additive manufacturing oriented design approach to mechanical assemblies. *Journal of Computational Design and Engineering*, *5*(1), 3–18.

Sossou, G., Demoly, F., Montavon, G., & Gomes, S. (2018b). Design for 4D printing: Rapidly exploring the design space around smart materials. *Procedia CIRP*, *70*, 120–125.

Spiegel, C. A., Hippler, M., Münchinger, A., Bastmeyer, M., Barner-Kowollik, C., Wegener, M., et al. (2020). 4D printing at the microscale. *Advanced Functional Materials*, *30* (26), 1907615.

Stapleton, S. E., Kaufmann, D., Krieger, H., Schenk, J., Gries, T., & Schmelzeisen, D. (2019). Finite element modeling to predict the steady-state structural behavior of 4D textiles. *Textile Research Journal*, *89*(17), 3484–3498.

Tawk, C., & Alici, G. (2021). A review of 3D-printable soft pneumatic actuators and sensors: research challenges and opportunities. *Advanced Intelligent Systems*, 2000223.

Tawk, C., Gillett, A., in het Panhuis, M., Spinks, G. M., & Alici, G. (2019). A 3D-printed omni-purpose soft gripper. *IEEE Transactions on Robotics*, *35*(5), 1268–1275.

Tawk, C., in het Panhuis, M., Spinks, G. M., & Alici, G. (2018). Bioinspired 3D printable soft vacuum actuators for locomotion robots, grippers and artificial muscles. *Soft Robotics*, *5* (6), 685–694.

Wu, L., de Andrade, M. J., Saharan, L. K., Rome, R. S., Baughman, R. H., & Tadesse, Y. (2017). Compact and low-cost humanoid hand powered by nylon artificial muscles. *Bioinspiration & Biomimetics*, *12*(2), 026004.

Yousuf, M. H., Abuzaid, W., & Alkhader, M. (2020). 4D printed auxetic structures with tunable mechanical properties. *Additive Manufacturing*, *35*, 101364.

Zolfagharian, A., Durran, L., Gharaie, S., Rolfe, B., Kaynak, A., & Bodaghi, M. (2021). 4D printing soft robots guided by machine learning and finite element models. *Sensors and Actuators A: Physical*, *328*, 112774.

Zolfagharian, A., Kaynak, A., Khoo, S. Y., & Kouzani, A. Z. (2018). Polyelectrolyte soft actuators: 3D printed chitosan and cast gelatin. *3D Printing and Additive Manufacturing*, *5*(2), 138–150.

Zolfagharian, A., Kaynak, A., & Kouzani, A. (2019). Closed-loop 4D-printed soft robots. *Materials & Design*, *108411*.

Zolfagharian, A., Kaynak, A., & Kouzani, A. (2020). Closed-loop 4D-printed soft robots. *Materials & Design*, *188*, 108411.

Zolfagharian, A., Kaynak, A., Noshadi, A., & Kouzani, A. Z. (2019). System identification and robust tracking of a 3D printed soft actuator. *Smart Materials and Structures*, *28*(7), 075025.

Zolfagharian, A., Kouzani, A. Z., Khoo, S. Y., Moghadam, A. A. A., Gibson, I., & Kaynak, A. (2016). Evolution of 3D printed soft actuators. *Sensors and Actuators A: Physical*, *250*, 258–272.

Zolfagharian, A., Kouzani, A. Z., Khoo, S. Y., Nasri-Nasrabadi, B., & Kaynak, A. (2017). Development and analysis of a 3D printed hydrogel soft actuator. *Sensors and Actuators A: Physical*, *265*, 94–101.

Zolfagharian, A., Kouzani, A. Z., Khoo, S. Y., Noshadi, A., & Kaynak, A. (2018). 3D printed soft parallel actuator. *Smart Materials and Structures*, *27*(4), 045019.

Zolfagharian, A., Kouzani, A. Z., Maheepala, M., Khoo, S. Y., & Kaynak, A. (2019). Bending control of a 3D printed polyelectrolyte soft actuator with uncertain model. *Sensors and Actuators A: Physical*, *288*, 134–143.

Zolfagharian, A., Kouzani, A., Moghadam, A. A. A., Khoo, S. Y., Nahavandi, S., & Kaynak, A. (2018). Rigid elements dynamics modeling of a 3D printed soft actuator. *Smart Materials and Structures*, *28*(2), 025003.

Zolfagharian, A., Mahmud, M. P., Gharaie, S., Bodaghi, M., Kouzani, A. Z., & Kaynak, A. (2020). 3D/4D-printed bending-type soft pneumatic actuators: Fabrication, modelling, and control. *Virtual and Physical Prototyping*, *15*(4), 373–402.

4D printing electro-induced shape memory polymers

2

Rytis Mitkus, Ferdinand Cerbe, and Michael Sinapius
Institute of Mechanics and Adaptronics, TU Braunschweig, Braunschweig, Germany

Introduction

Current three-dimensional (3D) printing technologies mostly provide static printing, while four-dimensional (4D) printing aims to create structures that can transform or change their properties over time when exposed to environmental stimuli. The fourth dimension is time. The status of 4D-printed structures can be achieved by using different smart materials and combining them with appropriate design options for specific applications. This technology could also potentially lead to improved material savings, complex 3D geometries, and enable several new applications in different fields such as metamaterials, biomedicine, sensors, and actuators.

Our study presents shape-memory polymer (SMP) structures, printed initially flat with a fused deposition modeling (FDM)-type 3D printer, made of off-the-shelf material polylactic acid (PLA). In printed flat, the structure bends itself into a U-shape when a temperature higher than glass transition temperature (T_g) is reached (Fig. 2.1). In the study presented in this chapter, the T_g of PLA is approximately 60°C. However in U-shape, i.e., above T_g, SMP structures can be manually deformed or twisted into nearly any arbitrary shape and can then be cooled down to retain that new shape (programming step). Repeated heating above T_g causes the SMP structure to deform back to a U-shape (recovery step). Fig. 2.1 depicts a shape-morphing cycle of a SMP structure.

The shape-morphing of the structure is determined by varying printing parameters. PLA is a one-way SMP (Xu, Song, & Yahia, 2015), and the performance of shape-morphing is dependent on printing parameters and printing conditions (An et al., 2018; Kačergis, Mitkus, & Sinapius, 2019; Manen, Janbaz, & Zadpoor, 2017; Noroozi, Bodaghi, Jafari, Zolfagharian, & Fotouhi, 2020). Even though in this study the geometry is limited to bending to clarify the effective relationships, extremely complex structures and behaviors are possible by a specific combination of printing paths and printing parameters (Gu et al., 2019; Noroozi et al., 2020; Wang, Tao, Capunaman, Yang, & Yao, 2019).

We also report in this chapter the integration of conductive PLA (CPLA) into the SMP structure, which allows the deformation of the SMP structure to be activated by Joule heating using electric current (see Fig. 2.2). CPLA is encapsulated inside the structure and is covered with nonconductive PLA. The SMP structures reported here with conductive traces can be activated with 60 V and offer full shape-morphing in a few minutes, where the speed of deformation depends on the electric current applied.

Smart Materials in Additive Manufacturing, Volume 2: 4D Printing Mechanics, Modeling, and Advanced Engineering Applications
https://doi.org/10.1016/B978-0-323-95430-3.00002-6
Copyright © 2022 Elsevier Inc. All rights reserved.

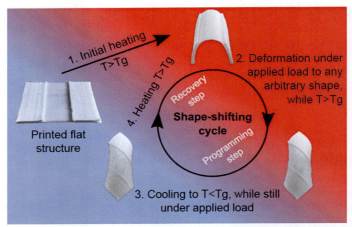

Fig. 2.1 A shape-shifting cycle of the SMP structure. Red (right-top) and blue (left-bottom) colors indicate a hot (above T_g) and cold (below T_g) environment.
Data from Kačergis, L., Mitkus, R., & Sinapius, M. (2019). Influence of fused deposition modeling process parameters on the transformation of 4D printed morphing structures. *Smart Materials and Structures, 28*(10), 105042. https://doi.org/10.1088/1361-665x/ab3d18.

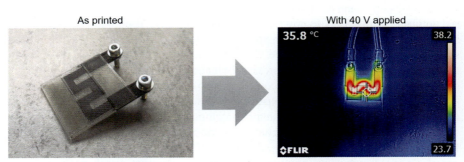

Fig. 2.2 Fully 3D-printed SMP structure with conductive traces (left) and an example of heating performance by applying voltage (right).
No permission required.

The current main drawback of the SMP structures investigated is that they cannot regain their flat form without straightening them manually. However, the SMP structure can be deformed into any shape by a programming step (see Fig. 2.1). Both shapes at the bottom of Fig. 2.1 represent a rather arbitrarily chosen deformation. It retrieves its U-shape after heating above T_g (step 4). A further drawback of the SMP structure is its very low Young's modulus above T_g and thus quite a low blocking force.

Since conductive filaments usually have very high electrical resistance, experiments are conducted to find printing parameters that would produce the lowest resistance possible for enhanced heating performance. To avoid any heating at the cable connection points, various connection methods are experimentally investigated. Furthermore, multiple designs of SMP structure with heating elements are created and are investigated, regarding maximum deformation angle, change in resistance, and blocking force.

SMP structures presented in this chapter are the initial step to understand the future possibilities of fully 4D-printed, complex, thermoplastic structures with embedded electric activation and sensing capabilities. The structures can be fully printed with a single machine and do not require any specific tools or devices. Embedding of electronics opens new possibilities for controlled deformation and sensing for various applications, such as flexible grippers (Shao, Zhao, Zhang, Xing, & Zhang, 2020; Yang, Chen, Li, Wang, & Li, 2017; Yang, Wang, Li, & Tian, 2017). SMP structures presented here are one of the first of such kind that use printable conductive material for heating purposes to activate the morphing of the structure.

Materials, working principle, and similar structures in 4D printing

Shape-morphing of various 4D-printed structures is usually based on different physical principles such as residual stresses (prestrain), energy release, or capillary forces (Zhang et al., 2017). For example, some materials such as hydrogels can absorb water and thus expand. This results in a deformation of the structure, which can be reversed. Various environmental stimuli are reported by researchers for active material activation, including humidity (Nishiguchi et al., 2020; Raviv et al., 2014; Tibbits, 2014), light (Jeong, Woo, Kim, & Jun, 2020), temperature (Ding et al., 2017; Kačergis et al., 2019), pH, magnetic field (Kim, Yuk, Zhao, Chester, & Zhao, 2018), combination of stimuli (Mao et al., 2016), and others (Hager, Bode, Weber, & Schubert, 2015; Kuang et al., 2018; Pei & Loh, 2018; Zafar & Zhao, 2020; Zolfagharian, Denk, Bodaghi, Kouzani, & Kaynak, 2020; Zolfagharian, Kouzani, Maheepala, Yang Khoo, & Kaynak, 2019). Most active materials used in 4D printing can be split into four classes: hydrogels, liquid crystalline elastomers (LCE), surface-tension-driven materials, and SMPs (Ionov, 2015).

As an example, hydrogels have a reversible volumetric expansion, can achieve high swelling ratios (Ionov, 2015), but their mechanical properties are quite low (Jeon, Bouhadir, Mansour, & Alsberg, 2009) and usually have slow response rates (hours to minutes) (Jeon et al., 2009). However, hydrogels are quite promising for biomedical applications because they can be used as a temporary substitute or other regenerative therapies (Lendlein & Shastri, 2010). LCE material deforms anisotropically, under a change in temperature or other environmental conditions (Ionov, 2015), and produces reversible deformation. Surface-tension-driven materials can be employed only on very small scales, have single-directional shape-variability, and behave quite similar to SMPs (Ionov, 2015). SMPs, however, have much faster response rates (seconds to minutes) and higher elastic modulus compared with the other classes, making them more suitable for various applications (Bodaghi et al., 2020; Ge et al., 2016; Manen et al., 2017), but mostly they offer only single-directional shape-morphing. Another huge advantage of SMPs is the possibility to mix them with particles (e.g., conductive, magnetic) to create composite materials that can be activated in specific conditions or in specific applications (Flowers, Reyes, Ye, Kim, & Wiley, 2017; Kim et al., 2018; Kwok et al., 2017; Shao et al., 2020). Similar trend with particle filled hydrogels is reported in literature (Jeon, Hauser, & Hayward, 2017). Hydrogels and SMPs are currently the most used active materials (or smart materials) in various 4D printing studies. An extensive review of printable active materials used in 4D printing can be found (Ryan, Down, & Banks, 2020).

4D printing with FDM

In the last decade, researchers investigated FDM type of 3D printing with known SMPs (Bodaghi et al., 2020; Bodaghi, Damanpack, & Liao, 2017; Garcia et al., 2018; Garcia et al., 2018b), while other researchers found possibilities to use common polymers printed with FDM to achieve shape memory effect (SME). Both single (Zhang et al., 2017) and multimaterial 4D structures using the SME have been successfully printed using FDM technology and its common materials (An et al., 2018; Elgeneidy, Neumann, Jackson, & Lohse, 2018; Kačergis et al., 2019). These polymers, such as the widely used PLA, are commercially available amorphous polymers that have one-way SME and whose behavior can be influenced by FDM printing processes (An et al., 2018; Bodaghi, Noroozi, Zolfagharian, Fotouhi, & Norouzi, 2019; Goo, Hong, & Park, 2020; Manen et al., 2017; Noroozi et al., 2020; Wang et al., 2018, 2019).

PLA is one of the most popular thermoplastics used in 3D printing. It can be prestressed during printing as the temporary shape is formed by heating the filament above its melting temperature, extruding along the set path, and cooling it. Melting and extruding the filament stretches and aligns the polymer chains along the directions of the extrusion path and fast cooling locks them inside the material (An et al., 2018; Noroozi et al., 2020; van Manen, Janbaz, & Zadpoor, 2018; Wang et al., 2018). The extruded material is constrained by depositing it on the build plate or the previous layer. The residual stresses (or prestrains) generated by the stretching are stored in the material until it is heated above T_g. When the SMP structure is removed from the printer and is heated over its T_g, the material softens and stretched polymer chains return to strain-/stress-free, low-energy mode (An et al., 2018) and the material slightly shrinks along the stretched path. While above T_g (soft material), the material can be stretched or deformed and then cooled down to retain the new shape (rigid material). Sequent heating above T_g would cause the material to deform back to its chaotic polymer chain configuration.

This prestressing of the material can be achieved using an FDM printer. Even though most of the 3D printing technologies are suitable for use in 4D printing processes, in principle, however, not all of them can be used equally. The FDM printer used in this study extrudes material through a nozzle, where this extrusion process can be precisely controlled. These 3D printers are cheap, easy to use, and easily modifiable. Furthermore, they allow combining deposition of pastes or inks with 3D printing by adding additional equipment. A wide choice of materials is available for FDM, and materials are relatively cheap compared with other additive manufacturing technologies. Most polymers can be easily recycled. Moreover, the SME exists in a large group of thermoplastics (An et al., 2018; Goo et al., 2020), and various common types of filaments can be used to print shape-morphing structures. Furthermore, polymer/solid particle blends can be used in FDM to increase or alter the properties of the polymers so that they would become active under specific conditions, for example, magnetic fields (Kim et al., 2018). Polymers used in FDM qualify for applications better than, for example, photocurable polymers, which are used in PolyJet or stereolithography/digital light processing (SLA/DLP) processes and become brittle over time. Several researchers show interest in 4D printing based on FDM machines.

Printing parameters and their influence on deformation of PLA

The behavior of PLA depends on multiple printing parameters, material thermal history, and geometrical configuration of the material itself. The latter refers to a combination of rasters of extruded PLA in a single layer. Moreover, some of our early experiments showed that even the geometry of the extruder itself (long or short heat transition zone) influences the degree of prestrain introduced into the material. The exact combination of parameters that would provide the highest prestrains is still not yet fully understood.

Our recent study clarified the influence of printing speed, the temperature of the build plate, and the number of active layers in the structure on the deformation of the layered bi-material structure, composed of PLA and thermoplastic polyurethane (TPU) (Kačergis et al., 2019). The investigation of the bi-material structure serves to understand and maximize the influence of printing parameters on the shrinkage of PLA. The investigations reveal an almost linear relationship between printing speed and structure deformation (Fig. 2.3), where the highest printing speed produces the

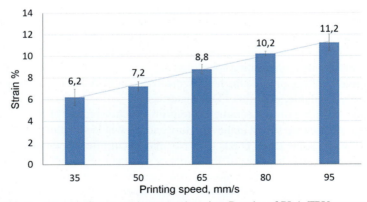

Fig. 2.3 Printing speed influence on generated strains. Results of PLA-TPU structure. Strain generated by the bending of the SMP structures, printed at different printing speeds.
Data from Kačergis, L., Mitkus, R., & Sinapius, M. (2019). Influence of fused deposition modeling process parameters on the transformation of 4D printed morphing structures. *Smart Materials and Structures, 28*(10), 105042. https://doi.org/10.1088/1361-665x/ab3d18.

highest shrinkage of PLA and thus the highest deformation of the bi-material structure. The same results are reported by a number of researchers (An et al., 2018; Bodaghi et al., 2017, 2019; Noroozi et al., 2020; Zhang, Yan, Zhang, & Hu, 2015). Recently, Goo et al. (2020) report a very similar behavior of acrylonitrile butadiene styrene (ABS), revealing that ABS also has an SME.

Manen et al. (2017) conduct extensive research on the influence of various FDM printing parameters on multilayer structures made of single PLA material. Lower extrusion temperature together with lower layer height (0.05 mm) produces the highest prestrains in the structures (Manen et al., 2017). Higher stored prestrain inside the material during printing on lower extruder temperature is reported for polyurethane-based printable SMP material as well (Bodaghi et al., 2017). The increase in prestrain due to printing at lower temperatures might be explained by the degree of crystallinity in the material because prestrain in the material originates from its amorphous phase. According to Wittbrodt and Pearce (2015), the degree of crystallinity in the material depends on the material color and extrusion temperature, where higher extrusion temperature produces higher crystallinity, thus less amorphous content in the polymer. However, change in crystallinity is nonlinear, while the decrease in structural deformations with decreasing extrusion temperature is linear (Manen et al., 2017). This indicates that crystallinity might be not the only influencing factor, but a faster cooling rate and the thermal history can play a significant role. We have found that the structures consisting of more layers take more time to reach their final shape (Kačergis et al., 2019). Although it may look like a drawback at a first glance, it can be treated as a feature. The number of active layers can be utilized as a mechanism for a sequential folding of a complex layout SMP structure. Furthermore, the maximum deformation and its speed can be controlled by the number of active layers (Fig. 2.4).

The temperature of the build platform also has a significant influence on the prestrain. Structures printed on colder build plates deform to a higher degree (Fig. 2.5). Other researchers reported similar results (Bodaghi et al., 2017). Furthermore, as reported by some researchers (Gu et al., 2019; Manen et al., 2017), a lower layer height also leads to higher strains, most likely due to a faster cooling rate (Bodaghi et al., 2017). Summarizing, high printing speed, low layer height, low extruder temperature, and cold build platform provide the highest prestrains in PLA and most likely in other commercially available, printable FDM materials.

By using constant printing parameters to print structures made of PLA, for example, in-plane stresses are stored in the structure and the whole structure mainly shrinks or expands, while some small bends or twists may occur after heating above T_g. For many applications, the bending of the structures instead of shrinking is favorable. Some studies report printing PLA directly onto a paper surface so that in-plane strains of PLA could be converted to bending motion when PLA is heated above its T_g (Bodaghi et al., 2019; Wang et al., 2018; Zhang, Zhang, & Hu, 2016). Using paper, self-straightening structures are possible instead of single-directional bending at temperatures higher than T_g (Bodaghi et al., 2019; Wang et al., 2018). However, paper is not the best option, because PLA does not form a strong bond with it, and paper is usually easy to peel off.

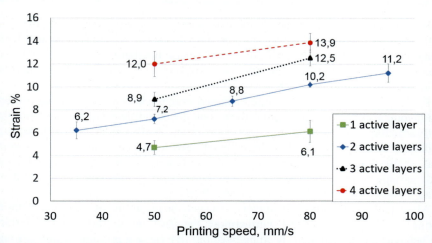

Fig. 2.4 Printing speed and number of active layer influence on generated strain. Results of PLA-TPU structure. Strain generated by the bending of the SMP structures, printed at different printing speeds and with the different number of active layers.
Data from Kačergis, L., Mitkus, R., & Sinapius, M. (2019). Influence of fused deposition modeling process parameters on the transformation of 4D printed morphing structures. *Smart Materials and Structures, 28*(10), 105,042. https://doi.org/10.1088/1361-665x/ab3d18.

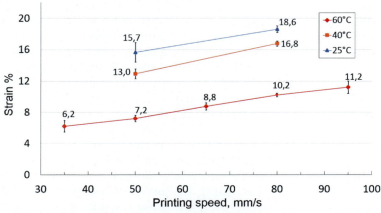

Fig. 2.5 Printing speed and build plate temperature influence on the generated strain. Results of PLA-TPU structure. Strain generated by the bending of the SMP structures, printed at different printing speeds and different build plate temperatures.
Data from Kačergis, L., Mitkus, R., & Sinapius, M. (2019). Influence of fused deposition modeling process parameters on the transformation of 4D printed morphing structures. *Smart Materials and Structures, 28*(10), 105042. https://doi.org/10.1088/1361-665x/ab3d18.

Researchers proposed an idea to use elastomer TPU as a passive layer (An et al., 2018). TPU is soft at room temperature because it is above its T_g. This passive layer made of TPU eliminates the need for paper because it is hardly compressible but is flexible at the same time and constrains the active layer of PLA so that bending motion would occur (Kačergis et al., 2019).

Design parameters of the structure, such as overall thickness, material composition, and layout, also have a significant influence on the performance and behavior of the SMP structures (Bodaghi et al., 2019; Goo et al., 2020; Gu et al., 2019; Janbaz, Hedayati, & Zadpoor, 2016; Kačergis et al., 2019; Manen et al., 2017; Noroozi et al., 2020; Wang et al., 2019). By properly combining layers and printing direction with varying printing parameters, sophisticated deformations can be achieved by using an only single material, for example, PLA.

Almost all main printing parameters influence the prestrains stored in the material during the 3D printing process. Furthermore, the direction of printed rasters defines the behavior of the structure (bending/twisting). By varying printing parameters and geometrical design, single-material or multimaterial morphing structures can be designed.

Integration of conductive PLA

Material activation by temperature or ambient medium as well as the activation method itself, whether internally or externally supplied energy, influences the achievable maximum bending/elongation and speed of deformation of 4D-printed structures (Zhang et al., 2016). Most 4D structures realized to date respond to either direct (An et al., 2018; Ge et al., 2016; Hu, Moqadam, & Liao, 2017) or indirect heating (Shao et al., 2020; Yang, Chen, et al., 2017). Some studies use hot water to transform SMP structures (Bodaghi et al., 2017; Kačergis et al., 2019). However, it is well known that polymers tend to absorb water. As the literature suggests (Ecker et al., 2019; Fernandes, Deus, Reis, Vaz, & Leite, 2018; Tham, Chow, Poh, & Ishak, 2016; Wahit, Hassan, Akos, Zawawi, & Kunasegeran, 2015), water absorption of PLA depends on the temperature of the water and the exposure time to water. Although the weight gain of PLA up to around 2% in 24 h in water at room temperature does not seem to have a significant effect on the morphing behavior of SMP structures, water or any other liquid is not suitable for most applications.

A possibility to actuate the SMP structure with electric current is a much more favorable approach. Wamg et al. successfully use CPLA to manufacture shape-morphing structures (Wang et al., 2018). The structure transforms from bend shape to flat using internal Joule heating, where the current is applied to 3D-printed CPLA material. It has quite high electrical resistance but can be heated above its glass transition temperature with 40–130 V, depending on the geometry printed.

A novel approach to internal heating the SMP structures is to integrate heating elements into the SMP structures made of PLA. First, different methods to attach cables to printed CPLA electrodes are investigated to find a connection with the lowest resistance. Second, the influence of printing speed and extrusion temperature on the resistance of the material is investigated. Finally, CPLA is investigated for its heating performance and influence to a maximum bending of SMP structures.

Conductive filaments for FDM printers are mostly used for low-current applications such as wiring (Kwok et al., 2017), shielding, or radiofrequency transmissions (Pizarro Torres et al., 2019; Roy, Qureshi, Asif, & Braaten, 2017). The main reason for low-current applications is their high resistance, which is only suitable for extremely low currents and relatively low resolution of printing process (short-circuiting possibility). The possibilities to 3D print capacitors, inductors, and high-pass filters with commercially available conductive filaments are reported (Flowers et al., 2017; Kim et al., 2019). By using modified filament, even batteries can be printed (Reyes et al., 2018).

A group of researchers reports 3D printing of strain gauge sensors by using conductive filament (Elgeneidy et al., 2018). The sensors exhibit a strong drift due to internal heating. However, it is possible to eliminate this by using specially designed electronics.

An interesting approach uses 3D printable conductive materials to change the stiffness of the structure, taking advantage of the Joule heating effect (Al-Rubaiai, Pinto, Qian, & Tan, 2019; Yang, Chen, et al., 2017). The use of this approach in soft robotics enables a change in the stiffness of the structure. At the same time, 3D-printed conductive traces can be used to determine the position of the actuator (bending angle) (Yang, Chen, et al., 2017).

The usage of conductive filaments allows the application of an electric current to generate Joule heating, thus changing the stiffness of the conductive material itself and of surrounding materials. Furthermore, the deformation of conductive traces changes their resistance, and they can be used for various sensing purposes.

Materials and equipment

Electrically conductive composite PLA (Proto-Pasta Conductive PLA, made by Proto Plant, USA) is used in this study. It is a compound of Natureworks 4043D PLA and about 21 wt% of electrically conductive carbon black Proto-Pasta Conductive PLA Safety Data Sheet. Its physical and mechanical properties of CPLA are displayed in Table 2.1. It should be noted that its Young's modulus at room temperature is only 1/3 of normal PLA.

Detailed results exist from several research groups on the same CPLA filament used in this study (Al-Rubaiai et al., 2017, 2019). A reduction of up to 98.6% in Young's modulus is reported, when a material is heated from room temperature to 80°C (Al-Rubaiai et al., 2017). This is beneficial for the SMP structures due to the low mechanical influence on their performance. Furthermore, an increase in the coefficient of thermal expansion with increasing temperature (Al-Rubaiai et al., 2017, 2019) can support the deformation of SMP structures. The main drawback of the used CPLA is an increase in electrical resistance with increasing temperature, which can be as high as 50% at 100°C (Al-Rubaiai et al., 2017; Kwok et al., 2017). Because a sharp increase in resistance is reported at about 100°C, it is recommendable to use lower temperatures being still sufficient to deform the studied SMP structures.

Table 2.1 Properties of CPLA.

Property	Value
Young's modulus (ASTM D638) at 20°C	1000 MPa
Young's modulus (ASTM D638) at 80°C	13.6 MPa
Thermal conductivity	∼0.366 W/(m K) (Al-Rubaiai, Pinto, Torres, Sepulveda, & Tan, 2017; Al-Rubaiai et al., 2019)
Resistance of a 10 cm length of 1.75 mm filament	2–3 kΩ
Volume resistivity along layers (*xy*-plane)	30 Ω cm (Product: Conductive PLA from Proto-Pasta, 2021), 6 Ω cm (FAQs of Multi3D. Contact Resistance Improvement, 2021)
Volume resistivity against layers (*z*-direction)	115 Ω cm (Product: Conductive PLA from Proto-Pasta, 2021), 10 Ω cm (FAQs of Multi3D. Contact Resistance Improvement, 2021)

All samples of this study are produced by a single 0.4-mm nozzle, FDM-type 3D printer Original Prusa i3 (Prusa Research, Czechia) equipped with MK3S Multi-Material Upgrade 2S unit (Prusa Research, Czechia). A spring steel sheet with a smooth polyetherimide (PEI) surface is used throughout this study. Multi-material upgrade unit allows to use of up to five materials in a single print, but all materials must be extruded through the same nozzle.

Because of the single nozzle printer used in this study, the printable material switching from CPLA to PLA causes the contamination of extruded PLA with a residue of CPLA. This contamination may cause short-circuit on the printed part and must therefore be avoided. Normally, material switching is done automatically by the printer by extruding a few millimeters of the subsequent material through the nozzle to clean the residue of the previously used filament, e.g., PLA is purged to clean the residue of CPLA. However, to remove the residue of CPLA, a large amount of PLA must be extruded, which creates additional material waste and increases printing time. To make the process faster and cheaper, the manufacturer of CPLA recommends purging the nozzle with a cleaning filament, which is typically polyamide-based (PA). Therefore, PA filament is extruded through the nozzle first to clean the residue of CPLA in the nozzle, and then a significantly smaller amount of PLA material is used to clean the PA filament from the nozzle. The material amount required to clean the nozzle has been decreased to 100 mm^3 of PA comparing to 1500–1700 mm^3 of PLA required to clean the same nozzle under the same conditions.

Electrical contacting of CPLA

Most literature sources do not give precise information about a proper electrical connection to the 3D-printed conductive filament. Some researchers used conductive inks to smooth the rough contact surfaces after FDM printing. While low currents do not require a particularly strong connection, higher currents demand a firm connection to avoid an increase of contact resistance, which could result in local heating. The connection of printed CPLA to the power cables should provide a mechanically reliable connection both at room temperature and at operating temperature ($<100°C$). Moreover, for most applications, an easily detachable connection is prioritized.

Contact resistance between the printed CPLA part and the cable is mainly influenced by the structure of surfaces and the applied load. The roughness of the surface causes a reduced contact area and thus an increase in the total resistance of contact. The higher total resistance in turn requires a higher input voltage to provide the constant current across the terminals required for the heating.

Reduction of contact resistance of the surfaces of printed conductive filaments can be achieved either by silver paint (FAQs of Multi3D. Contact Resistance Improvement, 2021; Flowers et al., 2017; Kwok et al., 2017) or by electroplating (Kim et al., 2018). However, the information in the literature is unfortunately limited because low temperatures and low currents are used or it is not specified how exactly the connections were made.

Low voltage limited to 120 V DC is recommended for experimental circuits without additional electrical protection against electric shocks. It carries a low risk of life-threatening electrical shock on contact (Schmolke, 2018). We limited in our study the maximum voltage to 80 V DC.

Crocodile clamps on exposed terminals and bolted connections are frequently used for temporary connections in laboratories or for semi-permanent connections. The literature states that silver paint enhances contact conductivity, but corresponding measurements on conductive filaments are not reported (FAQs of Multi3D. Contact Resistance Improvement, 2021; Flowers et al., 2017; Kwok et al., 2017). Therefore, our study investigates both crocodile clamp and bolted connections, with and without conductive silver ink. Furthermore, two permanent wiring methods have been added to the measurements, including cable gluing with silver ink and pushing soldering alloy preinfused wire with soldering iron into the CPLA structure. In early experiments, it was found that the end of cable coated with soldering alloy can be heated up and pushed inside the CPLA.

All samples used for contact resistance measurement are shown in Fig. 2.6. Individual specimens are combined into one part by adding a few layers of clear PLA on top of the CPLA. Due to their width, a physical end stop is produced by the PLA for the different wiring methods, enabling a constant measuring distance.

A thin layer of silver conductive lacquer (L204N from Ferro, USA) is applied to both sides of the respective test specimens and dried. The vertical sides and the inner surfaces of the holes of the test specimens remain free.

Silver-bonded cables (cross section of $0.5\,mm^2$) are produced by applying silver ink five times with intermediate drying. The silver ink is only applied to the top of the CPLA connectors. The screw connections consist of zinc-plated cylinder screws and are tightened to 0.40 Nm. Each specimen is 54 mm long, 7 mm wide with a height of 0.8 mm. The holes have a diameter of 3.1 mm and a distance between holes of 47 mm.

No separate infill printing parameters are used, and simply, the number of outside perimeters is set to 10 to achieve the toolpaths only parallel to the outside walls to

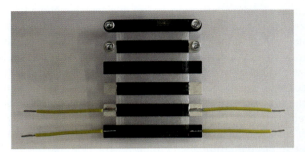

Fig. 2.6 Investigation of electrical connector types. Investigated connector types, from top to bottom: (1) screw connection; (2) screw connection on silver paint; (3) crocodile clips; (4) crocodile clips and silver paint; (5) silver paint as glue; (6) hot wire pushed in. Crocodile clips are not shown.
No permission required.

achieve the lowest resistance (FAQs of Multi3D. Contact Resistance Improvement, 2021). The manufacturer of CPLA recommends the use of standard PLA settings (or hotter) and no heated bed or special nozzles (Product: Conductive PLA from Proto-Pasta, 2021). Standard PLA printing settings were used (PrusaSlicer, print setting: "0.2 mm QUALITY," filament setting: "Das Filament PLA"). However, researchers in the literature report different printing speeds (15–90 mm/s) and printing temperatures (190–230°C). In the next experiment, we use a printing speed of 25 mm/s and an extrusion temperature of 215°C in the studies presented. An extrusion multiplier of 1.15 produces CPLA samples without gaps, while printing with the standard extrusion multiplier of 1.00 produces gaps between the grids of printed CPLA.

Since CPLA sticks very strongly to build plate surface (smooth PEI sheet, Prusa Research), a thin layer of school-grade paper glue stick is applied on the surface and is spread over the surface with glass cleaner containing alcohol to reduce adhesion. The resulting residue on CPLA specimens is cleaned with isopropyl alcohol.

All printing settings of CPLA for a current experiment are summarized in Table 2.2. Electrical resistance is measured at room temperature on the 10 prepared specimens. The sample with glued-in cables turns out to be extremely brittle, and cables detached after measuring. Some specimens with cables welded in also disconnected during measurements. While the resistance of these two types of connections can successfully be measured, the brittle behavior of them limits their use for applications. The measurement results are presented in Fig. 2.7.

The results show that silver ink significantly reduces resistance in all cases, and this is extremely important to achieve a nonheating connection. Furthermore, the standard deviation is slightly reduced when using silver ink, especially for bolted connections. The use of silver ink reduces the roughness of printed CPLA surfaces and provides a bigger contact area to the bolt and thus reduced resistance. Methods to glue the cable with silver ink and pushing the cable inside with soldering iron provided average resistance results and at the same time brittle, unusable connections. Overall, the bolted connection shows the lowest resistance, while this connection type is the strongest.

Table 2.2 Printing parameters used to print CPLA filament.

Printing parameter	Value
Printing speed (first layer)	20 mm/s
Printing speed	25 mm/s
Number of vertical shells (perimeters)	10
Layer height	0,2 mm
Extrusion width	0,45 mm
Nozzle diameter	0,4 mm
Extrusion multiplier	1,15
Extruder temperature	215°C
Build plate temperature	60°C
Active extrusion cooling (first layer)	0%
Active extrusion cooling	100%

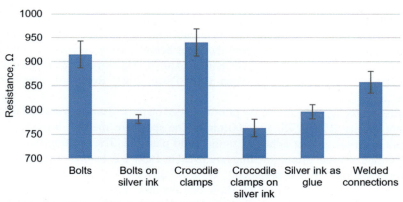

Fig. 2.7 The electrical resistance of different connector types. No permission required.

Printing parameter influence on resistance

Although websites already report that printing parameters influence the mechanical and electrical properties of the printed parts (FAQs of Multi3D. Contact Resistance Improvement, 2021), no peer-reviewed literature is available on this topic. To find the printing parameters that produce the lowest resistance of 3D-printed electric traces inside SMP structures, the influences of extrusion temperature and printing speed on the resistance are investigated below.

Three different extrusion temperatures are selected: 195°C, 215°C, and 233°C. For each temperature, 10 specimens are prepared. All other manufacturing parameters presented in Table 2.2 remain constant.

Fig. 2.8 reveals a strong influence of the printing temperature on the resistance. The resistance is lower at higher printing temperatures, and at the same time, the scatter of

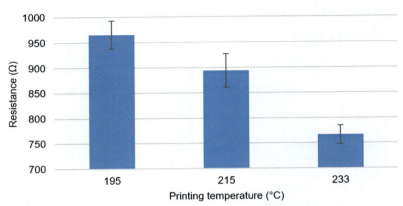

Fig. 2.8 Electrical resistance dependency from printing temperature. No permission required.

the results is smaller. Extruder temperature is reported to affect the mechanical properties, especially layer adhesion. At higher temperatures, the extruded filament contains more thermal energy to heat the layer below and the filament around thus creates stronger inter-and intra-layer bonds (Attoye, Malekipour, & El-Mounayri, 2019; Product: Conductive PLA from Proto-Pasta, 2021; Redwood, Schöffer, & Garret, 2017).

Furthermore, the influence of the printing speed on the resistance is studied for four different speeds: 25, 50, 75, and 100 mm/s. The speed for the first layer is always 20 mm/s to ensure proper adhesion to the build platform. The printing temperature is 215 °C, and all other printing settings are listed in Table 2.2. In total of 10 specimens, for every printing, speed is prepared.

The results presented in Fig. 2.9 show that the resistance slightly decreases with increasing printing speed up to 50 mm/s. However, at printing speeds 75 mm/s and

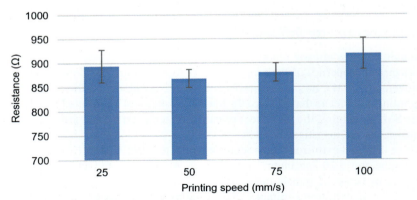

Fig. 2.9 Electrical resistance dependency from printing speed. No permission required.

higher, resistance starts to increase again. The reason for this unexpected behavior is still unexplained. Interestingly, the resulting resistance has a smaller scatter at printing speeds around 50–75 mm/s. Possibly, the best printing quality is achieved and thus the lowest porosity in the sample at these printing speeds. For comparison, standard PLA printing settings use a printing speed of 50–60 mm/s for most of the printing moves.

The highest performance of SMP structures requires the lowest possible electrical resistance of conductive CPLA traces, both in the element itself and at the connection point. The connection point itself must be sturdy up to temperatures <100°C. The method of contacting cables to the printed elements by applying conductive silver ink and using bolted connection together with ring-style crimp connectors provides a sturdy connection while producing the lowest resistance. From the CPLA experiments and the findings in the literature, the lowest resistance can be achieved by using high printing temperatures and medium printing speeds. The literature reports an almost linear increase of resistance over temperature up to 100°C for the same CPLA filament as used in our study (Al-Rubaiai et al., 2017). Furthermore, another literature source reports the same trend of increase of resistance with increasing temperature up to temperatures of 130°C from another conductive filament also made with the same filler—carbon black (Kwok et al., 2017).

Investigation of SMP structures

The SMP structure investigated here is a hinge-type basic element. It consists of PLA and CPLA. To achieve bending, the SMP structure has layers printed with different printing parameters of PLA. The layup builds a bi-layer-like structure that contains residual stresses after printing. A layerwise combination of PLA, printed with different printing settings, allows converting compressive in-plane stresses of PLA into out-of-plane bending. By adding conductive CPLA traces to the PLA layer, the structure can be electrically heated, thus yielding the deformation when the structure exceeds the T_g temperature.

When 4D printing a SMP structure with single-stimulus responsive material, it is important to produce layers with different directional strains (van Manen et al., 2018; Wang et al., 2019). Layers with unidirectional filling patterns exhibit anisotropic deformation behavior (Zolfagharian, Kaynak, Khoo, & Kouzani, 2018) in contrast to layers where the material is deposited in multiple directions, resulting in a quasi-isotropic behavior. A combination of these layers leads to an out-of-plane bending, twisting, or curling, depending on the direction of layers with the unidirectional filling pattern. Specific deformation type, twisting, for example, might be used to design a locking mechanism to enable single-activation shape-shifting and thus increase the achievable complexity and structural rigidity of 4D-printed geometries. Furthermore, the shrinkage of PLA (or amount of residual stresses stored in the material during printing) depends highly on printing parameters, such as printing speed, extrusion temperature, and others (Kačergis et al., 2019).

In the following, various activation voltages are investigated and the resulting temperatures on the surfaces of SMP structures are measured. After finding the proper activation voltage, five different designs of the SMP structure are used to find the most suitable layer to place the conductive CPLA layer. The most appropriate place for the CPLA layer is determined by investigating maximum free bending of the structures and blocking forces, produced by the SMP structures.

Design of SMP structure

Fig. 2.10 shows the design and dimensions of the SMP structure used in this study, which focuses on pure bending. The bending angle theta (Fig. 2.10) is used to characterize the bending. The SMP structure has a hinge area in the middle, i.e., the thinner section, thus being responsible for the bending. The thicker sides are printed so that their deformations during activation are negligible, which is achieved by simply using active PLA printing settings (Table 2.3) for two bottom layers and two top layers at nondeformable ends (Fig. 2.10). These stiff sides allow the SMP structure to be fixed and measurements to be taken.

Fig. 2.11 (left) shows an exploded view of the actuator used for initial performance characterization. The material PLA or CPLA of each layer is shown for a better understanding of the structure. The CPLA layer is printed with offsets of the edges, and the printing toolpath can be seen in detail (Fig. 2.11, right). Voids in the CPLA layer are filled with PLA, with either passive or active printing settings, depending on in which layer uses CPLA.

The structure has in total of six layers, each with a layer height of 0.2 mm. Two materials are used, plain PLA and CPLA. The CPLA is used only for the conductive traces, while plain PLA is used for both active and nonactive layers.

Thermal activation is achieved through a single CPLA layer in the structure, as a third layer, counting from the bottom (Fig. 2.11). The CPLA layer is designed to have a larger electrode area at the nondeformable ends to reduce resistance and thinner rasters in the hinge area to increase resistance and generate heating. This allows to vary the electrical resistance of printed CPLA paths to provide very low resistance at the ends and higher resistance at the hinge, so that heating is achieved only in the hinge.

Manufacturing of SMP structures

All specimens are produced using an Original Prusa i3 equipped with an MK3S Multi-Material Upgrade 2S unit (Prusa Research, Czechia). The investigated printing process is a single-step approach, where the entire SMP structure is printed in a single step (Fig. 2.12). The printing parameters are presented in Table 2.3. The printing parameters of CPLA are determined from the previous experiments in a subchapter: Printing parameter influence on resistance, where extrusion temperature of 233°C and printing speed of 50 mm/s showed the lowest resistance of printed CPLA.

At the nondeformable ends, we add a thin single layer of conductive silver ink on the CPLA layer to reduce the contact resistance of the SMP structure by about 25%. By reducing the total resistance of the structure, higher heating performance can be achieved.

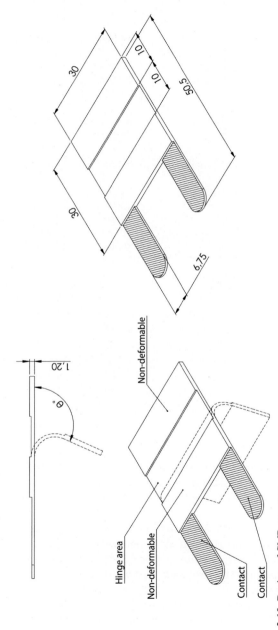

Fig. 2.10 Design of SMP structure. No permission required.

Table 2.3 Printing parameters used to print different segments of SMP structure.

Printing parameters	PLA (passive)	PLA (active)	CPLA
Printing speed (first layer)	n/a (no first layer)	80 mm/s	50 mm/s
Printing speed	20 mm/s	80 mm/s	50 mm/s
Acceleration	1000 m/s^2	1000 m/s^2	1000 m/s^2
Number of vertical shells (perimeters)	1	1	20
Infill fill pattern	Rectilinear	Rectilinear	Concentric
Infill/perimeters overlap	25%	25%	25%
Extrusion width	0.45 mm	0,45 mm	0.45 mm
Nozzle diameter	0.4 mm	0,4 mm	0.4 mm
Extrusion multiplier	1	1	1.2
Extruder temperature	215°C	215°C	235°C
Build plate temperature	30°C	30°C	30°C
Active extrusion cooling (first layer)	n/a (no first layer)	100%	100% (only relevant for CPLA 1)
Active extrusion cooling	0%	100%	100%

A rectilinear filling pattern with 100% of infill is used to create layers with no air gaps or any voids inside. However, micro-voids and very small air gaps are still possible due to the nature of the FDM process. Printing of each layer of the SMP structure starts with printing a single outer perimeter (vertical shells = 1) and then filling the layer (infill pattern = rectilinear, infill percentage = 100%). All active layers are filled parallel to bending orientation, all passive layers are printed perpendicular to bending orientation, and the CPLA layer is always printed as shown in Fig. 2.11.

Measuring setup

The maximum free bending angle and blocking force are measured in a specially designed test setup (Fig. 2.13). The test setup consists of a rotatable specimen holder, in which the SMP structure is clamped vertically during the tests to eliminate the influence of gravity.

The power is supplied with a Voltcraft VLP 2403pro power supply. The bending angle is extracted with GIMP software from photos taken from the above. A digital multimeter measures the current going into the SMP structure during activation.

The blocking force of SMP structures is measured using a test setup shown in Fig. 2.13 with a slight modification. A thin steel beam (cross section 25 mm × 0.25 mm) with a small steel dome attached at the bottom touches the bending SMP structure and deforms it back to the initial flat geometry. The force required to bend the specimen to its initial flat geometry is the blocking force. Once the SMP structure is activated (with an electric current) and starts to deform, it is rotated with the help of a rotatable specimen

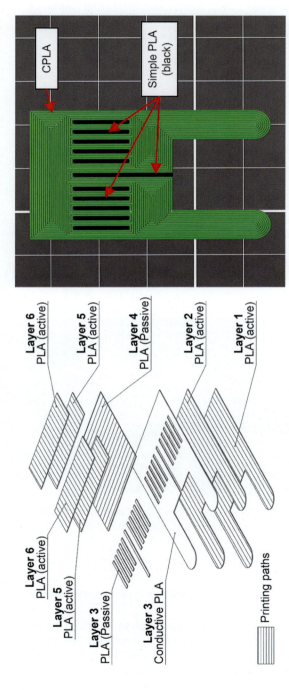

Fig. 2.11 Exploded view of SMP structure (left) and detailed composition of layer 3 (right). No permission required.

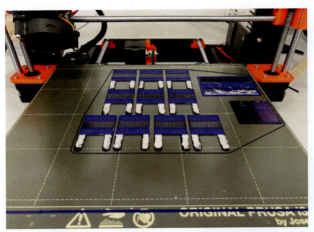

Fig. 2.12 Manufacturing of SMP structures. Blue material—PLA, black material—CPLA, gray material—conductive silver ink.
No permission required.

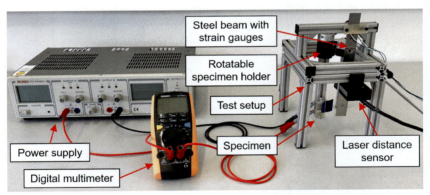

Fig. 2.13 Test setup for SMP structure investigation. Equipment used to activate the SMP structures and measure their deformation and bending force.
No permission required.

holder until the SMP structure touches the dome at the end of the steel beam (single point touch, see Fig. 2.14).

The force required to bend the specimen back to its flat geometry is calculated by measuring the strains on the steel beam with two strain gauges (1-LY11-3/120, HBK GmbH, Germany) and measuring the deflection of the beam with a help of a distance sensor (wenglorMel M7LL/10, 0.5 μm resolution, measuring range 10 mm, wenglorMEL GmbH, Germany) (Fig. 2.15). The force exerted by the SMP structure

4D printing electro-induced shape memory polymers

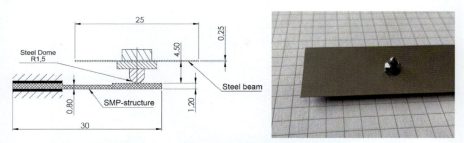

Fig. 2.14 Schematic of specimen interaction with steel dome during blocking force measurement (left) and photo of a steel dome (right).
No permission required.

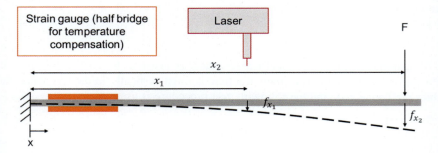

Fig. 2.15 Blocking force measuring principles. Measuring principle of steel beam to calculate blocking force.
No permission required.

at the bottom can be calculated with the beam stiffness and dimensions, the contact point of the SMP structure, and laser position. The force-deflection relation of the beam is calibrated with nine weights for a range between 10 and 50 mN before the experiment.

SMP structure behavior at different activation voltages

The initial experiments with the SMP structures investigate the heating performance of 3D-printed CPLA traces and the voltage required for activation. This measurement provides the lowest voltage required to activate the structures and the highest voltage, above which the structures are damaged. A total of five specimens are activated with voltages ranging from 50 to 65 V with 5 V increments. The current is not recorded in this experiment, but a readout from the power supply showed a current between 20 mA and 60 mA. A T-type thermocouple measures the temperature change on the surface of the SMP structure during heating and cooling in the center of the hinge area (Fig. 2.16).

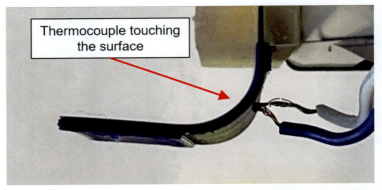

Fig. 2.16 Thermocouple attachment. Thermocouple touches the surface of the SMP structure to measure the change of temperature.
No permission required.

As an initial condition, the electrical resistance is measured for all test specimens in their flat form. Then one specimen at a time is fixed vertically in the test stand, the cables are connected, and the following test procedure is used:

1. Start temperature measurement;
2. Apply activation voltage (50 V) for 180 s;
3. Disable the power supply for 120 s;
4. Stop temperature measurement;
5. Take a photo from above to estimate the bending angle;
6. Measure resistance of the bent SMP structure, when the temperature reading of the thermocouple is below 23°C;
7. Starting over with *Step 2*, but using higher voltage (e.g., 55 V) until all voltages of interest are investigated.

The specimens are not deformed back to the initial shape (flat geometry) between measurements but rather left in their bent position. After cooling to 23°C, the resistance is measured. A higher voltage is then applied, and the process is repeated.

Fig. 2.17 shows the temperature development overtime at different voltages. As expected, the maximum temperature reached increases with higher voltage. Similar results are also reported in the literature (Garces & Ayranci, 2020). At each voltage studied, the SMP structures reach a temperature plateau after about 120 s of applied voltage, and the energy introduced by Joule heating and the energy dissipated into the environment form an equilibrium.

Fig. 2.18 shows the maximum free bending angle achieved and the electrical resistance of the SMP structures starting with the flat geometry, where the initial bending angle is zero. The bending angle increases with higher applied voltages. After Joule heating with 50 V, the bending angle is only 38°, whereas heating with 65 V gives the highest bending angle of 127°. The increase in bending is not proportional to the increase in applied voltage and reaches a plateau at higher voltages. This provides an indication of the limits of SMP structures investigated here. From the results, it

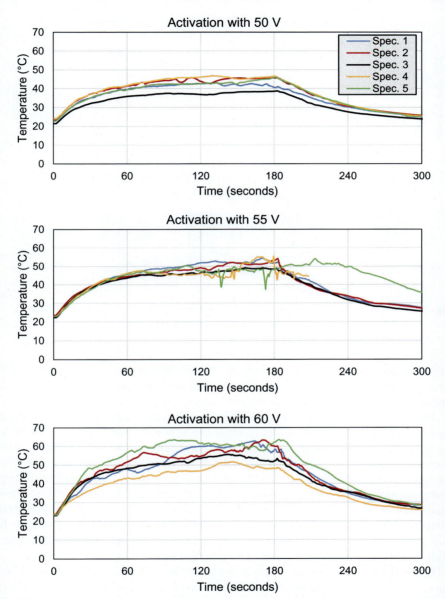

Fig. 2.17 Change of temperature in SMP structures over time by different activation voltages. No permission required.

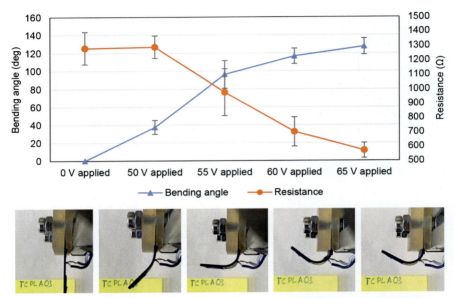

Fig. 2.18 Maximum bending angle and change in resistance of the structure at different activation voltages. Deformation after voltage application for 180 s.
No permission required.

can be concluded that activation with 60 V is the optimum voltage for these SMP structures investigated here, as it allows a high deformation of the structure in a few minutes and at the same time does not damage the structure.

Fig. 2.18 depicts the decrease in resistance over the increasing application voltage during the testing procedure. A similar trend can be found in the literature (Garces & Ayranci, 2020). The change in resistance is almost linear between 50 and 60 V. The decrease in resistance can be attributed to the partial melting of printed rasters and the change in geometry of the structure. The printed rasters touch each other after printing and are partially fused together. However, due to the nature of the FDM process, some small voids and air gaps can exist. Reheating with electric current results in additional partial melting occurs in the formation of a more uniform layer, reducing resistance. It might be that the SMP structures investigated here have to be activated multiple times until nearly constant resistance would be achieved.

Researchers also report a decrease in resistance due to bending of 3D-printed conductive filament layer (Elgeneidy et al., 2018). However, it is believed that the highest change in resistance originates from the effect of temperature and not from the deformation of the structure.

Conductive layer placement influence on the performance

The placement of the CPLA layer in the SMP structure decisively influences its shape variability, mainly because the CPLA layer has to replace one of the active or passive PLA layers.

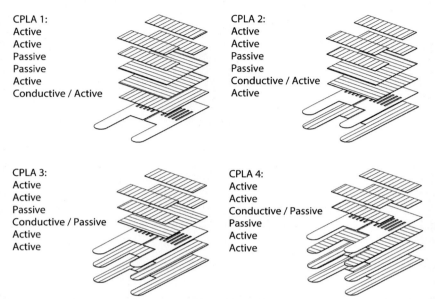

Fig. 2.19 Four different designs of the SMP structure. CPLA layer is varied from layer one to layer four. The fifth design investigated is not shown, because it is exactly as design number 3. No permission required.

To prove experimentally, a total of five different designs are investigated for their maximum free bending angle and blocking force. Four designs have different positions of the CPLA layer labeled 1 to 4 in Fig. 2.19, while the fifth design is exactly the same as CPLA 3, with the difference that CPLA is replaced by plain PLA. This design labeled PurePLA3 is activated in an oil bath with oil heated to 80°C. All other specimens are activated by 60 V for 90 s.

The performance characterization includes the measurement of the maximum free bending angle and the blocking force. For statistical validation, a total of 10 specimens per design are 3D-printed: five for measuring the maximum bending angle and five for measuring the blocking force. All ten specimens of the same design are manufactured in the same printing job. The current consumed by the samples during bending measurements is also recorded.

Free bending of SMP structures

Fig. 2.20 presents the results of the maximum free bending angle for different positions of the CPLA layer at the initial activation. The bending angle increases from CPLA position 1 to 3 and decreases again for position 4. On the other hand, the PurePLA3 design consisting only of plain PLA exhibits only average results, similar to the CPLA2 and CPLA4 designs.

PLA printed longitudinal to the bending direction and with high printing speeds is mainly responsible for the ability to change shape. This raster of shrinking PLA causes the bending of the structure and is referred to as active PLA. The middle section of the

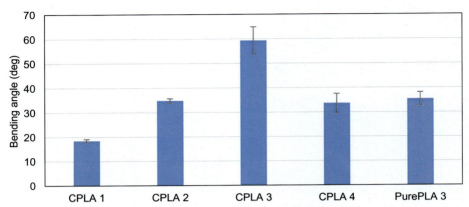

Fig. 2.20 Maximum free bending of different SMP structure designs. No permission required.

SMP structure contains PLA rasters printed transversely to bending direction and at lower printing speeds to have low or no prestress, here referred to as passive PLA. Such a configuration forms a bi-layer structure. The CPLA is not printed parallel to the bending direction, rather simply offsets the outside perimeter of the layer, with the raster parallel to the bending direction in some places, but perpendicular in others (see Fig. 2.11).

Since CPLA could also store and release residual stresses when heated above T_g, it must be printed parallel to the bending direction, with high printing speed and preferably with the lowest possible extrusion temperature. From previous studies, the printing parameters generating the highest prestrain in plain PLA do not produce the lowest resistance of CPLA. It can be assumed that CPLA does not release any prestress when heated due to the chaotic raster distribution in the layer and rather low printing speeds.

When the CPLA layer is at the bottom of the structure, the most prestrained plain PLA layer is exchanged with the nonprestressed CPLA layer, thus reducing the deformation of the whole SMP structure. By moving the CPLA layer to the second layer, the most prestrained plain PLA layer is on the bottom of the structure and generates a higher bending moment because it is the farthest from the neutral axis of the structure, which is in the center. On the other hand, when the CPLA layer is in the third position, it is close to the neutral axis, provides the biggest amount of heat to two bottom PLA layers with the highest prestress, and thus, the highest deformation of the structure is achieved. Moving the CPLA layer to the fourth layer position, the maximum deformation of SMP structure decreases because less heat reaches the lowest PLA layers.

Comparing results with the reference sample made completely of plain PLA, the performance increase between SMP structures of the same design can be observed. The internal heating indeed not only enables the activation of the structure with electric current but also increases the maximum free bending angle.

Fig. 2.21 shows the resistance of different SMP structure designs, before and after activation by 60 V for 90 s. The results show different resistance between the samples,

4D printing electro-induced shape memory polymers

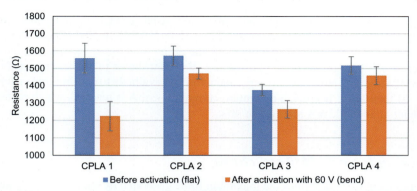

Fig. 2.21 Change in electrical resistance of different SMP structure designs. Before and after activation.
No permission required.

although all designs had the same electrode design and the same printing parameters. The only difference between designs is that the CPLA layer in design CPLA1 is printed directly on the build plate, instead of a previous layer. A slight miss-calibration of the first layer might change the height of it and thus influence adjacent raster bonding. A clear correlation between resistance and printing cannot be established at present.

Fig. 2.22 shows the current consumed by the CPLA3 design during activation. The red dashed line represents the change of theoretical resistance of the specimen CPLA3-9. This resistance is calculated from the measured current and applied voltage. When the current is applied to the structure, heating of the CPLA layer starts and in turn, the resistance of the CPLA layer starts to increase sharply over time. Other

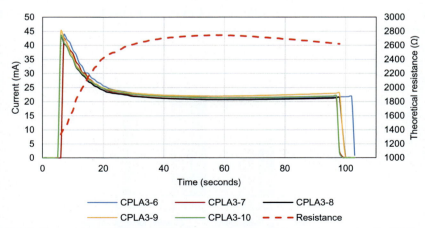

Fig. 2.22 Current consumption and change in resistance. Current consumption by SMP structure design CPLA3, when 60 V is applied for 90 s.
No permission required.

researchers have already reported the increase of resistance of CPLA with temperature (Al-Rubaiai et al., 2017; Kwok et al., 2017). Interestingly, the resistance starts to decrease after about 50–60 s of activation, and at the same time, the specimens start to deform. This slight decrease in resistance can be attributed to the bending angle of the SMP structure. However, no experiments are yet available to confirm this assumption. At the start of heating, a maximum current of 45 mA is reached, which remains constant at about 22 mA after about 40 s.

Blocking force of SMP structures

The blocking force describes the maximum force generated by the actuator. To measure it, the deformation of the sensor is completely blocked and the resulting force is measured. Besides the free stroke, it is the second central parameter for characterizing actuators (Sinapius, 2020).

To measure the blocking force, the respective SMP structure is activated by 60 V for 90 s to fully deform. Then, the specimen is rotated against a dome attached to a steel beam (Fig. 2.13) to measure the force required to push the SMP structure back to its flat position. The following sequence is used to measure the blocking force:

1. Apply activation voltage (60 V) with a power supply for 90 s;
2. Start strain and steel beam deflection measurement at second 85;
3. Turn off power supply;
4. Engage the SMP structure with the steel beam until the hot specimen is fully flattened.

The blocking force is derived from the measured strain and deflection of the steel beam and its calibrated force-deflection relation. Fig. 2.23 shows an increasing blocking force when the CPLA layer is moved toward the neutral axis of the SMP structure. However, inserting CPLA at layer 4 produces a slightly lower blocking force. On the other hand, the SMP structure without the CPLA layer produces only an average blocking force.

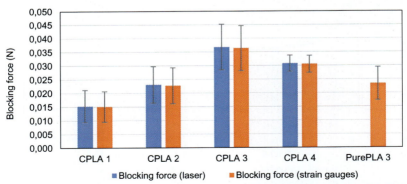

Fig. 2.23 Blocking force results. Blocking force generated by different designs of SMP structure. Measured by two measuring methods.
No permission required.

It must be noted that the results of blocking force are very similar to the maximum bending angle results (Fig. 2.20). Therefore, the results indicate that the addition of CPLA and application of Joule heating indeed produces SMP structures with higher blocking force and higher free bending angle. Unfortunately, blocking force for PurePLA3 specimens cannot be measured by laser because of its activation with hot oil. However, results show that both strain measurement and laser measurement produce very similar values, and therefore, only strain measurement to calculate blocking force is sufficient.

Conclusions

Recent advancements allow 3D printing of smart structures with shape-morphing capabilities. In this work, we present bi-layer SMP structures, factors influencing its morphing behavior, and methods to improve both its maximum capabilities and controllability. While the deformation of the SMP structures is currently only one way, the working principle of SMP structures produces self-healing behavior because, upon heating above T_g, it will always achieve a certain shape, defined by printing toolpaths and printing parameters.

The SMP structures consist of PLA material, where specific printing parameters of PLA introduce prestresses inside the material. High printing speed, low layer height, low extruder temperature, and cold build platform are just a few parameters of plain PLA that produce the highest prestressed and provide the highest deformation of SMP structures investigated here.

To activate the SMP structure, an environmental temperature above T_g (approx. 60°C) is required. In this study, we present a novel approach to add the CPLA layer into the structure to activate it with electric current by using the Joule heating principle, instead of other energy sources. With a help of CPLA, the application of 60 V heats the SMP structure and the shape-morphing of the structure occurs within a few minutes. Furthermore, structures with internal heating achieve higher performance than the same design structures made of a single, plain PLA material.

Extensive investigations of CPLA reveal that high printing temperatures and average printing speeds produce the lowest resistance for printed CPLA, which is important to achieve the maximum heating performance. Voltages lower than 60 V do not heat the structure enough to achieve maximum deformation, while higher voltages heat the structure too much until it starts to degrade/melt.

Different activation voltages applied on the same SMP structure produce a significant decrease in resistance. While from the first impression it could be attributed to changed geometry, most likely the activation of the structure with multiple voltages allows the CPLA to form a more uniform layer because of partial re-melting of extruded rasters. As a consequence, lower resistance is achieved. Furthermore, SMP structures with multiple activations at different voltages achieve almost twice the higher maximum free bending angle.

A study on CPLA layer position within the structure reveals the best performance of the SMP structure, both on a maximum free bending angle and on blocking force, with CPLA being in a third layer from six.

This study informs future investigators on selecting design and printing parameters of shape-morphing 4D printable structures. Moreover, insights reported here can help to understand the complex behavior of 3D-printed conductive filaments, especially for 3D printable internal heating and/or sensing applications.

Acknowledgments

The funding from German Research Foundation (DFG) for project "Printed Polymer Shape Memory Structures" under project number 447858794 is gratefully acknowledged.

References

Al-Rubaiai, M., Pinto, T., Qian, C., & Tan, X. (2019). Soft actuators with stiffness and shape modulation using 3D-printed conductive polylactic acid material. *Soft Robotics, 6*(3), 318–332. https://doi.org/10.1089/soro.2018.0056.

Al-Rubaiai, M., Pinto, T., Torres, D., Sepulveda, N., & Tan, X. (2017). *Characterization of a 3D-printed conductive PLA material with electrically controlled stiffness*. p. V001T01A003. https://doi.org/10.1115/SMASIS2017-3801.

An, B., Wu, H.-Y., Zhang, T., Yao, L., Tao, Y., Gu, J., et al. (2018). *Thermorph: Democratizing 4D printing of self-folding materials and interfaces* (pp. 1–12). https://doi.org/10.1145/3173574.3173834.

Attoye, S., Malekipour, E., & El-Mounayri, H. (2019). Correlation between process parameters and mechanical properties in parts printed by the fused deposition modeling process. In *Proceedings of the 2018 annual conference on experimental and applied mechanics* (pp. 35–41). https://doi.org/10.1007/978-3-319-95083-9_8.

Bodaghi, M., Damanpack, A. R., & Liao, W.-H. (2017). Adaptive metamaterials by functionally graded 4D printing. *Materials & Design, 135*. https://doi.org/10.1016/j.matdes.2017.08.069.

Bodaghi, M., Noroozi, R., Zolfagharian, A., Fotouhi, M., & Norouzi, S. (2019). 4D printing self-morphing structures. *Materials, 12*, 1353. https://doi.org/10.3390/ma12081353.

Bodaghi, M., Serjouei, A., Zolfagharian, A., Fotouhi, M., Rahman, H., & Durand, D. (2020). Reversible energy absorbing meta-sandwiches by 4D FDM printing. *International Journal of Mechanical Sciences, 173*. https://doi.org/10.1016/j.ijmecsci.2020.105451, 105451.

Ding, Z., Yuan, C., Peng, X., Wang, T., Qi, H., & Dunn, M. (2017). Direct 4D printing via active composite materials. *Science Advances, 3*. https://doi.org/10.1126/sciadv.1602890.

Ecker, J., Haider, A., Burzic, I., Huber, A., Eder, G., & Hild, S. (2019). Mechanical properties and water absorption behaviour of PLA and PLA/wood composites prepared by 3D printing and injection moulding. *Rapid Prototyping Journal, 25*. https://doi.org/10.1108/RPJ-06-2018-0149.

Elgeneidy, K., Neumann, G., Jackson, M., & Lohse, N. (2018). Directly printable flexible strain sensors for bending and contact feedback of soft actuators. *Frontiers in Robotics and AI, 5*. https://doi.org/10.3389/frobt.2018.00002.

FAQs of Multi3D. (2021). *Contact resistance improvement*. https://www.multi3dllc.com/faqs/.

Fernandes, J., Deus, A., Reis, L., Vaz, M. F., & Leite, M. (2018). Study of the influence of 3D printing parameters on the mechanical properties of PLA. In *2018. Progress in Additive Manufacturing (Pro-AM)* (pp. 547–552). Singapore: Nanyang Technological University. https://dr.ntu.edu.sg/handle/10220/45960.

Flowers, P., Reyes, C., Ye, S., Kim, M. J., & Wiley, B. (2017). 3D printing electronic components and circuits with conductive thermoplastic filament. *Additive Manufacturing, 18*. https://doi.org/10.1016/j.addma.2017.10.002.

Garces, I. T., & Ayranci, C. (2020). Active control of 4D prints: Towards 4D printed reliable actuators and sensors. *Sensors and Actuators A: Physical, 301*. https://doi.org/10.1016/j.sna.2019.111717, 111717.

Garcia, C., Kim, H., Duarte, M., Chavez, L., Castañeda, M., Tseng, B., et al. (2018a). Characterization of shape memory polymer parts fabricated using material extrusion 3D printing technique. *Rapid Prototyping Journal, 25*. https://doi.org/10.1108/RPJ-08-2017-0157.

Garcia, C., Kim, H., Duarte, M., Chavez, L., Tseng, B., & Lin, Y. (2018b). Toughness-based recovery efficiency of shape memory parts fabricated using material extrusion 3D printing technique. *Rapid Prototyping Journal, 25*. https://doi.org/10.1108/RPJ-09-2017-0188.

Ge, Q., Sakhaei, A. H., Lee, H., Dunn, C., Fang, N., & Dunn, M. (2016). Multimaterial 4D printing with tailorable shape memory polymers. *Scientific Reports, 6*, 31110. https://doi.org/10.1038/srep31110.

Goo, B., Hong, C.-H., & Park, K. (2020). 4D printing using anisotropic thermal deformation of 3D-printed thermoplastic parts. *Materials & Design, 188*. https://doi.org/10.1016/j.matdes.2020.108485, 108485.

Gu, J., Breen, D., Hu, J., Zhu, L., Tao, Y., Zande, T., et al. (2019). *Geodesy: Self-rising 2.5D Tiles by printing along 2D geodesic closed path* (pp. 1–10). https://doi.org/10.1145/3290605.3300267.

Hager, M., Bode, S., Weber, C., & Schubert, U. (2015). Shape memory polymers: Past, present and future developments. *Progress in Polymer Science, 49*. https://doi.org/10.1016/j.progpolymsci.2015.04.002.

Hu, G., Moqadam, A., & Liao, W.-H. (2017). Increasing dimension of structures by 4D printing shape memory polymers via fused deposition modeling. *Smart Materials and Structures, 26*. https://doi.org/10.1088/1361-665X/aa95ec.

Ionov, L. (2015). Polymeric actuators. *Langmuir, 31*(18), 5015–5024. https://doi.org/10.1021/la503407z.

Janbaz, S., Hedayati, R., & Zadpoor, A. (2016). Programming the shape-shifting of flat soft matter: From self-rolling/-twisting materials to self-folding origami. *Materials Horizons, 3*. https://doi.org/10.1039/C6MH00195E.

Jeon, O., Bouhadir, K., Mansour, J., & Alsberg, E. (2009). Photocrosslinked alginate hydrogels with tunable biodegradation rates and mechanical properties. *Biomaterials, 30*, 2724–2734. https://doi.org/10.1016/j.biomaterials.2009.01.034.

Jeon, S., Hauser, A. W., & Hayward, R. C. (2017). Shape-morphing materials from stimuli-responsive hydrogel hybrids. *Accounts of Chemical Research, 50*, 161–169. https://doi.org/10.1021/acs.accounts.6b00570.

Jeong, H., Woo, B., Kim, N., & Jun, Y. (2020). Multicolor 4D printing of shape-memory polymers for light-induced selective heating and remote actuation. *Scientific Reports, 10*, 6258. https://doi.org/10.1038/s41598-020-63020-9.

Kačergis, L., Mitkus, R., & Sinapius, M. (2019). Influence of fused deposition modeling process parameters on the transformation of 4D printed morphing structures. *Smart Materials and Structures, 28*(10). https://doi.org/10.1088/1361-665x/ab3d18, 105042.

Kim, M. J., Cruz, M. A., Ye, S., Gray, A. L., Smith, G. L., Lazarus, N., et al. (2019). One-step electrodeposition of copper on conductive 3D printed objects. *Additive Manufacturing, 27*, 318–326. https://doi.org/10.1016/j.addma.2019.03.016.

Kim, Y., Yuk, H., Zhao, R., Chester, S., & Zhao, X. (2018). Printing ferromagnetic domains for untethered fast-transforming soft materials. *Nature, 558*, 274–279. https://doi.org/10.1038/s41586-018-0185-0.

Kuang, X., Roach, D., Wu, J., Hamel, C., Ding, Z., Wang, T., et al. (2018). Advances in 4D printing: Materials and applications. *Advanced Functional Materials, 29*, 1805290. https://doi.org/10.1002/adfm.201805290.

Kwok, S. W., Goh, K. H. H., Tan, Z. D., Tan, S. T. M., Tjiu, W. W., Soh, J. Y., et al. (2017). Electrically conductive filament for 3D-printed circuits and sensors. *Applied Materials Today, 9*, 167–175. https://doi.org/10.1016/j.apmt.2017.07.001.

Lendlein, A., & Shastri, V. P. (2010). Stimuli-sensitive polymers. *Advanced Materials, 22*(31), 3344–3347. https://doi.org/10.1002/adma.201002520.

Manen, T., Janbaz, S., & Zadpoor, A. (2017). Programming 2D/3D shape-shifting with hobbyist 3D printers. *Materials Horizons, 4*. https://doi.org/10.1039/C7MH00269F.

Mao, Y., Ding, Z., Yuan, C., Ai, S., Isakov, M., Wu, J., et al. (2016). 3D printed reversible shape changing components with stimuli responsive materials. *Scientific Reports, 6*, 24761. https://doi.org/10.1038/srep24761.

Nishiguchi, A., Zhang, H., Schweizerhof, S., Schulte, M. F., Mourran, A., & Möller, M. (2020). 4D printing of a light-driven soft actuator with programmed printing density. *ACS Applied Materials & Interfaces, 12*(10), 12176–12185. https://doi.org/10.1021/acsami.0c02781.

Noroozi, R., Bodaghi, M., Jafari, H., Zolfagharian, A., & Fotouhi, M. (2020). Shape-adaptive metastructures with variable bandgap regions by 4D printing. *Polymers, 12*(3). https://doi.org/10.3390/polym12030519.

Pei, E., & Loh, G. H. (2018). Technological considerations for 4D printing: An overview. *Progress in Additive Manufacturing, 3*(1), 95–107. https://doi.org/10.1007/s40964-018-0047-1.

Pizarro Torres, F., Salazar, R., Rajo-Iglesias, E., Rodríguez, M., Fingerhuth, S., & Hermosilla, G. (2019). Parametric study of 3D additive printing parameters using conductive filaments on microwave topologies. *IEEE Access, 7*, 1. https://doi.org/10.1109/ACCESS.2019.2932912.

Product: Conductive PLA from Proto-Pasta. (2021). https://www.proto-pasta.com/products/conductive-pla.

Raviv, D., Zhao, W., McKnelly, C., Papadopoulou, A., Kadambi, A., Shi, B., et al. (2014). Active printed materials for complex self-evolving deformations. *Scientific Reports, 4*, 7422. https://doi.org/10.1038/srep07422.

Redwood, B., Schöffer, F., & Garret, B. (2017). *The 3D printing handbook: Technologies, design and applications*. 3D Hubs.

Reyes, C., Somogyi, R., Niu, S., Cruz, M., Yang, F., Catenacci, M., et al. (2018). 3D printing a complete lithium ion battery with fused filament fabrication. *ACS Applied Energy Materials, 1*. https://doi.org/10.1021/acsaem.8b00885.

Roy, S., Qureshi, M., Asif, S., & Braaten, B. (2017). *A model for 3D-printed microstrip transmission lines using conductive electrifi filament*. https://doi.org/10.1109/APUSNCURSINRSM.2017.8072592.

Ryan, K., Down, M., & Banks, C. (2020). Future of additive manufacturing: Overview of 4D and 3D printed smart and advanced materials and their applications. *Chemical Engineering Journal, 403*. https://doi.org/10.1016/j.cej.2020.126162, 126162.

Schmolke, H. (2018). *DIN VDE 0100*. VDE Verlag.

Shao, L.-H., Zhao, B., Zhang, Q., Xing, Y., & Zhang, K. (2020). 4D printing composite with electrically controlled local deformation. *Extreme Mechanics Letters, 39*. https://doi.org/10.1016/j.eml.2020.100793, 100793.

Sinapius, M. (2020). *Adaptronics—Smart structures and materials* (1st ed.). Springer Vieweg. https://doi.org/10.1007/978-3-662-61399-3.

Tham, W., Chow, W. S., Poh, B., & Ishak, Z. (2016). Poly(lactic acid)/halloysite nanotube nanocomposites with high impact strength and water barrier properties. *Journal of Composite Materials, 50*. https://doi.org/10.1177/0021998316628972.

Tibbits, S. (2014). 4D printing: Multi-material shape change. *Architectural Design*, *84*. https://doi.org/10.1002/ad.1710.

van Manen, T., Janbaz, S., & Zadpoor, A. A. (2018). Programming the shape-shifting of flat soft matter. *Materials Today*, *21*(2), 144–163. https://doi.org/10.1016/j.mattod.2017.08.026.

Wahit, M. U., Hassan, A., Akos, N., Zawawi, N., & Kunasegeran, K. (2015). Mechanical, thermal and chemical resistance of epoxidized natural rubber toughened polylactic acid blends. *Sains Malaysiana*, *44*, 1615–1623.

Wang, G., Cheng, T., Do, Y., Yang, H., Tao, Y., Gu, J., et al. (2018). *Printed paper actuator: A low-cost reversible actuation and sensing method for shape changing interfaces* (pp. 1–12). https://doi.org/10.1145/3173574.3174143.

Wang, G., Tao, Y., Capunaman, O., Yang, H., & Yao, L. (2019). *A-line: 4D printing morphing linear composite structures* (pp. 1–12). https://doi.org/10.1145/3290605.3300656.

Wittbrodt, B., & Pearce, J. (2015). The effects of PLA color on material properties of 3-D printed components. *Additive Manufacturing*, *8*. https://doi.org/10.1016/j.addma.2015.09.006.

Xu, J., Song, J., & Yahia, L. (2015). 10-Polylactic acid (PLA)-based shape-memory materials for biomedical applications. In *Woodhead publishing series in biomaterials* (pp. 197–217). Woodhead Publishing. https://doi.org/10.1016/B978-0-85709-698-2.00010-6.

Yang, C., Wang, B., Li, D., & Tian, X. (2017). Modelling and characterisation for the responsive performance of CF/PLA and CF/PEEK smart materials fabricated by 4D printing. *Virtual and Physical Prototyping*, 1–8. https://doi.org/10.1080/17452759.2016.1265992.

Yang, Y., Chen, Y., Li, Y., Wang, Z., & Li, Y. (2017). Novel variable-stiffness robotic fingers with built-in position feedback. *Soft Robotics*, *4*(4), 338–352. https://doi.org/10.1089/soro.2016.0060.

Zafar, M. Q., & Zhao, H. (2020). 4D printing: Future insight in additive manufacturing. *Metals and Materials International*, *26*(5), 564–585. https://doi.org/10.1007/s12540-019-00441-w.

Zhang, Q., Yan, D., Zhang, K., & Hu, G. (2015). Pattern transformation of heat-shrinkable polymer by three-dimensional (3D) printing technique. *Scientific Reports*, *5*(1), 8936. https://doi.org/10.1038/srep08936.

Zhang, Q., Zhang, K., & Hu, G. (2016). Smart three-dimensional lightweight structure triggered from a thin composite sheet via 3D printing technique. *Scientific Reports*, *6*, 22431. https://doi.org/10.1038/srep22431.

Zhang, Y., Zhang, F., Yan, Z., Ma, Q., Li, X., Huang, Y.-S., et al. (2017). Printing, folding and assembly methods for forming 3D mesostructures in advanced materials. *Nature Reviews Materials*, *2*, 17019. https://doi.org/10.1038/natrevmats.2017.19.

Zolfagharian, A., Denk, M., Bodaghi, M., Kouzani, A. Z., & Kaynak, A. (2020). Topology-optimized 4D printing of a soft actuator. *Acta Mechanica Solida Sinica*, *33*(3), 418–430. https://doi.org/10.1007/s10338-019-00137-z.

Zolfagharian, A., Kaynak, A., Khoo, S. Y., & Kouzani, A. (2018). Pattern-driven 4D printing. *Sensors and Actuators A: Physical*, *274*, 231–243. https://doi.org/10.1016/j.sna.2018.03.034.

Zolfagharian, A., Kouzani, A. Z., Maheepala, M., Yang Khoo, S., & Kaynak, A. (2019). Bending control of a 3D printed polyelectrolyte soft actuator with uncertain model. *Sensors and Actuators A: Physical*, *288*, 134–143. https://doi.org/10.1016/j.sna.2019.01.027.

4D printing modeling using ABAQUS: A guide for beginners

Hamid Reza Jarrah[a], Ali Zolfagharian[b], Bernard Rolfe[b], and Mahdi Bodaghi[a]
[a]Department of Engineering, School of Science and Technology, Nottingham Trent University, Nottingham, United Kingdom
[b]School of Engineering, Deakin University, Geelong, VIC, Australia

Introduction

Typically, smart materials, such as shape memory polymers (SMP)s (Bodaghi, Damanpack, & Liao, 2016, 2017; Bodaghi et al., 2020) and hydrogels (Zolfagharian et al., 2016a, 2019), are used to impart additional functionality to three-dimensional (3D) structures, such as shape change in response to a stimulus such as temperature, light, or electricity (Zolfagharian et al., 2016b). As a result, many elements of the four-dimensional (4D) printing continue to be studied at a greater degree of study via sophisticated multiphysic modeling principles, finite element method (FEM) simulations, and experimental testing.

4D printing is a potential technique for regulating self-morphing in custom-designed structures (Bodaghi et al., 2019; Zolfagharian et al., 2017, 2018). This chapter aims to provide a realistic demonstration of 4D printing control parameters such as geometrical design and stimulus effects on the bending angle of the 4D-printed structures, including physics principles, and a self-learning steps of FEM simulations procedure.

The liquid crystal gel-phase transition, thermal expansion coefficient, thermal conductivity discrepancies, and different swelling and de-swelling ratios of bi-layer or composite beams may all be utilized in 4D printing to create spatial and temporal changes (Kamal & Park, 2014; Zeng et al., 2017). The direct and indirect use of active materials, such as SMPs, is yielding encouraging results. Numerous stimuli, including electrical, magnetic, heat, light, pH, and moisture, cause these materials to exhibit elastic deformation (de Haan et al., 2014; Tolley et al., 2014; Zarek et al., 2015).

Light stimulation is a useful way to generate heat via photothermal conversion, as it can be controlled remotely, instantly switches on and off, and utilizes clean and sustainable energy. Sunflowers are an excellent example of light-activated 4D-printed soft actuators, since they fold and unfold in response to sunlight (Atamian et al., 2016). Long-range irradiation enables bending without human intervention, which is advantageous for a variety of applications, including medicinal release, packaging, and assembly of manufactured items (Nakata et al., 2012; Sumaru et al., 2013).

In this case study, a simple 4D printing approach is introduced to transform a flat sheet into a spatial structure using near infrared (NIR) light. An FEM modeling of the SMP substrate and 4D-printed patterns has been developed to predict the final shape of the 4D-printed structures. The study provides guidelines for the learning principles of 4D printing, design, modeling, and FEM simulation using ABAQUS software. The polystyrene (PS) film, commercially known as Shrink Film, is used to demonstrate the thermoresponsive properties of SMP as a substrate material here. The active hinges are supposed to be dark hydrogel paste, which is printed in a variety of designs on PS film. When the NIR lamp emits light, the printed black pattern absorbs more light than the clear PS substrate, which acts as a heat source, resulting in localized heating. This results in the underlying printed region heating up quicker than the unprinted area, which results in actuation owing to thermal stress gradients along the height of the film. As a result, the film shrinks locally at active hinges where the temperature exceeds that of the glass transition, resulting in bending. Bending measurements are shown to be influenced by 3D printing characteristics, such as geometrical parameters.

The rest of this chapter is organized as follows. "Methodology" section describes the methodology, including the mechanism and design of the 4D-printed structure. The following section explains the modeling principles and details of FEM simulation steps in ABAQUS. "Results and discussions" section presents the simulation results and discussions. "Conclusion" section concludes this chapter.

Methodology

4D printing mechanism and design

The 4D printing modeling is decided on prestrained PS film, Shrink Film, since it is easily accessible, inexpensive, translucent, and has the necessary stiffness. Prestrained Shrink Film is an extruded polystyrene sheet that has been chilled below its glass transition temperature (Tg) while being stretched above it, and then annealed. These materials release their internal tension when heated over Tg. According to this study, shrinkage may reach up to 60% when the Tg is at 102°C (Liu et al., 2012). A simple method of heat absorption is to employ dark-colored patterns that absorb NIR and serve as actuator hinges. As a consequence, a material capable of high light absorption and heat conversion is needed to serve as a hinge and initiate actuation. The internal temperature of printed materials is increased due to the nonradiative decay of light absorption. In other words, raising the local temperature over the glass transition temperature in the printed region of the backbone polymer causes the actuator to bend gradually. The temperature responses of the polymer substrate and the hinge are mismatched, resulting in structural morphing. Bending of the PS film occurs because of heat absorption-induced shrinking at Tg. When polymer chains in amorphous regions are heated to the glass transition temperature, the specific heat capacity increases, but the elastic modulus falls substantially.

Earlier studies (Zolfagharian et al., 2017) showed that different thicknesses and widths of hinges lead to sequential bending in 4D-printed structure under even light intensity. In this study therefore, the effects of different 4D printing patterns and the

4D printing modeling using ABAQUS: A guide for beginners

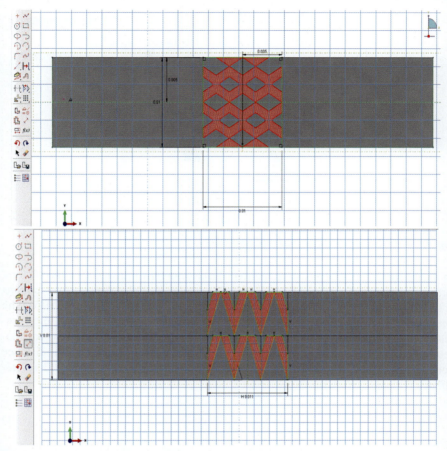

Fig. 3.1 The printed honeycomb (top) and zigzag (bottom) hinge patterns with same surface area.

number of layers for controlling the bending angle are focused while keeping the amount of printed hinge material same for a fair comparison's sake. Therefore, the same surface area with different patterns, namely, honeycomb and zigzag, is drawn from the center with assumed equal extrudate thicknesses as shown in Fig. 3.1. First, the PS films are assumed in the same dimensions of 50 mm × 10 mm. Then the center of the film is marked. Then, several hinge designs with the same quantity of printed ink and strand thicknesses are designed in a 10 mm × 10 mm square on the PS film.

Modeling of thermo-mechanical 4D-printed structure

The 4D printing behavior in this work is studied in terms of the bending angle control of the structure, when exposed to the IR light, in terms of thickness of the printed pattern, representing the number of layers, and the pattern designs on the PS film's center.

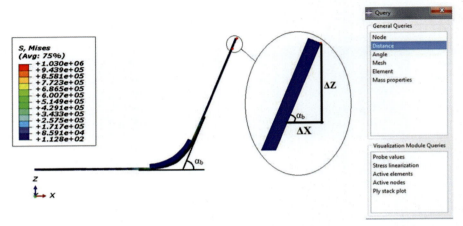

Fig. 3.2 Calculation of bending angle in ABAQUS.

In establishing the analysis of actuation, the assumption is made that the binding between the printed layer and PS film is strong enough such that the interface remains at the same length while the assembly is just being curved (Deng & Chen, 2015). The bending angle will be measured in the FEM software package of ABAQUS after simulation. Fig. 3.2 shows the bending angle definition in ABAQUS. Two points are chosen from PS film in the software, and then with using "Query" command, the distance between these points is measured. The simple trigonometric equation, Eq. (3.1), calculates the angle as:

$$\tan \alpha_b = \frac{\Delta Z}{\Delta x} \tag{3.1}$$

Heat generation and temperature rise due to NIR light

It is critical to know the temperature of the actuator surface since that controls the deflection of the actuator. When the polymer film is believed to be thin enough, heating it from one side with an IR light may result in a uniform temperature increase. Assuming all heating power of IR lamp (P_{IR}) is passed on to the polymer surface, the amount of heating power (q), which is able to pass through the polymer surface with the coefficient of absorbance α that is transmitted to the polymer surface and yields an increase in temperature as:

$$q = P_{IR}\alpha = \frac{m}{A} c_p \frac{dT}{dt} \tag{3.2}$$

where the mass, area, temperature, and the specific heat of PS film are m, A, T, and c_p, respectively. Substituting the $m = \rho h A$, where ρ and h are the density and thickness of film, respectively, in Eq. (3.2), and integrating from both sides result in the total heating time (t_{IR}) required for the polymer surface temperature to reach to T_f, where T_0 and T_f are initial and final temperatures of film, respectively:

$$\int_0^{t_{IR}} dt = \frac{\rho h c_p}{P_{IR}\alpha} \int_{T_0}^{T_f} dT \tag{3.3}$$

$$t_{IR} = \frac{\rho h c_p}{P_{IR}\alpha}(T_f - T_0). \tag{3.4}$$

A model that shows the temperature fluctuation throughout the thickness of the PS layer over time provides further insight into actuation control. This leads to the development of a model influenced by (Deng et al., 2017). The thermal conduction differential equation is restricted only on XZ plane governed by Eq. (3.5) in which T is the temperature field and r' is the distance between any point and heating source on the surface, t is the time, and a is the thermal diffusivity of the film. Assuming the surface temperature is constant and ignoring the surface diffusivity as it is far lower than the thermal conduction through the Z-axis (thickness direction) (Zolfagharian et al., 2017). In small-width patterns, solving Eq. (3.5) results in the temperature distribution function that equals to Eq. (3.6), where L is the printed pattern length, Q_0 is the absorbed heat from IR irradiation on surface as a single point in middle of printed pattern, and λ is the thermal conductivity of the film:

$$\frac{dT_{r'}}{dt} = a\left(\frac{\partial^2 T}{\partial x^2} + \frac{\partial^2 T}{\partial z^2}\right), \tag{3.5}$$

$$T_{r'} = \frac{Q_0}{4\pi\lambda L t} \exp\left(-\frac{r'^2}{4at}\right) \quad \text{where } \left(r' = \sqrt{x^2 + z^2}\right) \tag{3.6}$$

The radial position of the same temperatures might be derived on basis of Eq. (3.6) illustrated in Fig. 3.3. The irradiation heat, $P_{IR}\alpha \times t$ could be included into the Eq. (3.6) as:

$$T_{r'} = \frac{P_{IR}\alpha}{4\pi\lambda L} \exp\left(-\frac{r'^2}{4at}\right). \tag{3.7}$$

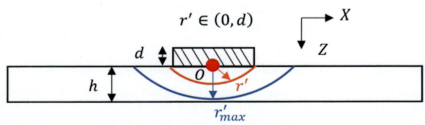

Fig. 3.3 The PS film surface temperature characteristics during IR irradiation.

Table 3.1 Specimen's physical and material properties of PS film.

Property	Value
Thermal conductivity (W/m K)	0.14
NIR intensity (mW/mm^2)	3.50
Glass temperature (K)	375.15
Density (kg/m^3)	1050
Distance of IR source to surface of PS (mm)	250
Onset temperature (K)	323.15
Thickness of PS (mm)	0.30
Width of PS (mm)	10
Length of PS (mm)	50
Specific heat (J/kg K)	1300

In the single-point model, the maximum temperature occurs on the surface as $T_{max} = \frac{P_{IR}\alpha}{4\pi\lambda L}$. The time necessary to achieve a certain temperature for the lower surface with position ($x = 0$, $z = h$) may also be computed as:

$$t = \frac{-h^2}{4a \times \ln\left(\frac{4\pi\lambda L T_{z'}}{P_{IR}\alpha}\right)}. \tag{3.8}$$

In determining the temperature distribution of PS film over time, the following assumptions are taken into account: (1) the black printed hinge pattern is a heat source comparable to the lamp's local intensity, (2) the absorption of the light by the printed pattern is the sole heat source, (3) the Shrink Film's exposed surface absorbs light by 15% pattern lines of absorption, (4) initial polymer film temperatures are the same with starting temperatures or ambient temperatures, and (5) the polymer film thermal conversion and heat capacity are indicated in Table 3.1. At 250 mm from the middle point at the top of actuators, light intensity was fixed at 3.50 mW/mm^2. The comparison results from an increase in temperature owing to NIR irradiation from the starting temperature (323.15 K) are presented in Fig. 3.3. Results show that the center point at above the actuator in the square pattern reached to the transition glass temperature of (375.15 K) quicker than the other two. The maximum temperature should reach the film transitional glass temperature to cause bending behavior (i.e., $T_{max} > T_g = 102\,°C$). To control the final shape accurately, it is critical to control the irradiation duration to attain a temperature of glass above the actuator, while the bottom of the actuator is more than 120°C.

FEM model of the thermal-mechanical coupling in ABAQUS

In ABAQUS, the FEM models are generated. PS film and 3D-printed design geometry are designed in CAD, then exported to ABAQUS. The default physical heat transfer model in solids is used, and boundary conditions and sources are set. Based on the parameters in Table 3.2, the material library is utilized and updated. Initial

Table 3.2 Material properties of PS film and printed hydrogel.

E (MPa)	Conductivity (W/m K)	Density (g/cm³)	Specific heat (J/kg K)
PS			
Fig. 3.5	0.14	1050	1300
Hydrogel			
0.7	1.7	1200	1050

temperature is set at 323.15 K, as the ambient temperature of the PS film is on the hotplate, at the time of beginning of irradiation of NIR source. A 3D model and a free tetrahedral mesh are used for the simulations.

There is a coupling between thermal and mechanical load in modeling of thermal-responsive 4D printing. For the case of thermo-elastic problems, Hook's law can be written for an isotropic material as:

$$\sigma = -\beta \Delta T a + D \varepsilon \tag{3.9}$$

where ΔT, D, β, and ε are temperature change, matrix of elastic constants, thermal stress modulus, and strain, respectively. Also, a and β are given by:

$$a^T = [1, 1, 1, 0, 0, 0] \tag{3.10}$$

$$\beta = \frac{E\alpha}{(1 - 2\nu)} \tag{3.11}$$

that E, ν, and α are Young's modulus, Poisson's ratio, and the thermal expansion coefficient, respectively.

The transient heat conduction equation for a 3D model in Cartesian coordinate can be written according to

$$\frac{\partial}{\partial x}\left(k_x(T)\frac{\partial T}{\partial x}\right) + \frac{\partial}{\partial y}\left(k_y(T)\frac{\partial T}{\partial y}\right) + \frac{\partial}{\partial z}\left(k_z(T)\frac{\partial T}{\partial z}\right) + G = \rho C_P \frac{\partial T}{\partial t} + \beta T \varepsilon_V \tag{3.12}$$

where $k_x(T)$, $k_y(T)$, and $k_z(T)$ are the temperature-dependent thermal conductivities in the x, y, and z directions, respectively. G, T, t, ρ, C_P, and ε_V, are the heat generation per unit volume, temperature, time, density, specific heat, and volume strain, respectively. Boundary and initial conditions should be considered for defining of problem. Additionally, thermoelastic deformations of interest occur during the time interval $[0, t]$.

The FEM is a numerical tool for determining approximate solutions to a large class of engineering problems. In this method, the solution region divides to non-overlapping elements. Each element is formed by the connection of a certain number of nodes. In the next step, the matrix equations that express the properties of the individual elements will be determined. These matrices involve element left-hand side (LHS[k]) matrix and load vector ([f]). The global matrix for whole region should be assembled with considering LHS and load vector for each element as:

$$[K][T] = [f] \tag{3.13}$$

where [T] is the global unknown vector and

$$T(x, y, z, t) = \sum_{i=1}^{n} N_i(x, y, z) T_i(t) \tag{3.14}$$

in which N_i are the shape functions, n is the number of nodes in an element, and T_i shows the temperature in each node of an element. There are different methods for solving this equation, and Galerkin method is chosen in this study. The Galerkin method applied to Eq. (3.12) could be expressed as:

$$-\int N_i \left[\frac{\partial}{\partial x} \left(k_x(T) \frac{\partial T}{\partial x} \right) + \frac{\partial}{\partial y} \left(k_y(T) \frac{\partial T}{\partial y} \right) + \frac{\partial}{\partial z} \left(k_z(T) \frac{\partial T}{\partial z} \right) + G - \rho C_P \frac{\partial T}{\partial t} - \beta T \varepsilon_V \right] d\Omega$$

$$+ \int N_i k_x(T) \frac{\partial T}{\partial x} l d\Gamma_q + \int N_i k_y(T) \frac{\partial T}{\partial y} m d\Gamma_q + \int N_i k_z(T) \frac{\partial T}{\partial z} n d\Gamma_q = 0 \tag{3.15}$$

where Ω and Γ are the domain of integration and boundary of domain, respectively. Also l, m, and n are the direction cosines of the appropriate outward surface normals. Additionally, the material is assumed isotropic so thermal conductivities where $k_x = k_y = k_z$. Incorporating the temperature relation in Eq. (3.14) into Eq. (3.15) yields

$$-\int_\Omega k \left[\frac{\partial N_i}{\partial x} \frac{\partial N_j}{\partial x} T_j(t) + \frac{\partial N_i}{\partial y} \frac{\partial N_j}{\partial y} T_j(t) + \frac{\partial N_i}{\partial z} \frac{\partial N_j}{\partial z} T_j(t) \right] d\Omega$$

$$+ \int \left[N_i G - N_i \rho C_P \frac{\partial N_j}{\partial t} T_j(t) - N_i \beta T_j(t) \varepsilon_V \right] d\Omega = 0 \tag{3.16}$$

where i and j represent the nodes in x and y directions, respectively. The final form of Eq. (3.16) can be written as:

$$[C_{ij}] \left\{ \frac{\partial T_j}{\partial t} \right\} + [K_{ij}] \{T_j\} = \{f_i\} \tag{3.17}$$

where

$$[C_{ij}] = \int \rho C_P N_i N_j d\Omega \tag{3.18}$$

$$[K_{ij}] = k \left[\frac{\partial N_i}{\partial x} \frac{\partial N_j}{\partial x} T_j(t) + \frac{\partial N_i}{\partial y} \frac{\partial N_j}{\partial y} T_j(t) + \frac{\partial N_i}{\partial z} \frac{\partial N_j}{\partial z} T_j(t) \right] d\Omega \qquad (3.19)$$

$$\{f_i\} = \int N_i \left[G - \beta T_j(t) \varepsilon_V \right] d\Omega \qquad (3.20)$$

The solution trend in the finite element software is shown in Fig. 3.4 flowchart. This loop repeats as per the flowchart until the desired result is achieved.

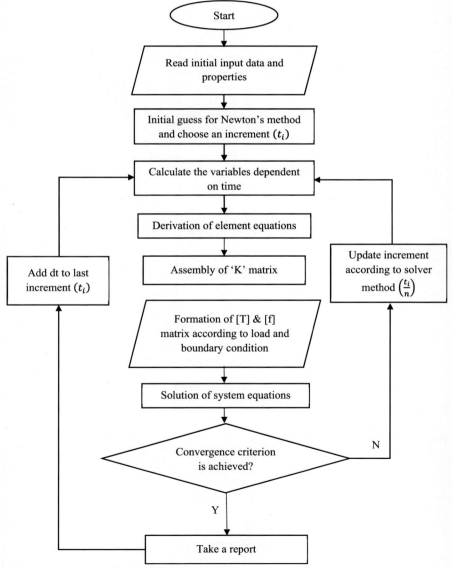

Fig. 3.4 Solution procedure in the FEM software.

The simulation procedures for 4D printing with various designs will be given here. The simulation of a model in the ABAQUS software consists of several steps. These procedures are summarized as follows:

1. *Create geometry of model*
   This section will create the geometry of the model based on the dimensions of the samples. The study made use of a three-dimensional model and a solid element. Once the method for model creation is selected, the sketch module is enabled till the sketch is completed. This task will be carried out in the part module. Fig. 3.1 illustrates the tools required to create geometry.
2. *Define property and assign it to the model*
   This section defines the material properties in ABAQUS. Table 3.2 shows the material property for both. These data will be input in the property module. Then, the specified properties will be assigned to the models. Some properties are temperature dependent. Since in this model the heating is not negligible, the relation between these properties and temperature should be defined. Fig. 3.5 demonstrates the relation between Young's modulus and temperature for PS film (Mailen et al., 2017). The most important of this step is defining thermal expansion coefficient for both materials. Already, the material has positive thermal expansion coefficient, but the film has some prestrain. The prestrain is reflecting the negative thermal expansion coefficient. Hence, a temperature-dependent negative thermal expansion is defined as shown in Table 3.3 for both hydrogel and PS. Fig. 3.6 shows the environment of this module in ABAQUS.

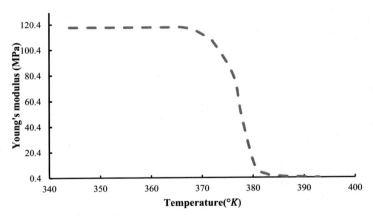

Fig. 3.5 Young's modulus of PS film.

Table 3.3 Thermal expansion coefficient.

	Temperature (K)				
		273	376.5	395	410	420
Thermal expansion coefficient (α) ($10^{-4} \times 1/K$)	PS					
		0	0	−4.5	−4.5	−4.5
	Hydrogel					
		0	0	−8	−8	−8

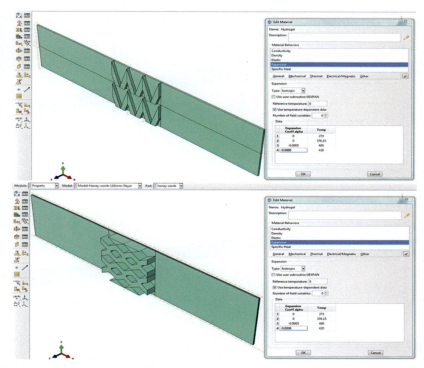

Fig. 3.6 Setting the materials properties.

3. *Define the solver*

Since in this study the mechanical load is the consequence of heating, a solver coupling mechanical loading and heat transfer is required. The best solver for this problem is "coupled temp-displacement." As shown in Fig. 3.7, a transient solution considering the time-dependent of temperature is chosen.

4. *Define loads and boundary conditions*

Defining right boundary conditions and loads are conducted in this section. The type of loading is heat flux. The major part of heat flux will be absorbed by the hydrogel, and only 15% of it will be absorbed by the film as per assumptions. Thus, there are two surfaces defined for applying heat: (1) hydrogel and (2) PS film. Two types of boundary for thermal and mechanical conditions are set. For the thermal one, the initial temperature sets to 323.15 K. For the mechanical one, the displacement in Z direction is constrained. Fig. 3.8 demonstrates the applied load surface and constraints that are defined for this work.

5. *Create seeds and mesh*

In this step, the size of seeds and mesh elements are specified. Indeed, the model needs a mesh refinement. This is a must for the finite element method. A parameter should be considered as a criterion for mesh refinement. The bending angle is selected as a parameter for this issue. First, big seed size is chosen for measuring bending angle and this amount is

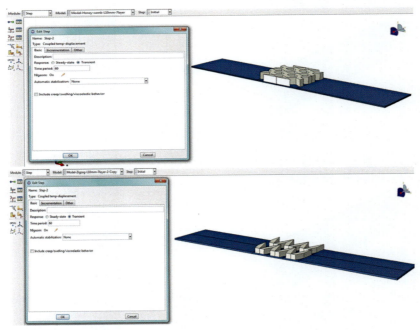

Fig. 3.7 Setting of the solver.

decreased until the bending angle does not change with seed size. This procedure is conducted for each of patterns with 3 layers and the results is demonstrated in Fig. 3.9. The zigzag and honeycomb patterns have acute angles, so the shape of element in these patterns is tetrahedral but hexahedral elements is used for the square pattern.

The approximate size of seed for the zigzag, square and honeycomb are 0.5, 0.9, and 0.5 mm respectively. Additionally, there are 3540 nodes and 2492 C3D8T elements for the creation of the mesh model as shown in Fig. 3.10. This element is an 8-node thermally coupled brick, trilinear displacement, and temperature. Also, the shape of element is Hex-structural.

Results and discussions

The relation of bending angles to the 4D printing parameters is investigated by simulation runs. The ambient temperature is supposed to be 323.15 K, all around the PS film. A NIR lamp source is placed at a suitable distance (250 mm) to irradiate the IR light evenly on the actuator with the maximum irradiation intensity of 3.5 mW/mm^2. Once the maximum deformation has been recorded before the softening temperature, the bending angle of each sample is determined. When the zigzag is the smallest, the honeycomb pattern exhibits the greatest bending angles (Figs. 3.11–3.14).

4D printing modeling using ABAQUS: A guide for beginners

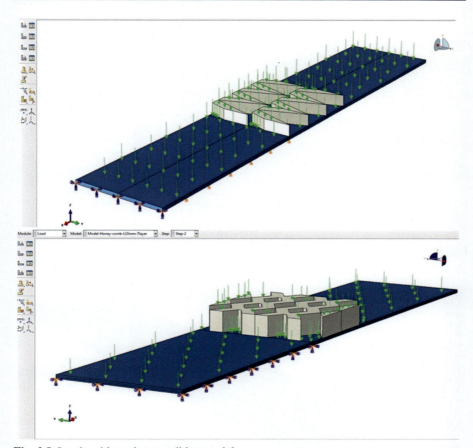

Fig. 3.8 Load and boundary condition module.

The simulation and experiment are conducted for evaluating the different layers effects on 4D-prited structures' bending angles. Furthermore, the comparison results of temperature rise in the middle of the PS film and the resultant bending angles of patterns are illustrated in Fig. 3.15 and Table 3.4. The findings demonstrate a 4D printing simulation model with various patterns and layer numbers. It can be noted that the maximum bending angle decreases with the number of printed layers, while the actuations are delayed correspondingly. The research successfully produced 4D-printed structures based on the same amounts of materials but varied patterns and layer control. When exposed to NIR light, the 4D-printed element absorbs the light and heats up, causing the prestrained SMP underneath to reach its glass transition temperature, resulting in shrinking of the prestrained film PS's top surface. Meanwhile, a temperature and therefore a shrinkage gradient is created between the top and bottom surfaces, allowing for controlled bending of the film through 4D printing.

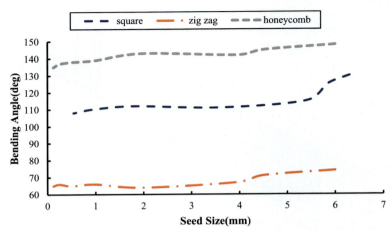

Fig. 3.9 Mesh refinement.

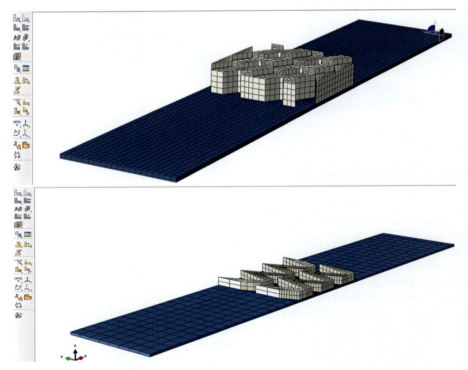

Fig. 3.10 The settings of mesh module.

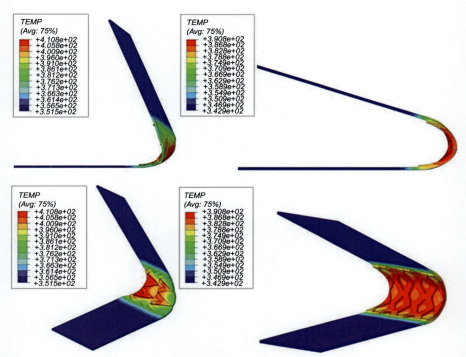

Fig. 3.11 Bending angles simulation results of 1-layer thickness zigzag (left) and honeycomb (right) designs.

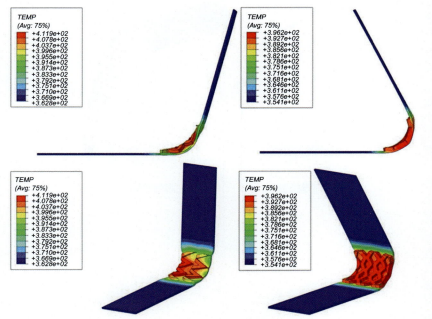

Fig. 3.12 Bending angles simulation results of 3-layer thickness zigzag (left) and honeycomb (right) designs.

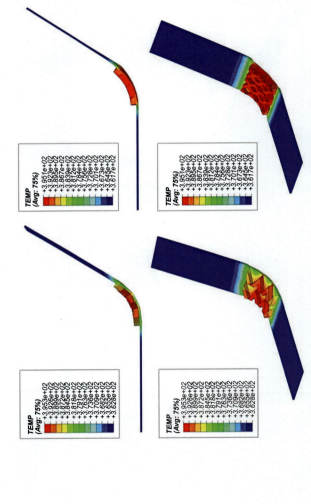

Fig. 3.13 Bending angles simulation results of 5-layer thickness zigzag (left) and honeycomb (right) designs.

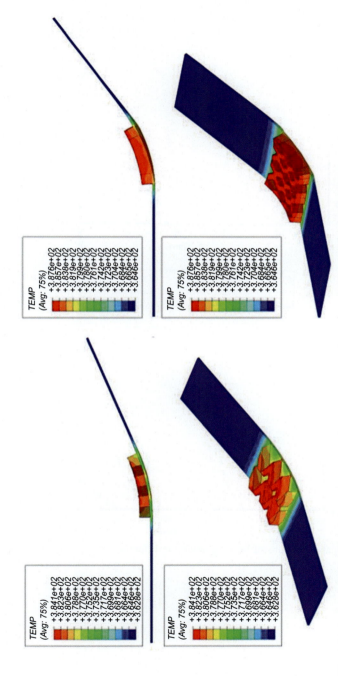

Fig. 3.14 Bending angles simulation results of 7-layer thickness zigzag (left) and honeycomb (right) designs.

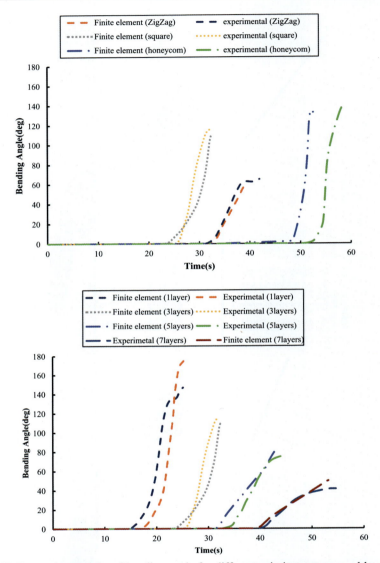

Fig. 3.15 Comparison results of bending angle for different printing patterns and layers.

Table 3.4 Simulation results bending angles.

Printed pattern	No. of layers	L (mm)	Bending angle (degrees)		Finite element	Error % (Exp)	Error % (FEM)
			Estimated	Experimental			
Honeycomb	1	2.5	139	147	141	5.44	4.08
	3	5	128	120	115	6.67	4.16
	5	7.5	71	67	70	5.97	4.47
	7	10	57	62	67	8.06	8.06
Zigzag	1	2.5	88	95	91	7.37	4.21
	3	5	79	72	75	9.72	4.16
	5	7.5	41	45	48	8.89	6.66
	7	10	33	30	33	10.00	10
Square	1	2.5	113	106	104	6.60	1.88
	3	5	98	91	88	7.69	3.29
	5	7.5	49	53	56	7.55	5.66
	7	10	39	36	38	8.33	5.55

Conclusion

With the increasing use of additive manufacturing and smart materials in industry, industries are looking to academics with interdisciplinary skills and expertise. This chapter presented an easy-to-implement and follow approach for students and early career researchers new to the field of 4D printing modeling and simulation, providing them a significant edge in an increasingly competitive job market and higher degree research. A commercial thermoplastic PS film and dark color patterns replicating active hinge on the SMP substrates are used to model 4D printing in this chapter. A photo-thermal stimulus was used to generate asymmetric shrinkage throughout the thickness of the SMP hinge. The temperature response of the printed actuator hinge to NIR irradiation was evaluated using a finite element model (FEM) in ABAQUS. A thermal-structural FEM model was built to predict the bending behavior of the 4D-printed structure. The simulation results showed the effects of the printed components' patterns and thickness on the bending angles. The present study could be utilized in different interdisciplinary engineering courses and classes to teach and practice 4D printing theoretical modeling and simulation.

References

Atamian, H. S., et al. (2016). Circadian regulation of sunflower heliotropism, floral orientation, and pollinator visits. *Science*, *353*(6299), 587–590.

Bodaghi, M., Damanpack, A., & Liao, W. (2016). Self-expanding/shrinking structures by 4D printing. *Smart Materials and Structures*, *25*(10), 105034.

Bodaghi, M., Damanpack, A., & Liao, W. (2017). Adaptive metamaterials by functionally graded 4D printing. *Materials & Design, 135*, 26–36.

Bodaghi, M., et al. (2019). 4D printing self-morphing structures. *Materials, 12*(8), 1353.

Bodaghi, M., et al. (2020). Reversible energy absorbing meta-sandwiches by FDM 4D printing. *International Journal of Mechanical Sciences, 173*, 105451.

de Haan, L. T., et al. (2014). Accordion-like actuators of multiple 3D patterned liquid crystal polymer films. *Advanced Functional Materials, 24*(9), 1251–1258.

Deng, D., & Chen, Y. (2015). Origami-based self-folding structure design and fabrication using projection based stereolithography. *Journal of Mechanical Design, 137*(2), 021701.

Deng, D., et al. (2017). Accurately controlled sequential self-folding structures by polystyrene film. *Smart Materials and Structures, 26*(8).

Kamal, T., & Park, S.-Y. (2014). Shape-responsive actuator from a single layer of a liquid-crystal polymer. *ACS Applied Materials & Interfaces, 6*(20), 18048–18054.

Liu, Y., et al. (2012). Self-folding of polymer sheets using local light absorption. *Soft Matter, 8*(6), 1764–1769.

Mailen, R. W., et al. (2017). A fully coupled thermo-viscoelastic finite element model for self-folding shape memory polymer sheets. *Journal of Polymer Science Part B: Polymer Physics, 55*(16), 1207–1219.

Nakata, K., et al. (2012). Bending motion of a polyacrylamide/graphite fiber driven by a wide range of light from UV to NIR. *Materials Letters, 74*, 68–70.

Sumaru, K., et al. (2013). Photoresponsive polymers for control of cell bioassay systems. In *Polymeric biomaterials* (3rd ed., pp. 683–708). Boca Raton: CRC Press.

Tolley, M. T., et al. (2014). Self-folding origami: Shape memory composites activated by uniform heating. *Smart Materials and Structures, 23*(9), 094006.

Zarek, M., et al. (2015). 3D printing of shape memory polymers for flexible electronic devices. *Advanced Materials, 28*, 4449–4454.

Zeng, H., et al. (2017). Self-regulating iris based on light-actuated liquid crystal elastomer. *Advanced Materials, 29*.

Zolfagharian, A., et al. (2016a). 3D printed hydrogel soft actuators. In *2016 IEEE region 10 conference (TENCON)*IEEE.

Zolfagharian, A., et al. (2016b). Evolution of 3D printed soft actuators. *Sensors and Actuators A: Physical, 250*, 258–272.

Zolfagharian, A., et al. (2017). 3D printing of a photo-thermal self-folding actuator. *KnE Engineering, 2*(2), 15–22.

Zolfagharian, A., et al. (2018). Pattern-driven 4D printing. *Sensors and Actuators A: Physical, 274*, 231–243.

Zolfagharian, A., et al. (2019). Topology-optimized 4D printing of a soft actuator. *Acta Mechanica Solida Sinica*, 1–13.

4D printing modeling via machine learning

Ali Zolfagharian[a], Lorena Durran[a], Khadijeh Sangtarash[a], Seyed Ebrahim Ghasemi[b], Akif Kaynak[a], and Mahdi Bodaghi[c]
[a]School of Engineering, Deakin University, Geelong, VIC, Australia
[b]Department of Engineering Sciences, Hakim Sabzevari University, Sabzevar, Iran
[c]Department of Engineering, School of Science and Technology, Nottingham Trent University, Nottingham, United Kingdom

Introduction

Robotics has always been synonymous with high precision and rigidity. However, in recent years, new technology has emerged that added the flexibility and adaptability that was not previously possible in rigid robots. Soft robots have extended the conventional robotics that has high precision in terms of movement and force but lack the ability to adjust to their dynamic environment situation (Yanlin, Lianqing, Guangkai, Mingxin, & Mingli, 2018). With the advancement of three-dimensional (3D) printing, their contribution to soft robotics has emerged as four-dimensional (4D) printing where the fourth dimension refers to the time-dependent response of the printed mechanism to varying stimuli, such as heat, magnetic, electricity, and pneumatic pressure.

The 4D-printed soft pneumatic actuator (SPA) focused in this study consists of finger-like structures with bellows on one longitudinal side, which inflate when pressurized causing it to extend and bend toward one side. It has a low material cost, lightweight, ease of manufacture, adaptability, flexibility, and deformability (Yanlin et al., 2018; Zolfagharian, Kaynak, et al., 2020; Zolfagharian, Mahmud, et al., 2020) compared to the rigid counterparts. Arguably, one of the most important benefits of this 4D-printed actuator is human safety, with their soft, flexible material properties enabling them to reduce impact and harm human skin or other delicate objects and surfaces. This type of soft robot actuator is utilized in the design of a soft surgical manipulator using an embedded optical fiber and FBG sensors to control bending in a closed-loop system (Yanlin et al., 2018). They also prove beneficial to applications where the objects to be manipulated may vary in size and shape, and fragile to handle. The potential for this technology also lies in its use for the rehabilitation of a wrist and finger exoskeleton to assist the movement of joints (Yap, Ng, & Yeow, 2016).

3D/4D printing is the method being studied due to its capability to produce soft robots and actuators with complex inner structures and without having seams that run the risk of tearing during actuation (Yanlin et al., 2018; Zolfagharian, Mahmud, et al., 2020; Zolfagharian et al., 2021). Ninjaflex material is shown as an ideal choice for 3D printing without creating any defects or air bubbles, which compromise airtightness and have the hyperelastic material properties that give soft actuators their flexibility and sensitivity to applied stress (Tawk, Gao, Mutlu, & Alici, 2019; Zolfagharian, Mahmud, et al., 2020).

Yet, one of the current challenging research in 4D printing of soft robots and actuators is the modeling and prediction of their motion particularly due to the nonlinearity of the materials. The studies have employed the hyperelastic material to handle this using the simulations in finite element method (FEM) platforms. However, this method has revealed difficulties where the different tasks and geometrical constraints apply during the design improvements (Tawk et al., 2019). A linear analytical model often fails to accurately predict actuation behavior in 4D printing while numerical simulation incorporating nonlinear material principles to improve accuracy. However, using the advantage of machine learning (ML) techniques trained via numerical results (Elgeneidy, Lohse, & Jackson, 2016) could help with saving time and effort during the design improvement and modifications (Runge, Wiese, & Raatz, 2017).

A pure data-driven modeling approach (Elgeneidy, Lohse, & Jackson, 2018) provides an opportunity to remove the need for accurate material models and material coefficients, and to predict actuation behavior under different operating conditions than originally trained for. A combination of the accuracy of FEM modeling and time efficiency of ML to formulate a neural network algorithm is introduced as a practical way (Runge et al., 2017). The FEM is employed for training the data, and the artificial neural network (ANN) is used to create the model, a method that will be utilized in this study to predict deflection angle in response to a varying input air pressure (Fig. 4.1). The nonlinear model is solved in ANSYS using Ogden constitutive model due to its simple form accuracy based on the uniaxial tensile test data (Cervenka, 2013; Zhang, Zhao, Wang, & Lu, 2020). The convergence of a nonlinear model is more difficult to achieve than linear models, due to its additional effects of material behavior approximation, large deformations, and time-dependent behavior or creep (Cervenka, 2013). This could be resolved by increasing the mesh size and therefore the stiffness of the geometrical model (Guachi et al., 2020).

Methodology

Fabrication

In this study, a 3D printing technique developed in Yap et al. (2016) is employed to fabricate the bellows-type SPA. This approach does not require support or other post-processing while accelerating customization of the 4D-printed SPA. Using this concept, a commercially available thermoplastic elastomer filament NinjaFlex (NinjaTek, PA), which was determined to be the most suitable for 4D printing the SPAs, was used to fabricate the actuators by a Flashforge FDM 3D printer. The printer has a 0.4-mm nozzle diameter and 0.1-mm layer resolution. An open-source 3D printing program, Cura, was used in combination with the 3D printer to change printing

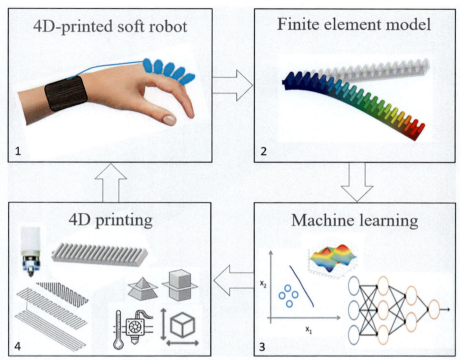

Fig. 4.1 4D printing modeling using machine learning.
The images are reproduced from Zolfagharian, A., Durran, L., Gharaie, S., Rolfe, B., Kaynak, A., & Bodaghi, M. (2021). 4D printing soft robots guided by machine learning and finite element models. Sensors and Actuators A: Physical, 328, 112774 with permission of Elsevier.

parameters. The CAD files are imported to Cura software, and the actuator layers were extruded sideways in a bottom-up layout. The parameters used for printing the SPAs in different sizes and geometries (Figs. 4.2 and 4.3) are expressed in Table 4.1. Note that the geometry of the actuator consists of a 2D shape seen from the side view, which is extruded to create width with a constant cross-section. It is also observed that the channel running between the bellows has the same wall thickness as the bellows, except for the thicker bottom layer of the actuator.

Analytical model

The analytical method outlined by (Alici, Canty, Mutlu, Hu, & Sencadas, 2018) estimates the bending angle with a given pressure input and dimensions. The derivation of this method stems from the idea that a difference between the center of pressure in an actuator and its centroid in the cross-section of the actuator, which is determined by its geometry, causes it to bend toward the bottom of the cross-section. When pressure is applied, this causes a tensile force to produce a bending moment that

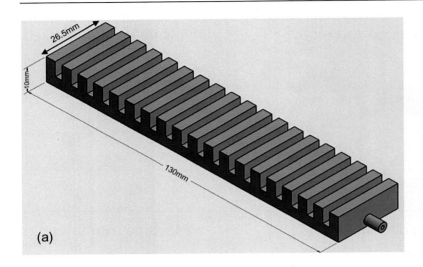

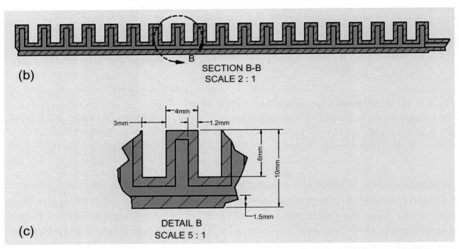

Fig. 4.2 The geometry of the soft actuator, including (A) outer geometry, (B) cross-section, and (C) detailed view.
The images are reproduced from Zolfagharian, A., Durran, L., Gharaie, S., Rolfe, B., Kaynak, A., & Bodaghi, M. (2021). 4D printing soft robots guided by machine learning and finite element models. Sensors and Actuators A: Physical, 328, 112774 with permission of Elsevier.

causes the deflection. It must be noted that this method takes a constant modulus of elasticity, E, as a generalization; however, in reality, this value would not remain constant, and therefore, we already introduce a margin for error in the results. In the method introduced in Alici et al. (2018), an effective modulus of elasticity is calculated using experimental results from a tensile test stress-strain data.

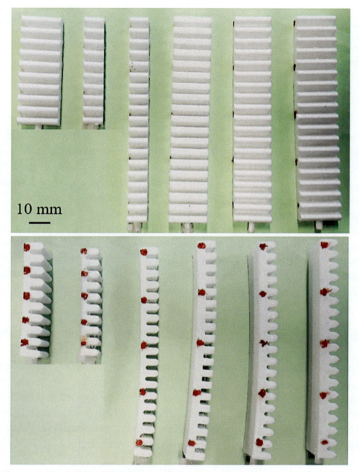

Fig. 4.3 The geometry of the 4D-printed soft actuators from the top (top) and (bottom) side views.
The images are reproduced from Zolfagharian, A., Durran, L., Gharaie, S., Rolfe, B., Kaynak, A., & Bodaghi, M. (2021). 4D printing soft robots guided by machine learning and finite element models. Sensors and Actuators A: Physical, 328, 112774 with permission of Elsevier.

Table 4.1 Printing parameters used for SPAs fabrication.

Parameters	Values
Printing speed (mm/s)	30
Nozzle size (mm)	0.4
Dual extrusion overlap (mm)	0.2
Initial layer thickness (mm)	0
Initial layer line width (%)	100
Platform temperature (°C)	0
Infill percentage (%)	100

The analytical expression derived for the steady-state bending angle of the actuator is then used to calculate the bending angle of the actuator as:

$$\theta(P) = \frac{L_i A^2 e}{A_w E^2 I} P^2 + \frac{L_i A e}{EI} P = CP^2 + DP \tag{4.1}$$

where the center of pressure, P, is taken to be at the centroid of the cross-section of the air chamber, e is the distance from the centroid of the actuator cross-section to the center of pressure, L_i is the initial length, A is the area of the cross-section of the chamber, and A_w is the area of the cross-section of the actuator. An assumed constant cross-sectional shape is used as the actuator is designed as a shape that is extruded along its length. Therefore, an average was taken between the maximum and minimum heights, and the bending angle was calculated using the method outlined in Alici et al. (2018). The bending angles calculated using the analytical model are compared to the experimental results. It is observed that the results of using an analytical approach did not closely match experimental results, and therefore are not feasible for modeling actuation behavior of hyperelastic materials; hence, a more complex model is required.

FEM modeling using hyperplastic material constitutive laws

The 4D printing performance in terms of the bending angle of SPAs of three different shapes and sizes was simulated using ANSYS Workbench. The geometries were modeled in the Design Modeler where parameters could be applied to their dimensions in order to enable simulation via parametrization. The bellow shapes analyzed in this project, as shown in Fig. 4.4, were rectangular, circular, and triangular, with the other input variables to the FEM model being the actuator width, height, bottom wall thickness, and air pressure.

The actuators were then modeled to reflect the 4D-printed SPAs experimental results of the bending angle in response to varying input air pressures. The behavior of hyperelastic materials could not be accurately modeled using a linear method due to their nonlinear stress-strain relationship. They can experience extension ratios up to 800% and are made up of long molecules between their cross-links that when under loading, can stretch to great lengths without permanent deformation. The cross-link density is a major influence on the mechanical response of hyperelastic materials (Elango, Faudzi, Rusydi, & Nordin, 2014). A nonlinear material model could be developed by fitting a curve to a set of experimental uniaxial or bi-axial stress-strain data using a suitable hyperelastic material constitutive model and then extracting the coefficients that are unique to that material. If the material is assumed to be isotropic, where its deformation behavior is uniform in all directions, only a uniaxial test is required, and a simplification of the model is possible (Elango et al., 2014). The strain energy density function in Mooney-Rivlin and Yeoh models is formed by strain invariants, whereas the Ogden model uses principal stretch ratios to represent the strain energy density function of the hyperelastic material (Elango et al., 2014). There are several models with their pros and cons, including Mooney-Rivlin, Ogden, Yeoh,

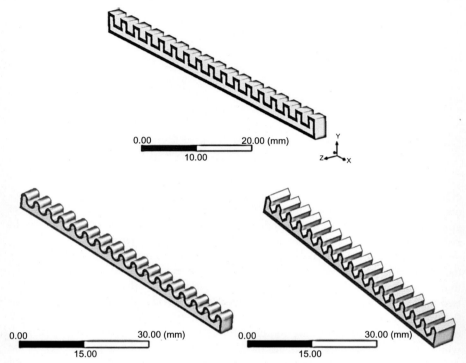

Fig. 4.4 Cross-sections of the rectangular, circular, and triangular shapes of actuators. The images are reproduced from Zolfagharian, A., Durran, L., Gharaie, S., Rolfe, B., Kaynak, A., & Bodaghi, M. (2021). 4D printing soft robots guided by machine learning and finite element models. Sensors and Actuators A: Physical, 328, 112774 with permission of Elsevier.

Neo-Hookean, Arruda-Boyce, and Blatz-Ko (Bodaghia et al., 2020; Zolfagharian, Mahmud, et al., 2020).

It is common for the specific geometry and function of the component to play a large role in the suitability of the various hyperelastic models, with the Mooney Rivlin model being unsuitable for capturing stiffening effects (Elango et al., 2014). Yet, in this study, it was found based on the tensile test on the printed sample that Ogden 3-parameter model (Yap et al., 2016) was the closest fitting model. The general form of the Ogden model for incompressibility is given by (4.2), where μ_i and α_i are the material constants, and λ_1, λ_2, and λ_3 are stretch ratio.

$$W(\lambda_1, \lambda_2) = \sum_{i=1}^{N} \frac{\mu_i}{\alpha_i} \left[\lambda_1^{\alpha_i} + \lambda_2^{\alpha_i} + \lambda_1^{-\alpha_i} \lambda_2^{-\alpha_i} - 3 \right] \tag{4.2}$$

Table 4.2 gives the hyperelastic model material coefficients.

A quadratic tetrahedral mesh was used for the models, and the settings were enabled to capture the curvature of the geometry for the circular actuator. A maximum

Table 4.2 Material coefficients for Ogden 3 parameter model.

Ogden parameter	Value
μ_1	−30.9212
α_1	0.5080
μ_2	10.3426
α_2	1.3755
μ_3	26.7912
α_3	−0.4827

element size of 1.4 mm was used, and no meshing methods were applied, which resulted in a relatively even mesh across the actuators. The number of nodes was approximately 90,000 for these models. In the meshing setup, the material is set to nonlinear material. Fig. 4.5 shows the mesh for the rectangular actuator.

The actuator was fixed at one end with the rest of it left free to deform. To simplify the model, air pressure was applied evenly to all inner surfaces of the actuator rather than introducing air through an inlet. The simulation was performed over 2.8 s using two-load steps, increasing the pressure linearly from 0 kPa to the set pressure from 0.4 s to 2.4 s. This ramping method was used due to the deformation not being an instantaneous effect and being required for successful convergence of the model, while also ensuring stability in the model. The number of substeps was set as the default value. Large deformations and weak springs were turned on to account for the comparatively large strains.

The solution setup included a total deformation plot and deformation probes on the end, the bottom edge of the actuator for the x- and the y-axes. The deformation probes

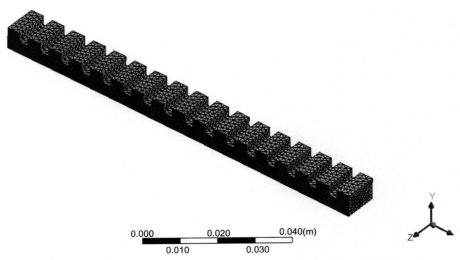

Fig. 4.5 Mesh for the rectangular actuator.
No permission required.

4D printing modeling via machine learning 81

enabled the extraction of the maximum endpoint deformation values, which were then used to calculate the bending angle. The simulation of the actuators using ANSYS validated the experimental results as shown in Fig. 4.6, and Fig. 4.7 shows that the rectangular-shaped actuator experiences the highest bending angles, followed by circular and then triangular. We can speculate that this is due to the larger surface area at the top of the bellows where the air pressure is applied, causing greater inflation of the bellows and therefore more bending.

Training data acquisition from FEM

In the process of dimensioning the geometry of the 4D-printed actuators, parameters were set to the dimensions associated with the width, height, and bellow thickness. The range of values used as the input parameters are outlined in Table 4.3. In

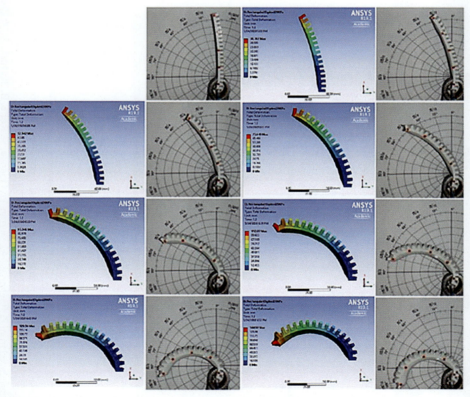

Fig. 4.6 Bending angle comparisons of FEM simulations and experimental results.
The images are reproduced from Zolfagharian, A., Durran, L., Gharaie, S., Rolfe, B., Kaynak, A., & Bodaghi, M. (2021). 4D printing soft robots guided by machine learning and finite element models. Sensors and Actuators A: Physical, 328, 112774 with permission of Elsevier.

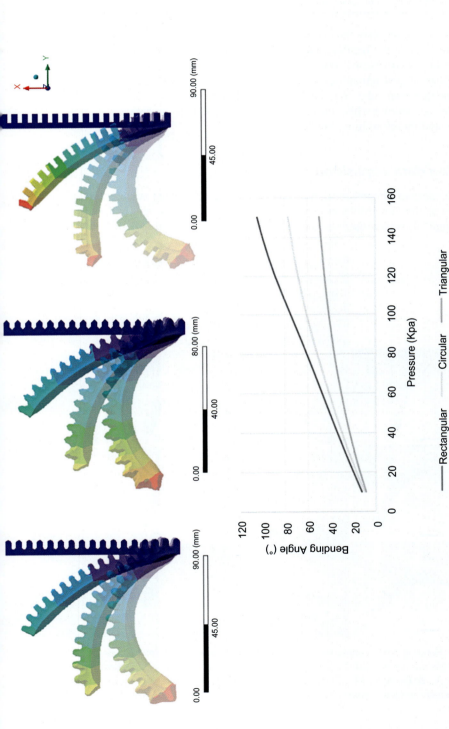

Fig. 4.7 FEM results comparing the bending sequence simulations of circular, triangular, and rectangular 4D-printed SPAs from left to right, respectively.

The images are reproduced from Zolfagharian, A., Durran, L., Gharaie, S., Rolfe, B., Kaynak, A., & Bodaghi, M. (2021). 4D printing soft robots guided by machine learning and finite element models. Sensors and Actuators A: Physical, 328, 112774 with permission of Elsevier.

Table 4.3 Variables and their value range for FEM simulations.

Variable	Minimum	Maximum	Interval
Height (mm)	8	10	1
Width (mm)	10	50	5
Bottom wall thickness (mm)	1.2	2	0.2
Air pressure (kPa)	10	150	10

Table of Design Points

	A	B	C	D	E
1	Name	P128 - Width	P132 - SmallWidth	P133 - BottomThickness	P134 - Height
2	Units	mm	mm	mm	mm
3	DP 0 (Current)	15.2	14	2	10
4	DP 1	5	3.8	1.2	8
5	DP 2	7.5	6.3	1.2	8
6	DP 3	10	8.8	1.2	8

Fig. 4.8 Setup of the parametrization process on ANSYS.
No permission required.

Fig. 4.8, the "width" parameter is the value of half of the actual width, which is due to the symmetric effect in the geometry setup.

In order to capture the bending angle at different input pressures, the time at which each pressure interval to be analyzed is first calculated. This was done by considering the pressure as a linear function over a time of 2 s. Position nodes positioned at the end of the actuator are then set as output parameters for each input pressure, which ranges from 10 kPa to 150 kPa at an interval of 10 kPa. The values of the input parameters are then set for the actuators, and a set of simulations are selected to be calculated automatically. This was done in batches of approximately 30 to allow for regular saving of the model and checking for issues or nonconvergence. The total number of data points to capture every combination of these parameters for all actuator shapes was 6027, with an average of 2009 data points required per actuator shape. These values meet the required minimum number of data points of 1000 in order to create a ML model that could accurately predict experimental bending angle results. Fig. 4.9 shows the components of the deformation vector used to calculate the bending angle θ.

Results and discussions

Initial analysis of the data

An analysis of variance (ANOVA) is needed in this study to disprove the null hypothesis, in which all groups are compatible, meaning that the spread of each of their means about the total mean is small enough to assume they all have the same mean

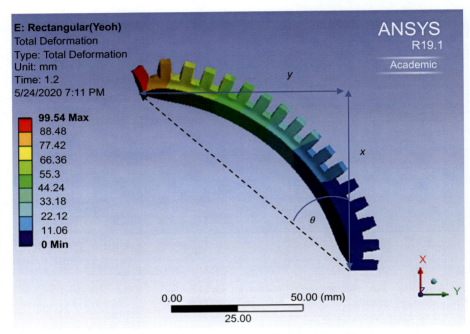

Fig. 4.9 x and y components of actuator bending angle.
No permission required.

(Barlow, 1993). In disproving the null hypothesis, the variability between the groups should be demonstrated (Lock, Lock, Morgan, Lock, & Lock, 2020). The F-statistic is the ratio of the mean square error of the groups (MSG) to the mean squared error within the groups (MSE) and is calculated if the null hypothesis is untrue. If it is true, the F-statistic will be close to 1 as the MSG and MSE will be (almost) equal. A P-value close to zero indicates there is evidence that there is a difference in the means among the groups (Lock et al., 2020). In an earlier study (Hu, 2019), the structural optimization of a SPA was analyzed for four variables and achieved a global ANOVA with a 95% confidence level, which identified and found the bottom layer thickness compared to bellow gap size and wall thickness had the highest influence on the bending angle. A global ANOVA was conducted here to determine the influence that each variable has on the bending angle of the 4D-printed actuator. In Fig. 4.10, the p-value is zero for all variables which disprove the null hypothesis and indicate that there is a difference in the means among the variable groups. In comparing the F-values for the different variables, we can see that the input pressure, which has an F-value of *1438.72*, has the most influence on the bending angle which is as expected. The geometrical variable with the next highest level of influence is the height with an F-value of 985.13. This may be due to the increased surface area of the bellow on which the air pressure is applied, causing greater inflation. This is followed by width with an F-value of 744.54 and then shape with an F-value of 580.56. The bottom wall

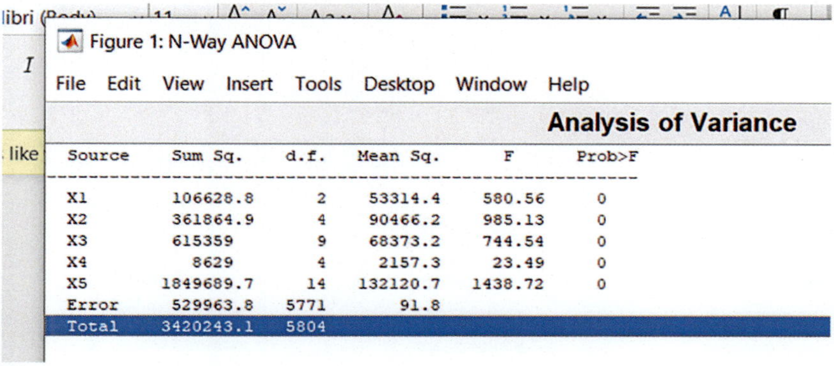

4D printing modeling via machine learning

Source	Sum Sq.	d.f.	Mean Sq.	F	Prob>F
X1	106628.8	2	53314.4	580.56	0
X2	361864.9	4	90466.2	985.13	0
X3	615359	9	68373.2	744.54	0
X4	8629	4	2157.3	23.49	0
X5	1849689.7	14	132120.7	1438.72	0
Error	529963.8	5771	91.8		
Total	3420243.1	5804			

Fig. 4.10 ANOVA results of 4D-printed SPA variables.
No permission required.

thickness, with an F-value of 23.49, has relatively little impact on the bending angle, which means future work could ignore this variable to simplify the model. This probably has such little influence due to the small range of values that this variable had, which was limited by a minimum of 1.2 mm due to printing requirements. As this part experiences less inflation and more bending than the rest of the actuator, it is not unsurprising that this would play a smaller role in the overall actuation. Insert: Loading….

The correlations between the input variables and the bending angle are plotted in Figs. 4.11 and 4.12 where Fig. 4.11 gives the correlations between height-width and bending angle, and Fig. 4.12 gives the correlations between bottom wall thickness-pressure and bending angle. Comparing the plots in confirmation to the ANOVA results, it is evident that the height and pressure show positive correlations with the bending angle. The width shows a positive correlation with the bending angle until a certain point for the circular and triangular actuators, after which it shows a downward trend. The bottom-layer thickness appears to have little influence on the bending angle, showing a negative correlation.

Linear regression

Linear regression is a method to create predictions based on historical data and consists of a dependent variable and one or more independent variables. This study uses four independent variables: height, width, bottom wall thickness, and air pressure, and therefore, a multiple linear regression is used. The general equation for multiple linear regression is given by (4.3), where Y is the dependent variable, β is a vector of regression coefficients, X is the independent variable, and the ϵ value is the error (Elgeneidy et al., 2016):

$$Y = \beta_0 + \beta_1 X_1 + \ldots \beta_k X_k + \epsilon \tag{4.3}$$

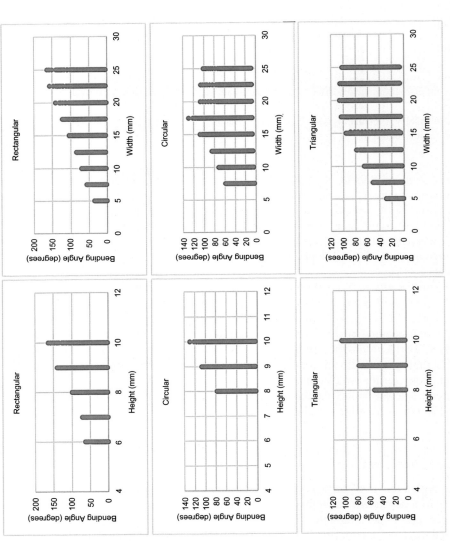

Fig. 4.11 Correlations between height and bending angle (left column) and between the width and bending angle (right column) for the rectangular, circular, and triangular actuators, from top to bottom, respectively. The images are reproduced from Zolfagharian, A., Durran, L., Gharaie, S., Rolfe, B., Kaynak, A., & Bodaghi, M. (2021). 4D printing soft robots guided by machine learning and finite element models. Sensors and Actuators A: Physical, 328, 112774 with permission of Elsevier.

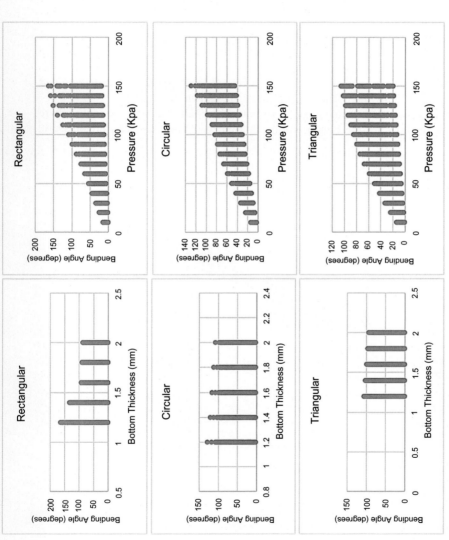

Fig. 4.12 Correlations between bottom wall thickness and bending angle (left) and between pressure and bending angle (right) for the rectangular, circular, and triangular actuators, from top to bottom, respectively.
The images are reproduced from Zolfagharian, A., Durran, L., Gharaie, S., Rolfe, B., Kaynak, A., & Bodaghi, M. (2021). 4D printing soft robots guided by machine learning and finite element models. Sensors and Actuators A: Physical, 328, 112774 with permission of Elsevier.

To evaluate the models, the R-squared and MSE are used. R-squared, or the coefficient of multiple determination, is a measure of how close the data are fitted to the regression line. It is the percentage of the response variable variation that is explained by a linear model, where the higher the R-squared, the better the model fits the data. However, there is a perfect fit to the data, and therefore, it is not possible to achieve 100% for such nonlinear problem.

The multiple linear regression models for each 4D-printed actuator shape are modeled using the regression tool on MATLAB, where the inputs and outputs are imported to the workspace as a numerical array with the variable categorized as the columns. The linear model is then trained, and the regression coefficients and residual plots are provided. The residual plots are shown in Fig. 4.13.

Table 4.4 gives the regression coefficients for the linear regression models, where X_1, X_2, X_3, and X_4 are the height, width, bottom wall thickness, and air pressure,

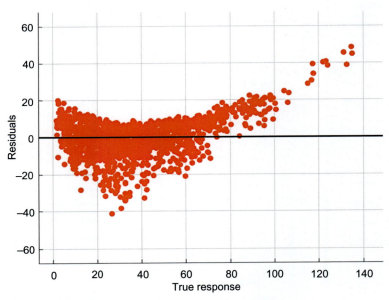

Fig. 4.13 Residual plots for all shapes linear regression models. No permission required.

Table 4.4 Regression coefficients for the linear regression model for the three individual actuator shapes.

	β_0	X_1	X_2	X_3	X_4
Rectangular	−75.108	8.205	1.576	−3.284	0.396
Circular	−63.348	6.002	0.580	−5.040	0.484
Triangular	−93.247	8.942	1.557	−2.991	0.383

Table 4.5 Regression coefficients for the linear regression model that encompasses all shapes.

	β_0	X_1	X_2	X_3	X_4	X_5	X_6
All shapes	−86.491	12.064	8.925	8.004	1.586	−3.708	0.413

respectively. Table 4.5 gives the regression coefficients for the linear regression model that encompasses all actuator shapes, where X_1, X_2, X_3, X_4, X_5, and X_6 are the rectangular variable, the circular variable, the height, width, bottom wall thickness, and air pressure, respectively. The linear regression models of the differently shaped actuators, as well as the model incorporating all shapes, show large residual error between the simulation results and the predicted values. High MSE values, as shown in Table 4.6, indicate that a linear representation of the data is not a good fit, which can be expected due to the nonlinear behavior of the material as it deforms.

Machine learning modeling using artificial neural network

The ML method in this project is constructed on the supervised learning ANN (Runge et al., 2017). Supervised learning is based on labeled training data consisting of input values and the corresponding desired outputs, known as target data, to solve complex differential equations while taking into consideration nonlinearities that would be otherwise ignored (Runge et al., 2017). Supervised learning deals with a set of inputs that correspond to a set of outputs that can be obtained experimentally, with the goal to create a function, $f: X \rightarrow Y$, to accurately predict the output given a new input. This differs from unsupervised learning where no outputs are given, and the algorithm identifies patterns in the input data. Each neuron in the network uses an activation function to determine its output, which is the input value with the weighting and bias applied, that is, then normalized into a range from 0 to 1 or −1 to 1. This value determines the influence that each neuron has on the neurons in the next layer, and therefore, its influence on the entire model (Dreyfus, 2005). A simple ANN has only one input layer, hidden layer, and output layer, in comparison with a more extended network with higher complexity and a greater computational requirement (Runge et al., 2017).

Table 4.6 MSE values for the linear regression models.

Shapes	MSE	R-squared
Rectangular	148.4003	0.74
Circular	54.586	0.90
Triangular	88.973	0.84
All shapes	108.4	0.81

The MSE loss function measures the error of the model and is used for regression problems (Fernandes de Mello & Antonelli Ponti, 2018). The general formula for this is given by (4.4) as:

$$\mathrm{MSE} = \frac{1}{n}\sum_{i=1}^{n}(y_i - \tilde{y}_i)^2 (Y - f(X)) \tag{4.4}$$

where y_i, $\tilde{y}_i$, and n are the network output, target data, and the number of training data, respectively. To optimize the algorithm, the MSE must be minimized using gradient-based method while updating the weights and biases (Dreyfus, 2005). Backpropagation is the term used to describe the refining of the weights and biases of the network to optimize the model (Rumelhart, Hinton, & Williams, 1986). It works backward from the output to make positive or negative changes to these values for each neuron using a solution method such as gradient descent, which moves the error of the model to a local minimum by following the gradient of the error function until it is close to zero. The learning rate determines how large steps are taken toward this local minimum in each iteration, with a balance being required to prevent overshooting or being unnecessarily time-consuming (Goodfellow, Bengio, & Courville, 2016).

In this study, the actuator variables and the bending configuration serve as input and target data, respectively. The accessible training data is a key point in ANN training. The data must cover the feasible state space of a 4D-printed system to achieve a model that can represent the whole state space. The FE-Model is simulated several times with randomly generated actuation profiles to acquire training results. The acquired data are divided into three sets: training set, validation set, and test set. The data are normalized in the first process of training. This gives the data a mean close to zero and places it within a specified range, which speeds up the training (Beale, Hagan, & Demuth, 2010). This is a default process when using MATLAB to train a neural network.

The parameters shown in Table 4.7 are set to study using ANN in this study adopted from a study on the closed-loop control system of a SPA (Runge et al., 2017). The bending curvature and angle were taken as outputs, and the model was deemed a good model for prediction, with small output errors. The Levenberg-Marquardt solution method is also used in regression analysis and ANN for modeling a soft finger with an embedded flex sensor (Alici et al., 2018). It was found that as additional inputs were added to the regression model, the R-value decreased and hence the accuracy of the model increased, with the ANN producing an even lower R-value. They found that the ANN is more computationally demanding but better captured the nonlinearity of the material compared to the linear regression model.

In this study to determine the most suitable machine learning model for predicting the bending angle of a 4D-printed soft pneumatic actuator, the multilayer perceptron (MLP) and radial basis function (RBF) neural networks have opted as the most feasible options for this problem. A classification model is not applicable as the output is not a category to be classified as, but rather, a value that can be any real, positive number. A recurrent neural network is also inapplicable as this problem does not

Table 4.7 Parameters used and errors associated with the neural network.

	Parameters	Values
ANN architecture		
Number of input neurons	n_{in}	3
Number of output neurons	n_{out}	5
Number of hidden neurons	n_{hid}	120
Activation function hidden layer	$f(v)$	$\tanh(v)$
Activation function output layer	$f(v)$	v (linear)
Training parameters		
Relative size training set	r_{tr}	70%
Relative size validation	r_{val}	15%
Relative size test set	r_{test}	15%
Maximum number of epochs	N_{epochs}	2000
Early stopping	N_{fail}	10
Optimization method Levenberg-Marquardt		
Initial value	μ_{init}	0.01
Increase factor	μ_{inc}	10
Decrease factor	μ_{dec}	0.1
Maximum value	μ_{max}	10

involve a sequence of values to be continued by the model and its output is influenced by a set of static input parameters. A radial basis function neural network uses Euclidean distances between the inputs and weights of a neuron, multiplied by the bias (Dreyfus, 2005). It uses the Gaussian activation function, which makes the neurons more locally sensitive (Dreyfus, 2005). Similar to the feedforward neural network, it consists of an input layer, hidden layer, and output layer and uses backpropagation to train; however, it typically has faster learning speed and is typically used for function approximation problems (Dreyfus, 2005).

The ANN is trained using the Neural Network Toolbox on MATLAB. The inputs and outputs are imported, and the training method is selected. The data are split and the number of neurons for the default one hidden layer is selected. After this initial creation of a neural network, it is imported to the workspace where the code can be directly altered to further make changes to the model. The Bayesian regularization method was found, through trial and error, to be the training algorithm that produced the most accurate results for these neural networks. Neural networks of this type are created for the rectangular-, circular-, and triangular-shaped actuators, as well as all shapes combined. These networks use 20 neurons in one hidden layer for the purpose of comparing each bellow shape, and the data were split into 70% training samples, 15% validation samples, and 15% testing samples. The seed used to assign the random weights and biases to the neural networks is fixed to enable an accurate comparison between the different types.

The model using all actuator shapes incorporates the three shapes using two Boolean variables. For this model, when the shape is rectangular, the "rectangular" variable is set to one and the "circular" variable is set to 0. When the shape is circular, the "rectangular" variable is set to 0 and the "circular" variable is set to 1. When the shape is triangular, both variables are set to 0. Table 4.8 gives the mean squared error and R-value for MLP neural networks.

The ANN model for the circular actuator was found to be the most accurate model, and the model using all shapes produced the least accurate results. This could be explained by a greater number of input variables and therefore a more diverse data set. Yet, the neural network model still demonstrates a sufficiently low mean squared error value and fit, and due it its ability to encompass all three actuator shapes, it was used in further refinement of the model.

Layer configurations analysis

The optimal number of layers and neurons is determined through an educated guess, to determine the configuration that results in the lowest error and best fit. MLP neural networks for the data set encompassing all three shapes are trained with single hidden layers of 5, 10, and 20 neurons each, two hidden layers with 5 and 10 neurons each, and 3 hidden layers with 2, 4, and 6 neurons each. The MSE and R values for each scenario are recorded in Table 4.9.

Table 4.8 MLP model results for actuators' shape comparison.

Shape	Training set MSE	Testing set MSE	Training set R-value	Testing set R-value
Rectangular	0.1673	0.4223	0.9999	0.9998
Circular	0.1080	0.1375	0.9999	0.9999
Triangular	0.1550	0.2052	0.9999	0.9999
All shapes	1.2060	1.4892	0.9992	0.9990

Table 4.9 Layer configurations analysis.

Configuration (layers, nodes)	Training MSE	Testing MSE	Training R-value	Testing R-value
(1,5)	5.8451	5.8769	0.9950	0.9949
(1,10)	3.6545	4.5215	0.9969	0.9964
(1,20)	0.3155	0.8382	0.9997	0.9993
(2,5)	3.8238	1.0036	0.9968	0.9991
(2,10)	0.1439	0.2477	0.9999	0.9998
(3,2)	52.3739	58.9353	0.9542	0.9509
(3,4)	3.7881	5.0175	0.9968	0.9956
(3,6)	0.5894	0.5766	0.9995	0.9995

The model with two hidden layers with 10 neurons each was found to perform the best, with an MSE of 0.1439. As expected, the MSE value improves with increasing the total number of neurons, and it is interesting to note that an increased number of layers with the same number of total neurons does not necessarily improve the MSE. It is therefore important to test different configurations of the neural network in order to determine the one most suited to the problem considering a balance to be struck between training time and accuracy. This, of course, could be automated using other heuristic optimization and deep learning algorithms too (Zolfagharian, Valipour, & Ghasemi, 2016).

Activation functions analysis

An analysis was performed on the influence of the activation function used at each layer on the accuracy of the output of the MLP model. A comparison of 5 different activation functions, including the tan-sigmoid function which has been used as the default, was conducted to determine that which produces the most accurate model. A neural network using all shapes listed in Table 4.10 with 2 layers with 10 neurons each is used, and the MSE and R values are given. The tan-sigmoid activation function was found to produce the lowest mean-squared errors for the model. This is an S-shaped function that outputs values within a range of -1 and 1. The function definition is given by (4.5) and is shown in Fig. 4.14.

$$\text{tansig} = f(x) = \frac{2}{1 + \exp^{-2x}} - 1 \tag{4.5}$$

An RBF was created with 2 hidden layers with 10 neurons each and was compared to the MLP in Table 4.11. The MLP was found to more accurately predict the bending angle than the radial basis function network.

The model that most accurately predicts the bending angle of the 4D-printed SPA was a MLP with two layers with 10 neurons each, using the tan-sigmoid activation function, as visually depicted in Fig. 4.14. As shown in Fig. 4.15, the neural network model is a near-perfect fit to the FEM simulation data, which means that the accuracy of the machine learning model is directly dependent on the accuracy of the FEM model.

The FEM simulations were conducted for only the rectangular actuator. The neural network was trained using the method obtained from the optimization of the ML

Table 4.10 Activation functions analysis.

Activation function	Training set MSE	Testing set MSE	Training set R	Testing set R
Tan-sigmoid	0.1439	0.2477	0.9999	0.9998
Triangular basis	3.5111	8.3595	0.9970	0.9930
Log-sigmoid	0.1996	0.2654	0.9998	0.9998
Positive linear	3.2666	7.1889	0.9972	0.9941
Normalized RBF	0.3732	0.5783	0.9997	0.9995

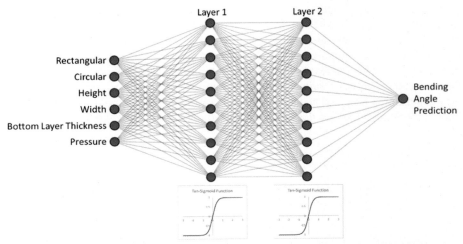

Fig. 4.14 Optimized ANN configuration for predicting the 4D-printed SPA bending angle. The images are reproduced from Zolfagharian, A., Durran, L., Gharaie, S., Rolfe, B., Kaynak, A., & Bodaghi, M. (2021). 4D printing soft robots guided by machine learning and finite element models. Sensors and Actuators A: Physical, 328, 112774 with permission of Elsevier.

Table 4.11 The MLP and RBF results comparisons.

ANN type	Training set MSE	Testing set MSE	Training set R-value	Testing set R-value
MLP	0.1439	0.2477	0.9999	0.9998
RBF	1.2227	2.6532	0.9990	0.9978

model and is compared to the experimental and FEM results in Fig. 4.16. The FEM results and ANN more accurately reflect the experimental results using this model; however, it is evident that there is some error at lower air pressures and less accurately reflects the FEM results. We observe that the training MSE value is better, probably due to the smaller number of data causing a better overall fit. Figure 4.16 shows the FEM model superimposed onto the actuator used in the experiment for air pressure of 120 kPa. It also shows the neural network output as well as the linear regression output.

4D-printed soft actuator shape classification using ML

The study so far achieved the initial objective of predicting the bending angle of 4D-printed SPA using ML; however, a complementary problem is presented here to demonstrate the capability of ML to predict the geometrical features of the 4D-printed constructs to obtain a required bending angle, given a set of design

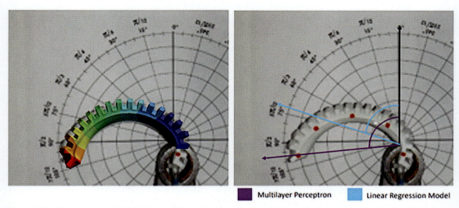

Fig. 4.15 Superimposed FEM model and neural network prediction on experiment for the rectangular actuator.
The images are reproduced from Zolfagharian, A., Durran, L., Gharaie, S., Rolfe, B., Kaynak, A., & Bodaghi, M. (2021). 4D printing soft robots guided by machine learning and finite element models. Sensors and Actuators A: Physical, 328, 112774 with permission of Elsevier.

constraints. Hence, a classification model was trained that can determine the 4D-printed SPA shape that would produce the required bending angle given a set geometry and input pressure. It was trained using the classifier wizard on MATLAB, which enables you to import input and output data and train and optimize a classification model. To begin with, all classification training methods were tested for the dataset containing all actuator shapes, with the best model being the medium Gaussian support vector machines (SVM) method. This model had an original accuracy of 82.8%, and as shown in Fig. 4.18, the model has the most difficulty distinguishing the circular shape from the others. An accuracy of 94.3% was then achieved with the optimization of the hyperparameters of the SVM optimizable classification model, which completed 30 iterations at the expense of an extended computation time. This indicates that the 4D-printed actuator shape for the given input parameters is able to be predicted with a high level of accuracy for design purposes (see Fig. 4.17).

Classification models were then created by comparing two shapes against one another in order to determine the confidence of distinguishing between each shape in Fig. 4.18. The circular versus triangular accuracy diagram model achieves 100% accuracy, meaning that there is little to no error when distinguishing between the circular and triangular shapes. The circular versus rectangular accuracy diagram model achieves 91.8% accuracy, having the lowest confidence level out of the three models comparing two shapes. This indicates that the highest error occurs when distinguishing between the rectangular and circular shapes. The triangular versus rectangular accuracy diagram model achieves 98.9% accuracy when distinguishing between the rectangular and triangular shapes.

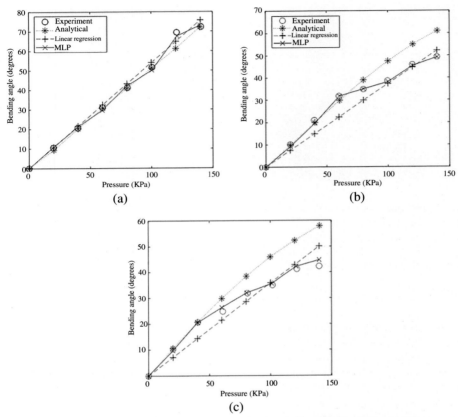

Fig. 4.16 Model comparison for the rectangular, circular, and triangular actuators from top to bottom, respectively.
The images are reproduced from Zolfagharian, A., Durran, L., Gharaie, S., Rolfe, B., Kaynak, A., & Bodaghi, M. (2021). 4D printing soft robots guided by machine learning and finite element models. Sensors and Actuators A: Physical, 328, 112774 with permission of Elsevier.

Discussions

The comparison of the different models used to predict the bending angle of the 4D-printed SPAs indicates that the MLP neural network successfully predicted the experiment results trained by FEM simulation data (see Fig. 4.19). The ineffectiveness of the analytical model stemmed from the utilization of a linear stress-strain relationship to characterize the material. This assumption failed to take into consideration the high levels of strain experienced by the hyperelastic material without plastic deformation, and the nonlinear characteristic of the stress as it deforms. It also assumes an isotropic material, which is not the case due to the manufacturing method. The orientation of the

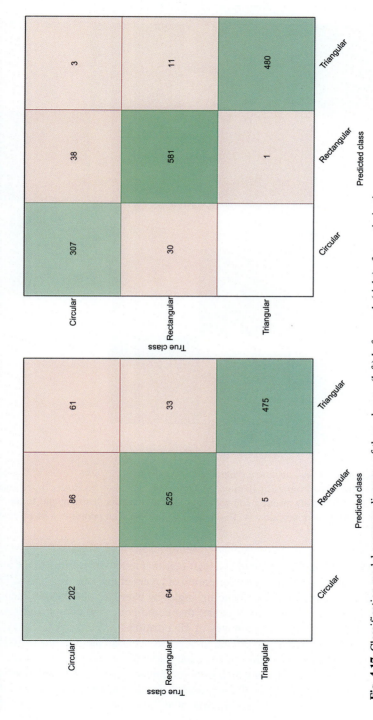

Fig. 4.17 Classification model accuracy diagrams of three shapes; (left) before and, (right) after optimization. The images are reproduced from Zolfagharian, A., Durran, L., Gharaie, S., Rolfe, B., Kaynak, A., & Bodaghi, M. (2021). 4D printing soft robots guided by machine learning and finite element models. Sensors and Actuators A: Physical, 328, 112774 with permission of Elsevier.

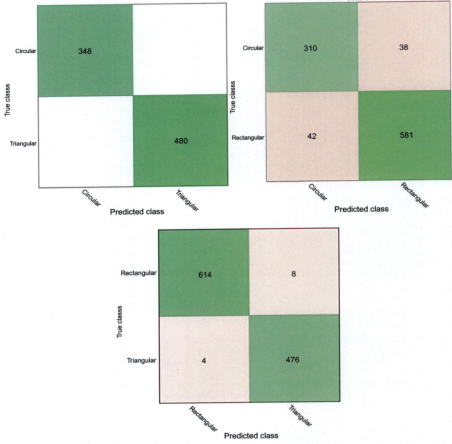

Fig. 4.18 Classification model accuracy diagrams of two shapes.
The images are reproduced from Zolfagharian, A., Durran, L., Gharaie, S., Rolfe, B., Kaynak, A., & Bodaghi, M. (2021). 4D printing soft robots guided by machine learning and finite element models. Sensors and Actuators A: Physical, 328, 112774 with permission of Elsevier.

4D-printed actuator means that it should be least resistant to tensile loading in the axis perpendicular to the print bed due to the formation of layers of material as it prints. The analytical method underpredicted the bending angles by over half of the experimental values, which validated the need for a model that could cover the complexities and nonlinearities of the material and geometry in 4D printing.

The linear regression model also proved to be a less accurate method of predicting the actuation. This model only allowed for each variable to influence the bending angle using a weighting assignment which is then summed up, with an additional bias value for the equation. The rigidity of this model is unsuited to the complexity of the problem and does not allow for the variation of influence of each variable as the

4D printing modeling via machine learning 99

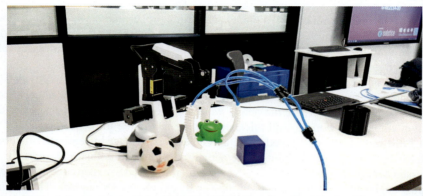

Fig. 4.19 Experimental tests of the 4D-printed SPAs based on the ML classification model to grasp objects with different bending and geometry specifications.
No permission required.

bending angle increases. The hyperelastic model with Ogden 3-parameter material coefficients found the most suitable constitutive model in ANSYS to fit the experimental data, validating the effectiveness of using FEM to simulate real actuation, resulting in an accurate ML model to accurately predict the bending angle of 4D-printed SPA.

The classification model for predicting shape is equally as useful for design purposes and enables the user to achieve their desired motion in 4D printing if, for example, there are geometrical requirements or limitations. The results from the ANOVA analysis enabled the identification of the variables that will maximize actuation, which can ultimately lead to efficiency in energy requirements and cost of the actuator.

Limitations of this project include the simulation time and processing power, which were quite large for the number of data points required. Access to a high-performance computer sped up the process, and once the simulations were completed, the ML training data were easily acquired. This approach of ML-based modeling could be tailored to the other types of 4D printing studies where the new design parameters could be defined first, and after validating the experimental results in FEM simulation, all the simulation data could be used to create a new ML model.

Conclusion

The outcome of this chapter is a model that can accurately predict the actuation of a 4D-printed, soft pneumatic actuator (SPA). It followed the development and simulation of a finite element method (FEM) model, which accurately reflects experimental results given a suitable hyperelastic model. An analytical model was developed using a constant elastic modulus, which was found to underpredict the bending angle and be insufficient for the nonlinearities and complexity of the problem. This validated the need for a machine learning (ML) model. The study proves that a ML model is more suited to a nonlinear problem than a linear analytical or regression model, and it follows the optimization of a multilayer perceptron (MLP) that achieves a reasonable error when predicting actuation behavior.

Models for predicting bending angle, pressure, and shape were successfully developed, and the results of a global ANOVA indicated the level of influence of each variable on bending angle where the height was found to have the highest influence on bending angle, with bottom wall thickness having negligible influence.

A FEM model was developed using an Ogden 3-parameter hyperelastic material model, which was used to simulate the actuation of the soft actuator as pressure is increased over time. The rectangular-shaped actuator proved to experience the highest bending angles for a given geometry, followed by circular and triangular. This model was used to produce the data required for training an artificial neural network (ANN). The variables used were the bellow shape, height, width, bottom layer thickness, and air pressure. Using this data, a ML learning model was developed and optimized to predict the bending angle given a certain set of inputs. The model achieved an MSE value of 0.1439 and proved to accurately predict the bending angle based on the simulation data. Besides, a classification model was developed by optimizing the hyperparameters to create a model with 94.3% accuracy of distinguishing between the three actuator shapes.

The importance of this work lies in the accurate modeling of nonlinear materials and complex geometries in 4D printing. The ML model accurately predicted experimentally validated FEM data and proved to be a viable solution for modeling such 4D-printed structures given an accurate FEM model. Future work may be conducted to analyze the blocking force. This work may be used to develop control systems utilizing this technology and design, or for design purposes of closed-loop 4D-printed soft robots and actuators with specific functional and geometrical requirements. External loads are not taken into account in this study. In real-world 4D printing problems, the impacts of these variables must be considered.

References

Alici, G., Canty, T., Mutlu, R., Hu, W., & Sencadas, V. (2018). Modeling and experimental evaluation of bending behavior of soft pneumatic actuators made of discrete actuation chambers. *Soft Robotics*, *5*(1), 24–35. https://doi.org/10.1089/soro.2016.0052.

Barlow, R. J. (1993). *Statistics: A guide to the use of statistical methods in the physical sciences. Vol. 29*. John Wiley & Sons.

Beale, M. H., Hagan, M. T., & Demuth, H. B. (2010). *Neural network toolbox user's guide* (pp. 77–81). The MathWorks.

Bodaghia, M., Serjoueia, A., Zolfagharianb, A., Fotouhic, M., Rahmana, H., & Durandad, D. (2020). Reversible energy absorbing meta-sandwiches by FDM 4D printing. *International Journal of Mechanical Sciences*, *173*. https://doi.org/10.1016/j.ijmecsci.2020.105451, 105451.

Cervenka, V. (2013). Reliability-based non-linear analysis according to fib model code 2010. *Structural Concrete*, *14*(1), 19–28. https://doi.org/10.1002/suco.201200022.

Dreyfus, G. (2005). Neural networks: Methodology and applications. In *Neural networks: Methodology and applications* (pp. 1–497). Berlin Heidelberg: Springer. https://doi.org/10.1007/3-540-28847-3.

Elango, N., Faudzi, A. A. M., Rusydi, M. R. M., & Nordin, I. N. A. M. (2014). Determination of non-linear material constants of RTV silicone applied to a soft actuator for robotic applications. In (Vols. 594–595). *Key engineering materials* (pp. 1099–1104). Trans Tech Publications Ltd. https://doi.org/10.4028/www.scientific.net/KEM.594-595.1099.

Elgeneidy, K., Lohse, N., & Jackson, M. (2016). Data-driven bending angle prediction of soft pneumatic actuators with embedded flex sensors. *IFAC-PapersOnLine*, *49*(21), 513–520. https://doi.org/10.1016/j.ifacol.2016.10.654.

Elgeneidy, K., Lohse, N., & Jackson, M. (2018). Bending angle prediction and control of soft pneumatic actuators with embedded flex sensors – A data-driven approach. *Mechatronics*, *50*, 234–247. https://doi.org/10.1016/j.mechatronics.2017.10.005.

Fernandes de Mello, R., & Antonelli Ponti, M. (2018). Statistical learning theory. *Machine learning* (pp. 75–128). Cham: Springer.

Goodfellow, I., Bengio, Y., & Courville, A. (2016). *Deep learning*. MIT press.

Guachi, R., Bini, F., Bici, M., Campana, F., Marinozzi, F., & Guachi, L. (2020). Finite element analysis in colorectal surgery: Non-linear effects induced by material model and geometry. *Computer Methods in Biomechanics and Biomedical Engineering: Imaging and Visualization*, *8*(2), 219–230. https://doi.org/10.1080/21681163.2019.1679669.

Hu, W. (2019). *Flexible fluidic actuators for soft robotic applications*. School of Mechanical, Materials, Mechatronic and Biomedical Engineering, University of Wollongong. https://ro.uow.edu.au/theses1/717. (Doctor of Philosophy thesis).

Lock, R. H., Lock, P. F., Morgan, K. L., Lock, E. F., & Lock, D. F. (2020). *Statistics: Unlocking the power of data*. John Wiley & Sons.

Rumelhart, D. E., Hinton, G. E., & Williams, R. J. (1986). Learning representations by back-propagating errors. *Nature*, *323*(6088), 533–536. https://doi.org/10.1038/323533a0.

Runge, G., Wiese, M., & Raatz, A. (2017). FEM-based training of artificial neural networks for modular soft robots. In *2017 IEEE International Conference on Robotics and Biomimetics (ROBIO)*.

Tawk, C., Gao, Y., Mutlu, R., & Alici, G. (2019). In *Vol. 2019. Fully 3D printed monolithic soft gripper with high conformal grasping capability In IEEE/ASME International Conference on Advanced Intelligent Mechatronics, AIM* (pp. 1139–1144). Institute of Electrical and Electronics Engineers Inc. https://doi.org/10.1109/AIM.2019.8868668.

Yanlin, H., Lianqing, Z., Guangkai, S., Mingxin, Y., & Mingli, D. (2018). Design, measurement and shape reconstruction of soft surgical actuator based on fiber bragg gratings. *Applied Sciences*, *1773*. https://doi.org/10.3390/app8101773.

Yap, H. K., Ng, H. Y., & Yeow, C. H. (2016). High-force soft printable pneumatics for soft robotic applications. *Soft Robotics*, *3*(3), 144–158. https://doi.org/10.1089/soro.2016.0030.

Zhang, Y., Zhao, W., Wang, N., & Lu, D. (2020). Development and performance analysis of pneumatic soft-bodied bionic basic execution unit. *Journal of Robotics*, *2020*. https://doi.org/10.1155/2020/8860550.

Zolfagharian, A., Durran, L., Gharaie, S., Rolfe, B., Kaynak, A., & Bodaghi, M. (2021). 4D printing soft robots guided by machine learning and finite element models. *Sensors and Actuators A: Physical*, *328*, 112774.

Zolfagharian, A., Kaynak, A., Bodaghi, M., Kouzani, A. Z., Gharaie, S., & Nahavandi, S. (2020). Control-based 4D printing: Adaptive 4D-printed systems. *Applied Sciences (Switzerland)*, *10*(9). https://doi.org/10.3390/app10093020.

Zolfagharian, A., Mahmud, M. A. P., Gharaie, S., Bodaghi, M., Kouzani, A. Z., & Kaynak, A. (2020). 3D/4D-printed bending-type soft pneumatic actuators: Fabrication, modelling, and control. *Virtual and Physical Prototyping*, *15*(4), 373–402. https://doi.org/10.1080/17452759.2020.1795209.

Zolfagharian, A., Valipour, P., & Ghasemi, S. E. (2016). Fuzzy force learning controller of flexible wiper system. *Neural Computing and Applications*, *27*(2), 483–493. https://doi.org/10.1007/s00521-015-1869-0.

4D-printed pneumatic soft actuators modeling, fabrication, and control

Charbel Tawk[a,b] and Gursel Alici[a]
[a]School of Mechanical, Materials, Mechatronic and Biomedical Engineering, Applied Mechatronics and Biomedical Engineering Research (AMBER) Group, ARC Centre of Excellence for Electromaterials Science, University of Wollongong, Wollongong, NSW, Australia
[b]Faculty of Engineering and Information Sciences, University of Wollongong in Dubai, Dubai, United Arab Emirates

Introduction

A soft actuator converts an input stimulus into a useful mechanical output. In general, soft actuators can be classified according to the input stimulus required for their activation. For example, chemical species are required for electroactive polymer actuators, air for pneumatic actuators, liquid for hydraulic actuators, heat for shape memory polymers and alloys, light for light-responsive actuators, electrical field for dielectric elastomers, and magnetic field for magnetorheological fluids and polymers (Alici, 2018; Tawk & Alici, 2021). Soft actuators that drive a soft robotic system are one of its main critical components (Fig. 5.1) (Tawk & Alici, 2021). Such soft actuators must be dexterous, fast, and reversible and should be able to contract, extend, bend, and/or twist with favorable relative precision while delivering sufficient output forces. The selection of the soft actuation concept for a soft robotic system and its realization directly affects the size, weight, performance, type of sensors and their location, control architecture, and power requirements of a soft robotic system. Therefore, the soft actuation concept must be designed based on the requirements and specifications of the soft robotic system and its main application.

There are several types of soft actuators and soft actuation concepts that can be directly used in soft robotic systems and devices, including smart materials and structures such as shape memory alloys (Jin et al., 2016; Laschi et al., 2012; Lin, Leisk, & Trimmer, 2011; Seok et al., 2012; She, Li, Cleary, & Su, 2015), dielectric elastomers (Gu, Zhu, Zhu, & Zhu, 2017; Kellaris, Gopaluni Venkata, Smith, Mitchell, & Keplinger, 2018), ionic polymer-metal composites (Yousef & Mohsen, 2014), coiled polymer fibers (Haines et al., 2014), hydrogels (Bakarich, Gorkin, in het Panhuis, & Spinks, 2015; Yuk et al., 2017), humidity-responsive materials (Shin et al., 2018), and magnetic-responsive structures (Hu, Lum, Mastrangeli, & Sitti, 2018). In addition, soft robotic systems and devices can be actuated using chemical reactions such as combustion (Bartlett et al., 2015), electrolysis (Suzumori, Wada, & Wakimoto, 2013),

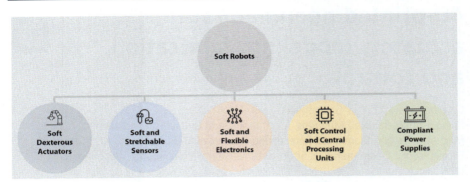

Fig. 5.1 The main components that constitute a soft robotic system.
No permission required.

and catalytic reactions (Wehner et al., 2016), and phase-change materials such as water (Miriyev, Stack, & Lipson, 2017) and wax (Lipton, Angle, Banai, Peretz, & Lipson, 2016) that can generate useful internal pressures. Moreover, electric motors can be used to drive and operate underactuated and adaptive tendon-driven soft grippers for grasping and manipulating delicate objects (Manti et al., 2015; Mutlu, Alici, in het Panhuis, & Spinks, 2016).

There is an additional type of soft actuators, commonly known as pneumatic soft actuators, that is widely used to drive soft robotic systems and devices. The most commonly used pneumatic soft actuators that require a positive pressure source to operate include McKibben actuators (Ching-Ping & Hannaford, 1996), fiber-reinforced actuators (Galloway, Polygerinos, Walsh, & Wood, 2013), and PneuNets (Keiko, Wakimoto, Suzumori, & Yasutaka, 2009; Mosadegh et al., 2014). Likewise, pneumatic soft actuators that require a negative pressure source (i.e., vacuum) to operate can also be used to develop and actuate a wide variety of soft robotic systems and devices (Li, Vogt, Rus, & Wood, 2017; Robertson & Paik, 2017; Yang et al., 2016; Yang, Chen, Sun, & Hao, 2017; Yang, Verma, Lossner, Stothers, & Whitesides, 2017), including soft grippers where particles jamming is required or used (Brown et al., 2010).

Pneumatic soft actuators, including positive and negative pressure actuators, are based on a volumetric expansion or contraction. The change in volume upon their actuation can be cleverly harnessed using a variety of geometric designs in their structures to generate multiple deformations such as extension, contraction, bending, twisting, or a combination of them. Pneumatic soft actuators are desired for soft robotic applications due to their high power density, high conformability, affordability, low weight, and ease of fabrication using various additive manufacturing technologies (Zolfagharian et al., 2020).

The development of novel soft, dexterous, and functional pneumatic soft actuators (El-Atab et al., 2020) that can be manufactured using additive manufacturing techniques is crucial for the progress in soft robotics and for developing functional, useful, and commercial soft robotic devices (Trimmer, 2013).

There is a type of soft actuators known as four-dimensional (4D)-printed soft actuators that are based on three-dimensional (3D)-printed passive deformable geometries (e.g., pneumatic soft actuators) or active smart and functional materials (Joshi et al., 2020) (e.g., shape memory alloys, shape memory polymers, hydrogels, variable stiffness materials, and others) that can change their shape with time in response to an external stimulus such as moisture, electric field, magnetic field, light, temperature, pH, chemical species, or pneumatic pressure (Zolfagharian, Mahmud, et al., 2020). 3D-printed structures are usually static, which means that they have a fixed shape that cannot be altered after their manufacturing. However, 4D-printed soft actuators can transform into or have different shapes when activated using an external stimulus. To this end, 4D-printed pneumatic soft actuators are defined as (3D)-printed soft actuators that change their shape (i.e., extend, contract, bend, and/or twist) after their fabrication using various additive manufacturing techniques upon their activation using an external stimulus (i.e., positive or negative air pressure).

In this chapter, we focus specifically on positive and negative pressure pneumatic soft actuators that are fabricated using additive manufacturing technologies. The 4D-printed pneumatic soft actuators are presented in terms of their types, modeling, materials, fabrication, sensing, control, capabilities, and applications in various functional soft robotic systems and devices. Also, a discussion is provided on the presented 4D-printable soft actuators along with a list of what is needed to develop robust and functional pneumatic soft actuators for soft robotic systems and devices.

In addition, this chapter emphasizes the importance of implementing a top-down approach to stimulate more interaction between materials scientists and robotic researchers so that 3D- and 4D-printable soft materials that meet the function and application requirements of a robotic system are developed to fully exploit such synthesized materials by roboticists and use them in practical robotic systems and devices (Tawk & Alici, 2021).

4D-printed pneumatic soft actuators

Types

Pneumatic soft actuators can be divided into two main categories depending on the type of input pressure used for their activation (Fig. 5.2). Such actuators can be usually designed to operate using positive pressure, negative pressure (i.e., vacuum), or a combination of those (Manti et al., 2015; Mutlu et al., 2016). Positive pressure soft pneumatic actuators inflate or expand in volume upon their activation due to the pumping of compressed air in their internal structure, whereas negative pressure soft pneumatic actuators shrink or contract in volume upon their activation due to the suction of air from their internal structure.

When it comes to using pneumatic soft actuators in soft robotic systems, positive pressure pneumatic soft actuators are usually more popular and preferred, or desired, for generating high output forces since the output force is directly related to the input pressure applied. However, positive pressure pneumatic soft actuators cannot

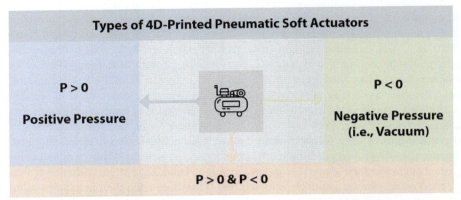

Fig. 5.2 Types of 4D-printed pneumatic soft actuators in terms of their actuation. No permission required.

withstand very high internal pressures, and usually, it is not recommended to use high pressures for activating such actuators (i.e., above 200 kPa) (Keong & Hua, 2018) due to the safety measures that are required to prevent the actuators from bursting and to improve their lifetime. For negative pressure pneumatic soft actuators, the possibility of exploding or burst is not present since they shrink in volume upon their activation and they are limited by the range of input negative pressures that can be applied to their internal structure, which makes them safer to operate in environments where humans are present (Tawk, in het Panhuis, Spinks, & Alici, 2018). However, this limitation in terms of input pressure range limits the output forces generated by such actuators compared to positive pressure actuators in most cases.

In some soft robotic applications, negative pressure soft actuators are preferred over positive pressure soft actuators due to this safety advantage and other additional advantages that they offer, including a fail-safe feature and improved lifetime (i.e., durability) (Tawk et al., 2018) due to the negative pressure input, they require to operate (Alici, 2018; Tawk & Alici, 2021). Fail-safe means that negative pressure soft actuators remain functional, under a continuous supply of vacuum, even after failure where their performance is not affected by minor air leaks or structural damage. Additionally, since negative pressure actuators shrink upon activation, they are ideal for use or deployment in confined spaces (Li et al., 2017).

Modeling

4D-printed pneumatic soft actuators can be modeled analytically and numerically to predict and optimize their performance before their fabrication (Tawk, Gao, Mutlu, & Alici, 2019; Xavier, Fleming, & Yong, 2021; Zolfagharian, Mahmud, et al., 2020). Analytical and numerical models are critical for the efficient design and rapid development of novel and functional pneumatic soft actuators for soft robotic systems and devices (Schmitt, Piccin, Barbé, & Bayle, 2018). Various analytical models and methods were developed to model pneumatic soft actuators such as fiber-reinforced actuators (Gu et al., 2017; Kellaris et al., 2018), PneuNets (Cao, Chu, & Liu, 2020; Gu, Wang, Ge, & Zhu, 2020; Zolfagharian, Mahmud, et al., 2020), and linear soft vacuum actuators (Tawk, Gao, et al., 2019; Tawk, Gillett, in het Panhuis, Spinks, & Alici, 2019; Tawk, in het Panhuis, Spinks, & Alici, 2019). Although various analytical models were developed, numerical methods are used and preferred instead (Schmitt et al., 2018) since it is very challenging to derive analytical models (Connolly, Walsh, & Bertoldi, 2016) for soft robotic actuators due to nonlinearities.

Finite element modeling (FEM) (Xavier et al., 2021) is one common numerical modeling approach to simulate 4D-printed pneumatic soft actuators (Tawk & Alici, 2020) to predict their behavior and performance (Bodaghi et al., 2020; Moseley et al., 2016) since their behavior can be accurately predicted under a specific input such as pressure or displacement.

FEM can be used to quickly predict and optimize accurately the performance of the pneumatic soft actuators before their fabrication, optimize and modify their geometry or topology by iterating rapidly and efficiently between several designs to meet certain design or performance requirements (Moseley et al., 2016; Tawk et al., 2018; Tawk, in het Panhuis, et al., 2019), and save huge amounts of time and potential fabrication resources.

For instance, 4D-printed linear vacuum soft actuators (Fig. 5.3) (Tawk, Gao, et al., 2019, Tawk, Gillett, et al., 2019) and positive pressure soft actuators (Fig. 5.4) (Drotman, Ishida, Jadhav, & Tolley, 2019; Tawk, Sariyildiz, et al., 2020, Tawk, Zhou, et al., 2020) can be modeled before their fabrication analytically and numerically using FEM to predict and optimize their performance in terms of output linear displacement and output blocked force. Similarly, 4D-printed bending vacuum (Fig. 5.3) (Tawk et al., 2018) and positive pressure bellow-like (Herianto, Irawan, Ritonga and Prastowo, 2019; Hu, Lum, et al., 2018) soft actuators can be modeled using FEM to predict their behavior in terms of bending deformation and blocked force.

Fig. 5.3 FEM of 4D-printed pneumatic soft actuators. FEM bending (*left*) and linear (*right*) deformation of negative pressure soft actuators.
Data from Tawk, C., & Alici, G. (2020). Finite element modeling in the design process of 3D printed pneumatic soft actuators and sensors. *Robotics*, *9*, 52. https://doi.org/10.3390/robotics9030052.

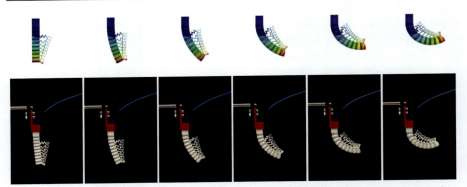

Fig. 5.4 FEM of 4D-printed pneumatic soft actuators. FEM bending deformation (*top*) of positive pressure soft actuators and its corresponding experimental bending deformation (*bottom*) at different input pressures.
From Tawk, C., Mutlu, R., & Alici, G. (2020). A 3D printed modular soft gripper for conformal grasping. In *2020 IEEE/ASME international conference on advanced intelligent mechatronics (AIM)* (pp. 583–588).

Highly nonlinear FE simulations can be performed to predict and optimize the performance of 4D-printed pneumatic soft actuators using commercial FEM software such as SOFA (Payan et al., 2012), Abaqus (Dassault Systèmes Simulia Corp) (Mosadegh et al., 2014), COMSOL Multiphysics® (Galley, Knopf, & Kashkoush., 2019), and ANSYS (ANSYS Inc.) (Tawk & Alici, 2020), that use a range of models for soft and 3D-printable materials such as hyperelastic materials (Xavier et al., 2021).

In addition to FEM, several other modeling methods were used to model soft actuators and robots, including lumped parameter modeling (Rieffel et al., 2009), point-lattices connected by linear beam elements (Hiller & Lipson, 2014), and Cosserat elements (Renda, Boyer, Dias, & Seneviratne, 2018).

For FEM, the FE modeling process (Tawk & Alici, 2020) starts with modeling the 3D computer-aided design (CAD) models of the 4D-printed pneumatic soft actuators (Fig. 5.5). The actuators are designed so that they can be 3D-printed using the desired 3D printing technology such as fused deposition modeling (FDM) (Tawk & Alici, 2020). Additionally, the soft materials used to 3D print the soft actuators are characterized to extract their stress–strain data, which can be used to develop hyperelastic material models for use in FE simulations. Various hyperelastic material models can be developed, including Mooney-Rivlin model, Ogden model, Yeoh model, and Neo-Hookean model, for use in FE simulations (Zolfagharian, Mahmud, et al., 2020). Afterward, the designed CAD models are imported to a FEM software along with the models of the materials developed to simulate the soft actuators to predict their behavior and optimize their performance. Based on the FEM results, the CAD models are modified, and their simulation is repeated accordingly until the desired performance is achieved before their fabrication. Once the desired or optimized performance is obtained, the 3D CAD models are sliced or imported to a 3D printing software specific to the 3D printing technology used where the 3D-printing parameters are

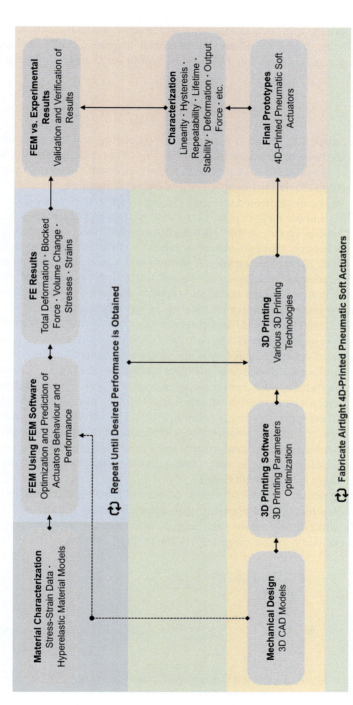

Fig. 5.5 FEM process or roadmap for 4D-printed pneumatic soft actuators. No permission required.

optimized to fabricate airtight and functional 4D-printed pneumatic soft actuators. Finally, the fabricated prototypes are characterized to assess their experimental performance and compare it with its FEM counterpart to verify and validate the simulation results.

Materials

4D-printed pneumatic soft actuators demand soft, stretchable, and flexible materials that are comparable with biological materials (Fig. 5.6) (Zhou et al., 2020). The development of novel 4D-printable soft materials for soft pneumatic actuators must be based on the function requirements and specifications of the robotic application where such actuators will be implemented and used. To fully exploit the synthesized materials by roboticists and use them in practical robotic systems and devices, a top-down approach must be considered so that 3D/4D-printable soft materials that meet the function and application requirements of a robotic system are developed (Carpi et al., 2016; Ewoldt, 2013). The integration of novel soft materials that are amenable to additive manufacturing techniques in various soft robotic systems and devices demands a collaboration between materials scientists and roboticists, which in turn accelerates and promotes the implementation of the top-down approach.

The synthesized materials should have programmable mechanical, rheological, and electrical properties and must be able to vary their overall compliance (i.e., stiffness) to deliver a range of desired output forces and interact safely with various environments that have different stiffnesses and contact conditions (Fig. 5.6). For instance, a soft actuator can distribute the contact forces on a larger contact area and therefore decrease the contact pressure to establish safe contact with an environment by changing its mechanical compliance actively (Majidi, 2013). The variable compliance of a soft actuator can be exploited to switch between different states, including a soft state where it is highly deformable and conformable and a rigid state where it can match the performance of conventional stiff actuators that are made of rigid materials with a bulk modulus of elasticity of at least 1000 MPa.

Pneumatic soft actuators are usually made of materials such as silicones and elastomers that have a modulus of elasticity that ranges between 100 kPa and 1 MPa (Majidi, 2019), while living organs such as skin, muscle tissue, and cartilage have a modulus of elasticity in the range of 100 Pa to 1 GPa (Ewoldt, 2013; Majidi,

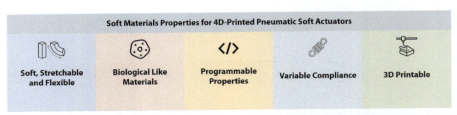

Fig. 5.6 The required properties of the materials used to fabricate 4D-printed pneumatic soft actuators for soft robotic systems and devices.
No permission required.

2013; Rus & Tolley, 2015). For instance, materials such as silicones (Schaffner et al., 2018; Sparrman, Kernizan, Laucks, Tibbits, & Guberan, 2019), thermoplastics polyurethane (TPUs) (Tawk et al., 2018), photocurable elastomeric resins (Bryan, Thomas, Huichan, & Robert, 2015; Ge, Dong, Wang, Ge, & Gu, 2018), and hydrogels (Mishra et al., 2020) can be used to fabricate 4D-printed pneumatic soft actuators.

Technically, the modulus of elasticity describes the softness or hardness of a uniform material that is fabricated in the shape of a prismatic bar to be subjected to an axial loading to generate solely elastic strains (Beer, Johnston, Dewolf, & Mazurk, 2009). Thus, the concept of modulus of elasticity should be cautiously used to describe the softness of a soft robotic material. However, it can be used to compare the relative softness of soft robotic systems with that of hard robotic systems.

There is an overlap between the range of modulus of elasticity of soft materials used for soft robotic actuators and the range of modulus of elasticity of living organs. This match in compliance makes it clear why soft actuators and soft robotic systems are desired to operate and interact with delicate environments compared to conventional rigid actuators. In addition, the programmable compliance of soft actuators makes them more desirable since they can span the whole range of the modulus of elasticity and therefore perfectly match the stiffness of the environment they are expected to interact with. Consequently, safe and adaptive interaction with environments that have different stiffnesses and contact conditions can be directly achieved without requiring complex control algorithms as in hard robotic systems.

Fabrication

4D-printed soft pneumatic actuators can be fabricated using various additive manufacturing techniques and technologies (Fig. 5.7) such as FDM (Elgeneidy, Neumann, Jackson, & Lohse, 2018; Hu & Alici, 2019; Keong & Hua, 2018; Stano, Arleo, & Percoco, 2020; Tawk, Spinks, in het Panhuis, & Alici, 2019a, Tawk, Spinks, in het Panhuis, & Alici, 2019b; Yap, Ng, & Yeow, 2016), stereolithography (SLA) (Bryan et al., 2015; Costas et al., 2018; Mishra et al., 2020; Shapiro, Wolf, & Gabor, 2011; Wallin et al., 2017; Yu, Xin, Du, Li, & Wang, 2019), digital light processing (DLP) (Ge et al., 2018; Patel et al., 2017; Thrasher, Schwartz, & Boydston, 2017; Zhang et al., 2019; Zhang et al., 2019; Zhang et al., 2019), selective laser sintering (SLS) (Grzesiak, Becker, & Verl, 2011; Laschi et al., 2017), material

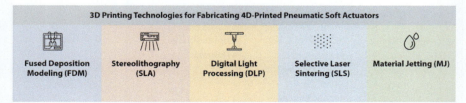

Fig. 5.7 The various 3D-printing technologies used to fabricate 4D-printed pneumatic soft actuators for soft robotic systems and devices.
No permission required.

jetting (MJ) (Dämmer, Gablenz, Hildebrandt, & Major, 2019; Manns, Morales, & Frohn, 2018), and rapid liquid printing (RLP) (Sparrman et al., 2021) using either a single soft material such as silicone (Bryan et al., 2015; Costas et al., 2018; Miriyev, Xia, Joseph and Lipson, 2019; Mishra et al., 2020; Shapiro et al., 2011; Wallin et al., 2017; Yu et al., 2019) or multiple soft materials (Skylar-Scott, Mueller, Visser, & Lewis, 2019) with multiple properties (Kalisky et al., 2017; MacCurdy, Katzschmann, Youbin, & Rus, 2016) and capabilities (Shih et al., 2019). The comparison between the different 3D-printing technologies with respect to the 4D-printed pneumatic soft actuators is presented in Table 5.1. The printing time

Table 5.1 3D printing technologies comparison with respect to 4D-printed pneumatic soft actuators.

3D printing technology	Printer cost	Availability	Layer resolution
Fused Deposition Modeling (FDM)	150–$5000 USD (depends on quality and print volume)	Commercially available models and laboratory-based modified models based on commercial models	100–200 µm (depends on airtightness requirements and exterior/surface quality)
Stereolithography (SLA)	~1500–$10,000 USD (depends on quality, print volume, and precision)	Laboratory-based models	50 µm
Digital Light Processing (DLP)	~3000–$20,000 USD (depends on quality, print volume, and precision)	Commercially available models and laboratory-based modified models based on commercial models	50–100 µm
Selective Laser Sintering (SLS)	~1000–$15,000 USD (expensive compared to the other technologies)	Commercially available	100–120 µm
Material Jetting (MJ)	>$40,000 USD (expensive compared to the other technologies)	Commercially available	16–30 µm
Rapid Liquid Printing (RLP)	Not Reported	Laboratory-based models	1000 µm (1 mm)

No permission required.

of each technology is not presented in Table 5.1 since it is highly dependent on many factors, including the size of the actuator, layer resolution, printing speed, infill settings, pre-processing time, post-processing time, and complexity of the geometry.

3D-printing technologies are highly desired for fabricating and prototyping pneumatic soft actuators since various designs and concepts can be rapidly and efficiently realized (Gul et al., 2018; Wallin, Pikul, & Shepherd, 2018; Zolfagharian et al., 2016). For instance, pneumatic soft actuators with programmable motion (Schaffner et al., 2018) and various capabilities such as locomotion and gripping (MacCurdy et al., 2016) can be directly developed in one manufacturing step. Also, through 3D-printing, the elasticity of soft actuators and robots can be tuned and controlled using complex soft structures (Schumacher et al., 2015). Additionally, their performance (Yirmibesoglu et al., 2018) and durability (Tawk et al., 2018) can be both significantly enhanced compared to the performance and durability of pneumatic soft actuators that are fabricated using time-consuming conventional manufacturing techniques that involve multiple laborious fabrication steps that limit the development of soft actuators with complex geometries and topologies (Marchese, Katzschmann, & Rus, 2015; Schmitt et al., 2018).

However, not all available and developed 3D-printing technologies are accessible and affordable. Accessibility and affordability are major limitations when it comes to considering 3D-printing technologies for developing soft actuators and specifically 4D-printed pneumatic soft actuators. For example, FDM 3D-printing is easily accessible due to the variety of affordable and commercial FDM 3D-printers in the market and the availability of numerous low-cost hard and soft materials that they use to 3D-print soft pneumatic actuators. However, the remaining additive manufacturing technologies are either too expensive or not readily available outside research laboratories. Moreover, the fabrication of functional airtight 4D-printed pneumatic soft actuators using FDM 3D-printing, for instance, requires essential skills including CAD design skills and 3D-printing skills (Tawk et al., 2018).

Each 3D-printing technology uses a range of soft materials for developing 4D-printed pneumatic soft actuators. For instance, FDM 3D printers use a range of commercially available materials, with different shore hardness ranging from 60A to 95A, including NinjaFlex (modulus of elasticity of 12 MPa, NinjaTek, USA) (Tawk et al., 2018), FilaFlex Shore 60A, 70A, and 82A (stress at 300% elongation is 10 MPa, Recreus Industries S.L.) (Anver, Mutlu, & Alici, 2017; Hu & Alici, 2019), eSUN TPU eFlex flexible (Shenzhen Esun Industrial Co., Ltd) (Salem, Wang, Wen, & Xiang, 2018), X60 Ultra-Flexible (stress at 300% elongation is 2 MPa, Diabase Engineering) (Ang & Yeow, 2020), Anycubic TPU (Anycubic 3D Corporation) (Zhou, Fu, & He, 2020), and Ultimaker TPU 95A (modulus of elasticity of 26 MPa, Ultimaker USA Inc.) (Scharff, Doornbusch, et al., 2019; Scharff, Wu, Geraedts, & Wang, 2019).

For SLA 3D printers, commercial resins are used, including Spot-E resin (modulus of elasticity of ~5.2 MPa, Spot-A Materials, Inc.) (Bryan et al., 2015), SL5180 resin (Huntsman LLC) (Kang, Lee, & Cho, 2006), and a mixture of hard and soft resins, including Formlabs Clear and Formlabs Flexible (Formlabs) (Costas et al., 2018). Also, synthesized resins (modulus of elasticity ranges between 6 and 287 kPa) that are based on commercially available materials (i.e., blend of poly(mercaptopropyl) methylsiloxane-co-dimethylsiloxane and bifunctional vinyl terminated PDMS) (Wallin et al., 2017) are developed to produce 4D-printed pneumatic soft actuators using SLA 3D-printers.

For DLP 3D-printers, commercial Tangoplus FLX930 (Stratasys) (Ge et al., 2018), with shore 26-28A hardness and elongation at break up to 220%, is used to develop bending 4D-printed pneumatic soft actuators. Additionally, various elastomeric photo resins are synthesized based on different formulations that result in elongations at break up to 472% (Thrasher et al., 2017) and 1100% (with a modulus of elasticity ranging between 0.58 and 4.21 MPa) (Patel et al., 2017) for use with DLP 3D-printers to fabricate soft pneumatic actuators.

For SLS 3D-printers, commercial nylon composite powder (Liu et al., 2020) and flexible TPU92A-1 material with shore 88A hardness (Materialise, Belgium) (Laschi et al., 2017) are used to develop 4D-printed soft pneumatic actuators.

For MJ 3D-printers, soft commercial materials, including TangoBlackPlus with shore 26-28A hardness (Stratasys, Ltd) (Dämmer et al., 2019; Drotman et al., 2019; Drotman, Jadhav, Karimi, deZonia, & Tolley, 2017; Lee, Kim, Choi, & Cho, 2016; MacCurdy et al., 2016; Shapiro et al., 2011; Shih et al., 2019; Wang, Chathuranga, & Hirai, 2016; Wang, Torigoe, & Hirai, 2017), Agilus30 with shore 30-35A hardness (Stratasys, Ltd) (Ge et al., 2018; Patel et al., 2017; Thrasher et al., 2017; Zhang, Ng, et al., 2019), FLX4670-DM Rubber-Like Digital Material (Stratasys, Ltd) (Ito, Kojima, Okui, & Nakamura, 2020), and FLX9070-DM Rubber-Like Digital Material (Stratasys, Ltd) (Kalisky et al., 2017), are used to fabricate 4D-printed soft pneumatic actuators.

For silicone 3D-printing and 3D-printers, commercial silicones, including Ecoflex 0010 (modulus of elasticity of 0.008 MPa at 100% strain, Smooth-On, Inc.) (Hamidi & Tadesse, 2020), ACEO Shore 20A, Shore 30A, and Shore 60A silicones (ACEO, Wacker Chemie AG, Burghausen, Germany) (Heung, Tang, Shi, Tong, & Li, 2020; Heung et al., 2019; Tang, Heung, Tong, & Li, 2020), Ecoflex 00-35 platinum-catalyzed silicone rubber (modulus of elasticity ranges between 90 and 105 kPa at 100% strain, Smooth-On, Inc.), Dragon Skin 10 Very Fast (Smooth-On, Inc.) (Byrne et al., 2018; Yirmibesoglu et al., 2018), EcoFlex 00-30 (Smooth-On, Inc.) (Morrow, Hemleben, & Menguc, 2017; Truby, Katzschmann, Lewis, & Rus, 2019; Truby et al., 2018), Dragon Skin 10 (DS10) Very Fast (Smooth-On, Inc.) with a modulus of elasticity of 147 kPa (Yirmibeşoğlu, Oshiro, Olson, Palmer, & Mengüç, 2019) and 133 kPa (Walker et al., 2019), Ecoflex 00-50 (stress of 0.4 MPa at 100% strain, Smooth-On, Inc.) (Miriyev et al., 2017), Dow Corning®737 with shore 33A hardness (Dow Corning, USA) (Plott & Shih, 2017), Sylgard 184 (S184; Dow Corning) (Skylar-Scott et al., 2019), and Moldstar 30 (modulus of elasticity of 0.68 MPa, Smooth-On, Inc.) (Sparrman et al., 2021) are used to 4D-print pneumatic soft actuators.

Sensing and control

The control of 4D-printed pneumatic soft actuators requires the integration of sensing elements or soft sensors in their body to sense themselves (e.g., mechanical strains) and their immediate environment (i.e., contact forces) for precise control applications.

Ideally, it is desired to seamlessly integrate the soft sensors in the body of the soft actuator or soft robotic system where the soft actuator or robotic system has a continuum topology embodying all its essential elements such as soft actuators, soft sensors, electronic skins (e-skins) (Shih et al., 2020), stretchable and flexible electronics, power sources, and passive and/or active soft and compliant structures (Fig. 5.1). In addition, it is desired and preferable that such soft sensors have a minimal impact on the compliance of the soft actuator or soft robotic system (Zolfagharian, Mahmud, et al., 2020). In other words, the material of the soft sensors should be softer (i.e., more compliant) than the surface of the soft actuator, in case the sensors are placed on it (e.g., e-skin), or its body in case the sensors are placed in it (Tawk & Alici, 2021). This is crucial for not creating mechanical resistance to actuation and for the sensors to stay stable over their full lifetime.

A range of soft sensors or soft sensing technologies can be used along with soft robotic actuators or in soft robotic systems (Fig. 5.8). For instance, resistive stretchable and/or flexible strain sensors such as flex sensors (Elgeneidy, Neumann, et al., 2018; Gerboni, Diodato, Ciuti, Cianchetti and Menciassi, 2017; Hu & Alici, 2019; Keong & Hua, 2018; Stano et al., 2020; Tawk, Spinks, et al., 2019a, 2019b; Yap et al., 2016), conductive inks (Kumbay Yildiz, Mutlu, & Alici, 2016; Muth et al., 2014; Yeo et al., 2016), ionic conductive liquids (Truby et al., 2018), liquid metals (Dickey, 2017; Park, Chen, & Wood, 2012; Xi, Yeo, Yu, Zhang, & Lim, 2017), fabrics and textiles (Lina & Alison, 2014; Stoppa & Chiolerio, 2014), highly compressive, sensitive, and biodegradable foam sensors (Sencadas, Tawk, & Alici, 2019, 2020), resistive 3D-printable thermoplastics (Elgeneidy, Neumann, et al., 2018), and ultrathin piezoresistive sensors (Mai, Mutlu, Tawk, Alici, & Sencadas, 2019; Sencadas, Mutlu, & Alici, 2017) can be used in soft robotic systems and devices. Other soft sensing technologies include soft capacitive pressure (Atalay, Atalay, Gafford, & Walsh, 2018; Sencadas, Tawk, Searle, & Alici, 2021; Viry et al., 2014), tactile (Bin, Yang, Adam, & Yon, 2016), and strain sensors (Frutiger et al., 2015), in addition to optical strain, curvature, texture, and force sensors (Zhao, O'Brien, Li, & Shepherd, 2016), optical tactile soft sensors (Xiang, Guo, & Rossiter, 2019), self-powered soft tactile sensors based on magnetoelectric materials (Ge et al., 2018; Patel et al., 2017; Thrasher et al., 2017; Zhang, Ng, et al., 2019), and acoustic soft sensors (Chossat & Shull, 2021). Finally, some soft sensing techniques are based on soft pneumatic sensing chambers that can be directly 3D-printed using FDM 3D-printers (Drotman et al., 2019; Tawk, Mutlu, & Alici, 2020), PolyJet (Kim, Alspach, & Yamane, 2015; Slyper & Hodgins, 2012; Vázquez, Brockmeyer, Desai, Harrison, & Hudson,

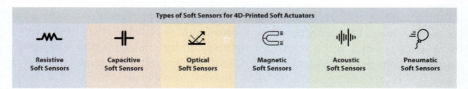

Fig. 5.8 The various types of sensors developed for integration in soft robotic systems and devices, and their use alongside 4D-printed pneumatic soft actuators.
No permission required.

2015), and SLA (Aulia Putra, 2018) to exploit the volume change of soft deformable hollow structures. Such soft sensing chambers can be used as human motion monitoring or tracking sensors (Kong & Tomizuka, 2009), touch sensors and skins (Aulia Putra, 2018; Kim et al., 2015; Tawk, in het Panhuis, et al., 2019), force sensors (Choi, Jung, Jung and Kong, 2017; Li et al., 2017; Robertson & Paik, 2017; Yang et al., 2016; Yang, Verma, et al., 2017), tactile sensors (Gong, He, Yu, & Zuo, 2017; Lu, He, Nanayakkara, & Rojas, 2020), bending sensors (Tawk, in het Panhuis, Spinks, & Alici, 2020; Tawk, Sariyildiz, et al., 2020; Tawk, Zhou, et al., 2020), linear displacement sensors, and torsional sensors (Tawk, in het Panhuis, et al., 2019).

Despite the numerous sensing technologies developed, the use of such technologies in closed-loop control systems for soft actuators and robotic systems is still limited mainly because such sensors were originally developed for state estimation in wearable soft robotic devices (Shih et al., 2020). However, in some cases, the soft sensors can be either purposefully 3D-printed (e.g., soft pneumatic sensing chambers) (Tawk, Sariyildiz, et al., 2020; Tawk, Zhou, et al., 2020) simultaneously with a soft body or actuator, or integrated manually in the body of the built soft actuator (e.g., a silicone matrix containing channels filled with either eutectic gallium–indium (eGaIn) or conductive carbon paste) (Morrow et al., 2016; Yildiz, Mutlu, & Alici, 2018) to achieve accurate and reliable position and force control.

In addition, the use of soft sensors in control applications for soft actuators and robotic systems is usually limited by the drawbacks of such sensors, including slow response, nonlinear response, hysteresis, lack of repeatability and reliability, drift over time, short lifetime, and limited stretchability (i.e., only small strains can be measured).

Despite all these challenges, there are several control methods and techniques that are implemented to control soft robotic actuators and systems such as 3D-printed soft monolithic robotic fingers with embedded soft pneumatic sensing chambers where proportional-integral-derivative (PID) controllers are implemented (C. Tawk, Zhou, et al., 2020), pneumatic soft skins for closed-loop haptic feedback control where PID controllers are used as well (Sonar, Gerratt, Lacour, & Paik, 2019), and pneumatic soft artificial muscles with biomimetic microfluidic sensors where a sliding mode controller is used (Wirekoh, Valle, Pol, & Park, 2019).

The information acquired from the soft sensors can be used to develop feedforward and feedback controllers (George Thuruthel, Ansari, Falotico, & Laschi, 2018) to achieve closed-loop control of 4D-printed pneumatic soft actuators (Zolfagharian, Kaynak, & Kouzani, 2020; Zolfagharian, Mahmud, et al., 2020).

The use of conventional PID controllers for controlling pneumatic soft actuators is not desired due to the highly nonlinear kinematics of such actuators in unstructured environments, which result from their hyperelastic material properties and design geometry (Zolfagharian, Mahmud, et al., 2020), and hysteresis that might result from the sensor or actuator material (Abbasi, Nekoui, Zareinejad, Abbasi, & Azhang, 2020). Alternatively, 4D-printed pneumatic soft actuators can be controlled using data-driven machine learning models and controllers (Elgeneidy, Lohse, & Jackson, 2018; George Thuruthel et al., 2017; Thuruthel, Shih, Laschi, & Tolley,

2019) that are suitable for such nonlinearities. Additionally, in some cases, FEM can be efficiently used to accurately control soft robotic systems (Bieze et al., 2018; Zolfagharian, Kaynak, & Kouzani, 2020).

Capabilities

4D-printed pneumatic soft actuators can be fabricated using smart, functional, and cleverly designed materials and structures that provide them with several capabilities and functionalities, including self-healing properties, fail-safe features, scalability, customizability, modularity, and multimodal programmable actuation (Fig. 5.9).

Self-healing properties and fail-safe features

One of the fascinating characteristics of biological muscles is their ability to self-heal after being repeatedly damaged or mechanically stressed and consequently becoming denser and stronger. Soft and 3D-printable materials with similar self-healing properties (Roels et al., 2020) are highly desired for 4D-printed positive and negative pressure pneumatic soft actuators because any damage or crack in their structure that leads to air leaks, and therefore, their failure can be automatically repaired to restore their original function (Yu et al., 2019).

For example, sunlight can be focused on the structures of soft bidirectional bending actuators that are 3D-printed using a stereolithography 3D-printer to rapidly self-heal their punctured structure and consequently restore their functionality (Wallin et al., 2017). In addition, fused filament fabrication (FFF) can be used to print self-healing materials in filament form to produce self-healing 4D-printed pneumatic soft actuators (Terryn et al., 2021), including fast healing capacity at room temperature.

In addition to being able to self-heal, soft and 4D-printable negative pressure actuators are characterized by their fail-safe feature since they remain functional, under a continuous supply of negative pressure, even after their rupture or any crack propagation in their structure (i.e., 3D-printed layers separation due to mechanical stresses) (Kalisky et al., 2017; Tawk, Spinks, et al., 2019a, 2019b).

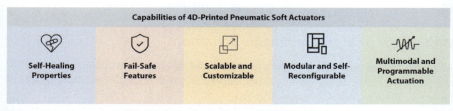

Fig. 5.9 The diverse capabilities and functionalities of 4D-printed pneumatic soft actuators. No permission required.

Scalability and customizability

4D-printed pneumatic soft actuators can be easily and rapidly scaled, in terms of their overall size and internal volume, and customized using additive manufacturing techniques to target specific applications with specific requirements.

For instance, miniature soft robotic systems and devices, including grippers, artificial muscles, locomotion robots, and camouflage robots, can be developed based on micro-sized 4D-printed pneumatic soft actuators (Kang et al., 2006; Zhang, Ng, et al., 2019). The size of such micro-sized actuators can be down to 2 mm in terms of overall size along with 150–350 μm in terms of voids (i.e., the width of channels or internal chambers) (Zhang, Ng, et al., 2019).

Likewise, 4D-printed macro-sized negative pressure soft actuators can be scaled either in terms of their internal volume or in terms of the number of actuators assembled in one single unit to amplify their output force and linear stroke (i.e., displacement) (Drotman et al., 2019; Tawk, Spinks, et al., 2019a, 2019b). The size of macro-sized 4D-printed pneumatic soft actuators can range from few centimeters (e.g., 1–20 cm) (Bryan et al., 2015; Charbel Tawk, in het Panhuis, et al., 2019; Tawk et al., 2018) up to large-scale silicone pneumatic soft actuators (Sparrman et al., 2021) with an overall size that is bigger than 20 cm (Sparrman et al., 2019), which is usually larger than the size of the print bed or print height (i.e., volume) of the majority of 3D printers used.

Modularity

Soft modular actuators allow soft robotic systems and devices to self-reconfigure so that they can disassemble and reassemble to form new morphologies to adapt to various environments and tasks. Also, the modularity of soft actuators allows the distribution of actuation and sensing and consequently improves the functionality and reliability of the actuators and leads to a decrease in their overall cost and maintenance costs (Yim et al., 2007). Distributed actuation and sensing means that their different configurations in the same soft robotic system or device can be considered to target specific requirements (Qi, Shi, Pinto, & Tan, 2020; Singh, Tawk, Mutlu, Sariyildiz, & Alici, 2019; Skylar-Scott et al., 2019).

For instance, 4D-printed pneumatic soft actuators with variable length and degrees of freedom (DoF) can be easily and rapidly built, based on modular pneumatic units or blocks that can be 3D-printed separately and assembled, for implementation in various robotic devices (Lee et al., 2016; Tawk et al., 2018).

Multimodal and programmable actuation

4D-printed pneumatic soft actuators that can bend, twist, contract, and extend simultaneously are essential for various robotic applications that require multiple modes of deformation to accomplish a specific task.

For instance, the gripping performance of a soft gripper can be enhanced by using soft helical actuators that wrap around grasped objects to realize a firm grip by generating bending and twisting motions simultaneously (Hu & Alici, 2019).

Likewise, the material properties and architectures of pneumatic soft actuators can be exploited and programmed through 3D-printing to develop soft actuators that can either perform a single motion or various motions simultaneously, which are required by specific soft robotic systems and devices (Byrne et al., 2018; Schaffner et al., 2018).

Moreover, soft manipulators that are based on a bundle of 4D-printed pneumatic soft actuators can perform a desired motion in a 3D space for pick and place tasks by activating a specific actuator from the actuation bundle (Tawk, Spinks, et al., 2019a, 2019b).

Applications

4D-printed pneumatic soft actuators that possess such various characteristics when it comes to their fabrication, materials, sensing, control, and capabilities are highly desired to develop a variety of functional soft robotic systems and devices such as locomotion robots, grippers, manipulators, artificial muscles, and assistive wearable and medical devices with haptic feedback capabilities (Fig. 5.10).

Soft locomotion robots

4D-printed pneumatic soft actuators are well suited to drive and actuate soft locomotion robots (Figs. 5.11 and 5.12) that can handle extreme conditions such as elevated temperatures, excessive deformations and pressures, impact forces, confined spaces (Tolley et al., 2014), and radiations (Yirmibeşoğlu et al., 2019) in extreme and harsh unstructured (i.e., dynamic) environments (Calisti, Picardi, & Laschi, 2017).

For instance, inspection, search, and rescue locomotion robots that are based on 4D-printed pneumatic soft actuators can be deployed to operate in confined spaces (Fig. 5.12) (Tawk, Spinks, et al., 2019a, 2019b) and dynamic terrains (Drotman et al., 2017) since they adapt to various environments, due to their inherent compliance, and navigate them by jumping (Bartlett et al., 2015), crawling (Keong & Hua, 2018; Lee et al., 2016; Miriyev et al., 2017), undulating (Qi et al., 2020), hopping (Fig. 5.11) (Tawk et al., 2018), and walking (Fig. 5.11) (Drotman et al., 2019; MacCurdy et al., 2016; Tawk et al., 2018).

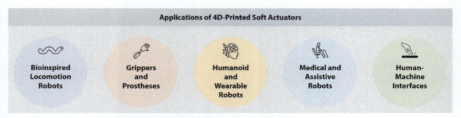

Fig. 5.10 The various soft robotic applications of 4D-printed pneumatic soft actuators. No permission required.

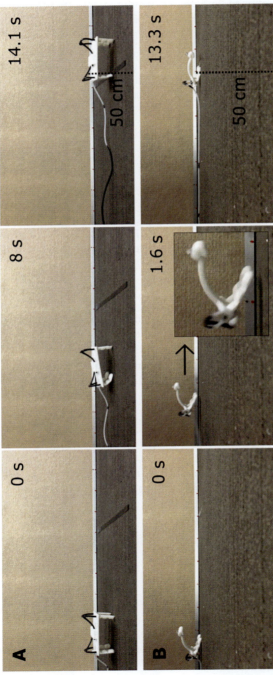

Fig. 5.11 Soft locomotion robots based on 4D-printed pneumatic soft actuators. A walking (*top*) and a hopping (*bottom*) robot based on 4D-printed negative pressure bending soft actuators. No permission required.

Fig. 5.12 Soft locomotion robots based on 4D-printed pneumatic soft actuators. A navigation robot based on 4D-printed negative pressure linear soft actuators.
From Tawk, C., Spinks, G. M., in het Panhuis, M., & Alici, G. (2019). 3D printable linear soft vacuum actuators: Their modeling, performance quantification and application in soft robotic systems. *IEEE/ASME Transactions on Mechatronics*, *24*(5), 2118–2129. https://doi.org/10.1109/TMECH.2019.2933027

Also, in addition to their softness and ability to handle extreme conditions, such soft locomotion robots can be modular (Lee et al., 2016) and scalable (Zhang, Ng, et al., 2019), which means they can be easily customized for various locomotion applications such as lifting and transporting moderate and heavy loads (Skylar-Scott et al., 2019; Yirmibesoglu et al., 2018).

Soft grippers and parallel manipulators

4D-printed pneumatic soft actuators can be directly used to develop soft, adaptive, and compliant grippers that are ideal for grasping, picking, manipulating, and placing delicate and fragile objects in unstructured environments (Figs. 5.13 and 5.14) (Shintake, Cacucciolo, Floreano, & Shea, 2018).

Such grippers can be used to grasp and pick a wide variety of heavy and lightweight objects (Fig. 5.13 and Fig. 5.14), including objects with irregular and various shapes (Drotman et al., 2019; Tawk, Spinks, et al., 2019a), fruits and vegetables (Tawk, Gillett, et al., 2019; Tawk et al., 2018), food (Wang et al., 2017), and packages (Wang et al., 2016). They are usually designed based on the human-hand model to realize dexterous anthropomorphic grippers and robotic hands (Laschi et al., 2017) that are built either using multiple and separate 4D-printed fingers or monolithically in one manufacturing step (Anver et al., 2017; Ge et al., 2018; MacCurdy et al., 2016; Patel et al., 2017; Thrasher et al., 2017).

Also, such grippers can possess various functions and capabilities, including variable stiffness (Mutlu, Tawk, Alici, & Sariyildiz, 2017; Zhang, Zhang, et al., 2019; Zhu, Mori, Wakayama, Wada, & Kawamura, 2019), modularity (Lee et al., 2016; Tawk, Mutlu, & Alici, 2020), scalability (i.e., micro-sized grippers) (Kang et al., 2006; Zhang, Ng, et al., 2019), pressure and/or position sensing (Bryan et al., 2015; Costas et al., 2018; Low et al., 2017; Mishra et al., 2020; Shapiro et al., 2011; Shapiro, Wolf, & Kósa, 2013; Wallin et al., 2017; Wei et al., 2019;

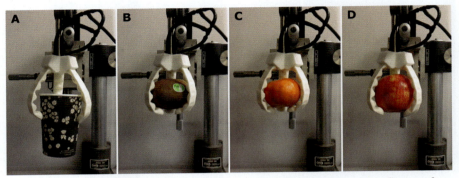

Fig. 5.13 Soft grippers based on 4D-printed pneumatic soft actuators. A three-finger soft gripper based on 4D-printed negative pressure bending soft actuators.
No permission required.

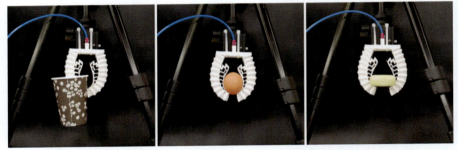

Fig. 5.14 Soft grippers based on 4D-printed pneumatic soft actuators. A monolithic soft gripper based on 4D-printed positive pressure bending soft actuators.
From Tawk, C., Gao, Y., Mutlu, R., & Alici, G. (2019). Fully 3D printed monolithic soft gripper with high conformal grasping capability. In *2019 IEEE/ASME international conference on advanced intelligent mechatronics* (AIM) (pp. 1139–1144).

Yang & Chen, 2018; Yu et al., 2019), and multiple modes of deformation such as bending and twisting simultaneously (Hu, Li, & Alici, 2018; Hu & Alici, 2019; Rosalia, Ang, & Yeow, 2018). Their grasping performance can be enhanced using 4D-printed pneumatic soft actuators with reduced out-of-plane deformation (Scharff, Wu, et al., 2019), and bioinspired compliant structures (Tawk, Gao, et al., 2019). Finally, soft robotic grippers can be based on hybrid structures that incorporate 4D-printed pneumatic soft actuators along with rigid, soft, or semi-soft materials (Miriyev et al., 2017; Tawk, Gillett, et al., 2019; Zhou, Chen, et al., 2020).

4D-printed soft pneumatic actuators can also be used to drive and build parallel manipulators with multiple DoF to guide soft robotic grippers or soft end-effectors (Kumar et al., 2017) in space to grasp, manipulate, or pick and place objects (Fig. 5.15). For instance, parallel manipulators can be based on a bundle of 4D-printed vacuum-powered soft actuators that are coupled with either soft suction cups for

Fig. 5.15 Soft manipulators based on 4D-printed pneumatic soft actuators. A soft manipulator based on 4D-printed negative pressure linear soft actuators.
From Tawk, C., Spinks, G. M., in het Panhuis, M., & Alici, G. (2019). 3D printable linear soft vacuum actuators: Their modeling, performance quantification and application in soft robotic systems. *IEEE/ASME Transactions on Mechatronics*, *24*(5), 2118–2129. https://doi.org/10.1109/TMECH.2019.2933027.

picking and placing applications (Drotman et al., 2019; Tawk, Spinks, et al., 2019a) or a laser pointer for handwriting applications (Kalisky et al., 2017). Also, such soft manipulators can move in space by bending, extending, and contracting either separately or simultaneously and therefore making use of the multimodal and programmable actuation of 4D-printed pneumatic soft actuators (Liu et al., 2020; Sparrman et al., 2019).

Soft artificial muscles

4D-printed pneumatic soft actuators can also be implemented as artificial muscles in various robotic applications, such as humanoid robots, to mimic the behavior and performance of natural muscles (Figs. 5.16 and 5.17) (Miriyev et al., 2017; Tawk et al., 2018) in terms of lifting heavy loads (Tawk, Spinks, et al., 2019a) and generating various modes of deformation (Schaffner et al., 2018).

Such artificial muscles are characterized by their high power density, high output force, fast response, compliance, durability, modularity, accessibility, affordability, scalability (Daerden, Lefeber, Verrelst, & Ham, 2001; Tawk, Spinks, et al., 2019a; Xie, Zuo, & Liu, 2020; Zhang, Ng, et al., 2019), and their ability to either expand (Ito et al., 2020) or contract in volume (Tawk et al., 2018) to generate the desired output displacement and force.

Soft assistive wearable and medical devices

4D-printed pneumatic soft actuators are highly desired for assistive and medical soft robotic devices such as wearable robots for human performance augmentation and rehabilitation applications (Cianchetti, Laschi, Menciassi, & Dario, 2018; Thalman & Artemiadis, 2020), again, due to their various capabilities and advantages such as high power density, high output force, fast response, compliance, durability, modularity, accessibility, affordability, and scalability.

For instance, wearable upper limb exoskeletons (Keong & Hua, 2018) and hands (Heung et al., 2020; Heung et al., 2019) for finger (Rosalia et al., 2018), wrist (Yap et al., 2016), elbow (Ang & Yeow, 2020), and hand rehabilitation can be worn and used by patients to assist them with recovering their functional motor skills and their

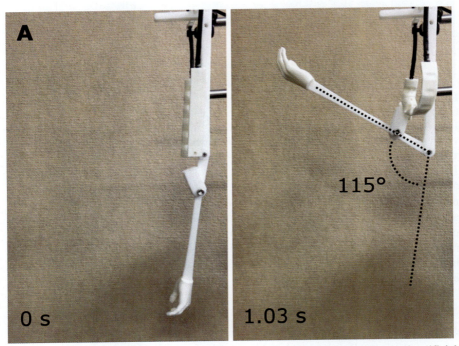

Fig. 5.16 Soft artificial muscles based on 4D-printed pneumatic soft actuators. A soft artificial muscle based on a pair of 4D-printed negative pressure bending soft actuators.
No permission required.

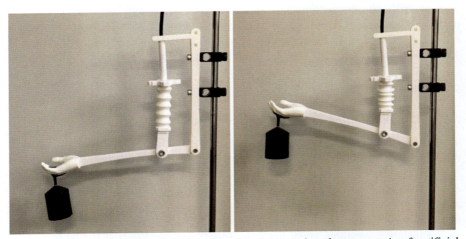

Fig. 5.17 Soft artificial muscles based on 4D-printed pneumatic soft actuators. A soft artificial muscle based on a single 4D-printed negative pressure linear soft actuator.
From Tawk, C., Spinks, G. M., in het Panhuis, M., & Alici, G. (2019). 3D printable linear soft vacuum actuators: Their modeling, performance quantification and application in soft robotic systems. *IEEE/ASME Transactions on Mechatronics, 24*(5), 2118–2129. https://doi.org/10.1109/TMECH.2019.2933027

daily activities. Additionally, assistive upper limb devices such as soft prosthetic hands can replace a lost upper limb and partially recover some of its functions (Laschi et al., 2017).

Similarly, wearable lower limb soft devices with haptic feedback capabilities can assist stroke patients by improving the biofeedback provided during their rehabilitation process (Singh et al., 2019, 2020).

Finally, such wearable upper and lower limbs soft robotic devices, for instance, can be used to augment human performance for certain tasks that require heavy load lifting (Xie et al., 2020).

Discussion

Challenges of 4D-printed pneumatic soft actuators

Although pneumatic soft actuators (Walker, Harbel, Strickland, & Shin, 2020), and specifically 4D-printed pneumatic soft actuators, are one of the most used actuators for building soft robotic systems and devices (Tawk & Alici, 2021), they still face some challenges that hinder their usage in portable soft robotic devices (Fig. 5.18).

Portability

4D-printed pneumatic soft actuators require additional bulky equipment to operate, including a pressure source along with power supplies, complex electric circuits, and pneumatic valves. Usually, the additional bulky equipment required limits the adoption of pneumatic soft actuators in portable soft robotic systems and devices since the overall weight of the device is a critical constraint in certain applications such as soft prosthetic hands (Belter, Segil, Dollar, & Weir, 2013; Zhou, Mohammadi, Oetomo, & Alici, 2019) in which the mobility and comfort of the user highly depend on the weight of the hand. However, this additional weight does not impede the usage of 4D-printed pneumatic soft actuators in other several applications such as soft wearable upper and lower limbs exoskeletons (Dickey, 2017; Park et al., 2012, 2014; Xi et al., 2017) and rehabilitation devices that can be used at home or in a healthcare facility (Shahid, Gouwanda, Nurzaman, & Gopalai, 2018).

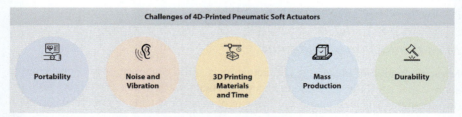

Fig. 5.18 The main challenges of 4D-printed pneumatic soft actuators in terms of their integration in various soft robotic systems and devices.
No permission required.

This additional equipment required by 4D-printed pneumatic soft actuators can be dramatically downsized as such actuators can be miniaturized. For instance, the downsizing of the pneumatic equipment is desired for haptic feedback applications (Sonar et al., 2019) where only small forces and mechanical deformations are required. However, lightweight pneumatic equipment along with miniature soft pneumatic actuators means that only small output forces can be generated. Thus, the choice of the pneumatic equipment highly depends on the requirements of the device and its functions in terms of size, output force, and deformation, primarily.

Noise and vibration

4D-printed pneumatic soft actuators are by themselves silent upon their actuation; however, since they are driven by air compressors or vacuum pumps (Dämmer et al., 2019; Drotman et al., 2019, 2017; Lee et al., 2016, 2017; MacCurdy et al., 2016; Shapiro et al., 2011; Shih et al., 2019; Wang et al., 2016, 2017), their operation generates some levels of noise and vibrations that are usually not desired, especially in wearable applications. In addition, such levels of noise and vibration limit the portability of soft robotic systems and devices based on 4D-printed pneumatic soft actuators when additional components that result in an additional weight are used to reduce or suppress the noise or vibration levels.

3D-printing materials and printing time

It is challenging to 3D/4D-print a variety of soft materials with elastic moduli matching the ones of soft tissues that range between 3 kPa for stromal tissue and 900 kPa for cartilage (Wallin et al., 2017, 2018). To address such a challenge, novel additive manufacturing technologies along with polymer chemistries must be developed (Wallin et al., 2018). For instance, recently developed 3D/4D-printable materials such as silicones (Wallin et al., 2020) and hydrogels (Cheng et al., 2019) can be used to develop soft robotic systems that address such a challenge.

In addition, some 3D-printing technologies such as FDM are time-consuming as they require multiple hours to produce a single airtight pneumatic soft actuator (Tawk, Spinks, et al., 2019a). However, the printing speed can be increased using novel 3D printing technologies to rapidly produce, for instance, silicone-based soft pneumatic actuators (Sparrman et al., 2019).

Mass production and lifetime

Mass production of pneumatic soft actuators is still a challenge or not considered in some cases as 4D-printed pneumatic soft actuators are meant for prototyping and developing new actuation concepts instead of mass-producing commercial soft actuators. The production of 4D-printed pneumatic soft actuators must take into consideration several factors, including production volume (i.e., number of printed actuators per hour or day), reliability and repeatability (i.e., consistency in terms of performance and dimensions), and most importantly durability (i.e., lifetime) since it is proved that

4D-printed pneumatic soft actuators can last longer compared to their molded counterparts (Tawk et al., 2018).

Requirements for 4D-printed pneumatic soft actuators

Here, we present a list of 15 requirements or suggestions that must be considered when developing functional 4D-printed pneumatic soft actuators. A pneumatic soft actuator should have one or more of the following characteristics depending on the application and requirements (Tawk & Alici, 2021).

1. Soft, flexible, and stretchable (i.e., should be able to exhibit strains between 10% and 100% (Madden et al., 2004; Polygerinos et al., 2017; Rus & Tolley, 2015) and their tangent modulus should range between 100 and 1000 MPa, which is consistent with the mechanical properties of biological materials ranging from body fat to tendon (Rus & Tolley, 2015)).
2. Its compliance should be below the compliance of the environment it is expected to directly interact with (e.g., humans and other delicate environments).
3. Amenable to additive manufacturing technologies where no or minimal post-processing should be required.
4. Scalable and customizable.
5. Ability to actively change its mechanical compliance (e.g., variable stiffness and damping constants), rheological properties (e.g., variable viscoelastic moduli, stress relaxation modulus, and shear viscosity), and electrical properties (e.g., resistance and capacitance).
6. Biocompatible so that no toxic or harmful materials are used in its structure and components.
7. Integrative and distributive like natural muscles that smoothly contain the structure, support, and actuation, and where 30%–80% of their fibers contribute to the generation of the force output.
8. Most of the actuator mass (i.e., active/total mass ratio of up to 80%) should contribute to the force output.
9. Ability to be integrated seamlessly into a continuum body with a low footprint along with soft sensors, flexible and stretchable electronics, and power sources.
10. Reversible and predictable with a reasonably short response time of less than 1 s.
11. Sensitive with a linear response and negligibly small hysteresis and creep.
12. Low power to be able to operate for multiple hours or days.
13. Low cost to increase its accessibility.
14. Reliable, durable (i.e., long lifetime), and resistant to fatigue under large and reversible strains.
15. Robust against external disturbances and excessive mechanical stresses.

Conclusion

This chapter presented 4D-printed pneumatic soft actuators that are fabricated directly using various additive manufacturing technologies. The actuators are presented in terms of their types, modeling, materials, fabrication, sensing, control, capabilities, and applications in diverse soft robotic systems. A discussion was provided on the presented 4D-printable soft actuators to highlight their major challenges and limitations along with a list of requirements or suggestions for developing robust and functional pneumatic soft actuators for soft robotic systems and devices. Finally, this chapter emphasized the importance of implementing a top-down approach to stimulate more interaction between materials scientists and robotic researchers so that 3D/4D-printable soft materials that meet the function and application requirements of a robotic system are developed to fully exploit such synthesized materials by roboticists and use them in practical robotic systems and devices.

References

Abbasi, P., Nekoui, M. A., Zareinejad, M., Abbasi, P., & Azhang, Z. (2020). Position and force control of a soft pneumatic actuator. *Soft Robotics*, *7*(5), 550–563. https://doi.org/10.1089/soro.2019.0065.

Alici, G. (2018). Softer is harder: What differentiates soft robotics from hard robotics? *MRS Advances*, *3*(28), 1557–1568. https://doi.org/10.1557/adv.2018.159.

Ang, B. W. K., & Yeow, C. (2020). Design and modeling of a high force soft actuator for assisted elbow flexion. *IEEE Robotics and Automation Letters*, 1.

Anver, H. M. C. M., Mutlu, R., & Alici, G. (2017). 3D printing of a thin-wall soft and monolithic gripper using fused filament fabrication. In *2017 IEEE international conference on advanced intelligent mechatronics (AIM)* (pp. 442–447).

Atalay, O., Atalay, A., Gafford, J., & Walsh, C. (2018). A highly sensitive capacitive-based soft pressure sensor based on a conductive fabric and a microporous dielectric layer. *Advanced Materials Technologies*, *3*(1), 1700237. https://doi.org/10.1002/admt.201700237.

Aulia Putra, H. (2018). Exploring air properties for fMRI-compatible interaction devices. *MATEC Web of Conferences*, *215*, 01001. https://doi.org/10.1051/matecconf/201821501001.

Bakarich, S. E., Gorkin, R., in het Panhuis, M., & Spinks, G. M. (2015). 4D printing with mechanically robust, thermally actuating hydrogels. *Macromolecular Rapid Communications*, *36*(12), 1211–1217. https://doi.org/10.1002/marc.201500079.

Bartlett, N. W., Tolley, M. T., Overvelde, J. T. B., Weaver, J. C., Mosadegh, B., Bertoldi, K., et al. (2015). A 3D-printed, functionally graded soft robot powered by combustion. *Science*, *349*(6244), 161–165. https://doi.org/10.1126/science.aab0129.

Beer, F. P., Johnston, E. R., Dewolf, J. T., & Mazurk, D. F. (2009). *Mechanics of materials* (5th ed.). McGraw-Hill Companies, Inc.

Belter, J. T., Segil, J. L., Dollar, A. M., & Weir, R. F. (2013). Mechanical design and performance specifications of anthropomorphic prosthetic hands: A review. *Journal of Rehabilitation Research and Development*, *50*(5), 599–618.

Bieze, T. M., Largilliere, F., Kruszewski, A., Zhang, Z., Merzouki, R., & Duriez, C. (2018). Finite element method-based kinematics and closed-loop control of soft, continuum manipulators. *Soft Robotics*, *5*(3), 348–364. https://doi.org/10.1089/soro.2017.0079.

Bin, L., Yang, G., Adam, F., & Yon, V. (2016). Soft capacitive tactile sensing arrays fabricated via direct filament casting. *Smart Materials and Structures*, *25*(7), 075009. http://stacks.iop.org/0964-1726/25/i=7/a=075009.

Bodaghi, M., Serjouei, A., Zolfagharian, A., Fotouhi, M., Rahman, H., & Durand, D. (2020). Reversible energy absorbing meta-sandwiches by FDM 4D printing. *International Journal of Mechanical Sciences*, *173*. https://doi.org/10.1016/j.ijmecsci.2020.105451, 105451.

Brown, E., Rodenberg, N., Amend, J., Mozeika, A., Steltz, E., Zakin, M. R., et al. (2010). Universal robotic gripper based on the jamming of granular material. *Proceedings of the*

National Academy of Sciences of the United States of America, *107*(44), 18809–18814. https://doi.org/10.1073/pnas.1003250107.

Bryan, N. P., Thomas, J. W., Huichan, Z., & Robert, F. S. (2015). 3D printing antagonistic systems of artificial muscle using projection stereolithography. *Bioinspiration & Biomimetics*, *10*(5), 055003. http://stacks.iop.org/1748-3190/10/i=5/a=055003.

Byrne, O., Coulter, F., Glynn, M., Jones, J. F. X., Ní Annaidh, A., O'Cearbhaill, E. D., et al. (2018). Additive manufacture of composite soft pneumatic actuators. *Soft Robotics*, *5* (6), 726–736. https://doi.org/10.1089/soro.2018.0030.

Calisti, M., Picardi, G., & Laschi, C. (2017). Fundamentals of soft robot locomotion. *Journal of the Royal Society Interface*, *14*(130), 20170101. https://doi.org/10.1098/rsif.2017.0101.

Cao, G., Chu, B., & Liu, Y. (2020). Analytical modeling and control of soft fast pneumatic networks actuators. In *IECON 2020 The 46th annual conference of the IEEE industrial electronics society* (pp. 2760–2765). https://doi.org/10.1109/IECON43393.2020.9254517.

Carpi, F., Alici, G., Mutlu, R., Melling, D., Jager, E. W. H., & Kaneto, K. (2016). Conducting polymers as EAPs: Device configurations. In *Electromechanically active polymers: A concise reference* (pp. 257–291). Springer International Publishing. https://doi.org/10.1007/978-3-319-31530-0_12.

Cheng, Y., Chan, K. H., Wang, X.-Q., Ding, T., Li, T., Lu, X., et al. (2019). Direct-ink-write 3D printing of hydrogels into biomimetic soft robots. *ACS Nano*, *13*(11), 13176–13184. https://doi.org/10.1021/acsnano.9b06144.

Ching-Ping, C., & Hannaford, B. (1996). Measurement and modeling of McKibben pneumatic artificial muscles. *IEEE Transactions on Robotics and Automation*, *12*(1), 90–102. https://doi.org/10.1109/70.481753.

Choi, H., Jung, P., Jung, K., & Kong, K. (2017). Design and fabrication of a soft three-axis force sensor based on radially symmetric pneumatic chambers. In *2017 IEEE international conference on robotics and automation (ICRA)* (pp. 5519–5524). https://doi.org/10.1109/ICRA.2017.7989650.

Chossat, J. B., & Shull, P. B. (2021). Soft acoustic waveguides for strain, deformation, localization, and twist measurements. *IEEE Sensors Journal*, *21*(1), 222–230. https://doi.org/10.1109/JSEN.2020.3013067.

Cianchetti, M., Laschi, C., Menciassi, A., & Dario, P. (2018). Biomedical applications of soft robotics. *Nature Reviews Materials*, *3*(6), 143–153. https://doi.org/10.1038/s41578-018-0022-y.

Connolly, F., Walsh, C. J., & Bertoldi, K. (2016). Automatic design of fiber-reinforced soft actuators for trajectory matching. *Proceedings of the National Academy of Sciences of the United States of America*. https://doi.org/10.1073/pnas.1615140114, 201615140.

Costas, A., Davis, D. E., Niu, Y., Dabiri, S., Garcia, J., & Newell, B. (2018). Design, development and characterization of linear, soft actuators via additive manufacturing. In *ASME 2018 conference on smart materials, adaptive structures and intelligent systems: Vol. Volume 1: Development and characterization of multifunctional materials; modeling, simulation, and control of adaptive systems; integrated system design and implementation*. https://doi.org/10.1115/smasis2018-8097.

Daerden, F., Lefeber, D., Verrelst, B., & Ham, R. V. (2001). Pleated pneumatic artificial muscles: Actuators for automation and robotics. In *Vol. 2. 2001 IEEE/ASME international conference on advanced intelligent mechatronics. Proceedings (Cat. No.01TH8556)* (pp. 738–743).

Dämmer, G., Gablenz, S., Hildebrandt, A., & Major, Z. (2019). PolyJet-printed bellows actuators: Design, structural optimization, and experimental investigation. *Frontiers in Robotics and AI*, *6*(34). https://doi.org/10.3389/frobt.2019.00034.

Dickey, M. D. (2017). Stretchable and soft electronics using liquid metals. *Advanced Materials*, *29*(27), 1606425. https://doi.org/10.1002/adma.201606425.

Drotman, D., Ishida, M., Jadhav, S., & Tolley, M. T. (2019). Application-driven design of Soft, 3-D printed, pneumatic actuators with bellows. *IEEE/ASME Transactions on Mechatronics*, *24*(1), 78–87.

Drotman, D., Jadhav, S., Karimi, M., deZonia, P., & Tolley, M. T. (2017). 3D printed soft actuators for a legged robot capable of navigating unstructured terrain. In *2017 IEEE international conference on robotics and automation (ICRA)* (pp. 5532–5538). https://doi.org/10.1109/ICRA.2017.7989652.

El-Atab, N., Mishra, R. B., Al-Modaf, F., Joharji, L., Alsharif, A. A., Alamoudi, H., et al. (2020). Soft actuators for soft robotic applications: A review. *Advanced Intelligent Systems*, *2*(10), 2000128. https://doi.org/10.1002/aisy.202000128.

Elgeneidy, K., Lohse, N., & Jackson, M. (2018). Bending angle prediction and control of soft pneumatic actuators with embedded flex sensors—A data-driven approach. *Mechatronics*, *50*, 234–247. https://doi.org/10.1016/j.mechatronics.2017.10.005.

Elgeneidy, K., Neumann, G., Jackson, M., & Lohse, N. (2018). Directly printable flexible strain sensors for bending and contact feedback of soft actuators. *Frontiers in Robotics and AI, 5* (2). https://doi.org/10.3389/frobt.2018.00002.

Ewoldt, R. H. (2013). Extremely soft: Design with rheologically complex fluids. *Soft Robotics*, *1*(1), 12–20. https://doi.org/10.1089/soro.2013.1508.

Frutiger, A., Muth, J. T., Vogt, D. M., Mengüç, Y., Campo, A., Valentine, A. D., et al. (2015). Capacitive soft strain sensors via multicore–shell fiber printing. *Advanced Materials*, *27* (15), 2440–2446. https://doi.org/10.1002/adma.201500072.

Galley, A., Knopf, G., & Kashkoush. (2019). Pneumatic hyperelastic actuators for grasping curved organic objects. *Actuators, 8*, 76. https://doi.org/10.3390/act8040076.

Galloway, K. C., Polygerinos, P., Walsh, C. J., & Wood, R. J. (2013). Mechanically programmable bend radius for fiber-reinforced soft actuators. In *2013 16th international conference on advanced robotics (ICAR)* (pp. 1–6). https://doi.org/10.1109/ICAR.2013.6766586.

Ge, L., Dong, L., Wang, D., Ge, Q., & Gu, G. (2018). A digital light processing 3D printer for fast and high-precision fabrication of soft pneumatic actuators. *Sensors and Actuators A: Physical, 273*, 285–292. https://doi.org/10.1016/j.sna.2018.02.041.

George Thuruthel, T., Falotico, E., Manti, M., Pratesi, A., Cianchetti, M., & Laschi, C. (2017). Learning closed loop kinematic controllers for continuum manipulators in unstructured environments. *Soft Robotics, 4*(3), 285–296. https://doi.org/10.1089/soro.2016.0051.

George Thuruthel, T., Ansari, Y., Falotico, E., & Laschi, C. (2018). Control strategies for soft robotic manipulators: A survey. *Soft Robotics, 5*(2), 149–163. https://doi.org/10.1089/soro.2017.0007.

Gerboni, G., Diodato, A., Ciuti, G., Cianchetti, M., & Menciassi, A. (2017). Feedback control of soft robot actuators via commercial flex bend sensors. *IEEE/ASME Transactions on Mechatronics, 22*(4), 1881–1888. https://doi.org/10.1109/TMECH.2017.2699677.

Gong, D., He, R., Yu, J., & Zuo, G. (2017). A pneumatic tactile sensor for co-operative robots. *Sensors, 17*(11), 2592. http://www.mdpi.com/1424-8220/17/11/2592.

Grzesiak, A., Becker, R., & Verl, A. (2011). The bionic handling assistant: A success story of additive manufacturing. *Assembly Automation, 31*(4), 329–333. https://doi.org/10.1108/01445151111172907.

Gu, G., Wang, D., Ge, L., & Zhu, X. (2020). Analytical modeling and design of generalized Pneu-net soft actuators with three-dimensional deformations. *Soft Robotics*. https://doi.org/10.1089/soro.2020.0039.

Gu, G.-Y., Zhu, J., Zhu, L.-M., & Zhu, X. (2017). A survey on dielectric elastomer actuators for soft robots. *Bioinspiration & Biomimetics, 12*(1). https://doi.org/10.1088/1748-3190/12/1/011003, 011003.

Gul, J. Z., Sajid, M., Rehman, M. M., Siddiqui, G. U., Shah, I., Kim, K.-H., et al. (2018). 3D printing for soft robotics—A review. *Science and Technology of Advanced Materials, 19*(1), 243–262. https://doi.org/10.1080/14686996.2018.1431862.

Haines, C. S., Lima, M. D., Li, N., Spinks, G. M., Foroughi, J., Madden, J. D. W., et al. (2014). Artificial muscles from fishing line and sewing thread. *Science, 343*(6173), 868–872. https://doi.org/10.1126/science.1246906.

Hamidi, A., & Tadesse, Y. (2020). 3D printing of very soft elastomer and sacrificial carbohydrate glass/elastomer structures for robotic applications. *Materials & Design, 187*. https://doi.org/10.1016/j.matdes.2019.108324, 108324.

Herianto, Irawan, W., Ritonga, A. S., & Prastowo, A. (2019). Design and fabrication in the loop of soft pneumatic actuators using fused deposition modelling. *Sensors and Actuators A: Physical, 298*. https://doi.org/10.1016/j.sna.2019.111556, 111556.

Heung, K. H. L., Tang, Z. Q., Ho, L., Tung, M., Li, Z., & Tong, R. K. Y. (2019). Design of a 3D printed soft robotic hand for stroke rehabilitation and daily activities assistance. In *2019 IEEE 16th international conference on rehabilitation robotics (ICORR)* (pp. 65–70).

Heung, H. L., Tang, Z. Q., Shi, X. Q., Tong, K. Y., & Li, Z. (2020). Soft rehabilitation actuator with integrated post-stroke finger spasticity evaluation. *Frontiers in Bioengineering and Biotechnology, 8*(111). https://doi.org/10.3389/fbioe.2020.00111.

Hiller, J., & Lipson, H. (2014). Dynamic simulation of soft multimaterial 3D-printed objects. *Soft Robotics, 1*(1), 88–101. https://doi.org/10.1089/soro.2013.0010.

Hu, W., & Alici, G. (2019). Bioinspired three-dimensional-printed helical soft pneumatic actuators and their characterization. *Soft Robotics, 7*(3), 267–282. https://doi.org/10.1089/soro.2019.0015.

Hu, W., Li, W., & Alici, G. (2018). 3D printed helical soft pneumatic actuators. In *2018 IEEE/ASME international conference on advanced intelligent mechatronics (AIM)* (pp. 950–955).

Hu, W., Lum, G. Z., Mastrangeli, M., & Sitti, M. (2018). Small-scale soft-bodied robot with multimodal locomotion. *Nature, 554*, 81. https://doi.org/10.1038/nature25443.

Ito, F., Kojima, A., Okui, M., & Nakamura, T. (2020). Proposal for 3D-printed pneumatic artificial muscles—Effect of leaf spring stiffness on contraction amount and contraction force. In *2020 IEEE/SICE international symposium on system integration (SII)* (pp. 1163–1167).

Jin, H., Dong, E., Xu, M., Liu, C., Alici, G., & Jie, Y. (2016). Soft and smart modular structures actuated by shape memory alloy (SMA) wires as tentacles of soft robots. *Smart Materials and Structures, 25*(8). https://doi.org/10.1088/0964-1726/25/8/085026, 085026.

Joshi, S., Rawat, K., Karunakaran, C., Rajamohan, V., Mathew, A. T., Koziol, K., et al. (2020). 4D printing of materials for the future: Opportunities and challenges. *Applied Materials Today, 18*. https://doi.org/10.1016/j.apmt.2019.100490, 100490.

Kalisky, T., Wang, Y., Shih, B., Drotman, D., Jadhav, S., Aronoff-Spencer, E., et al. (2017). Differential pressure control of 3D printed soft fluidic actuators. In *2017 IEEE/RSJ international conference on intelligent robots and systems (IROS)* (pp. 6207–6213). https://doi.org/10.1109/IROS.2017.8206523.

Kang, H.-W., Lee, I. H., & Cho, D.-W. (2006). Development of a micro-bellows actuator using micro-stereolithography technology. *Microelectronic Engineering, 83*(4), 1201–1204. https://doi.org/10.1016/j.mee.2006.01.228.

Keiko, O., Wakimoto, S., Suzumori, K., & Yasutaka, N. (2009). Micro pneumatic curling actuator—Nematode actuator. In *2008 IEEE international conference on robotics and biomimetics* (pp. 462–467). https://doi.org/10.1109/ROBIO.2009.4913047.

Kellaris, N., Gopaluni Venkata, V., Smith, G. M., Mitchell, S. K., & Keplinger, C. (2018). Peano-HASEL actuators: Muscle-mimetic, electrohydraulic transducers that linearly contract on activation. *Science Robotics*, *3*(14), eaar3276. https://doi.org/10.1126/scirobotics.aar3276.

Keong, B. A. W., & Hua, R. Y. C. (2018). A Novel fold-based design approach toward printable soft robotics using flexible 3D printing materials. *Advanced Materials Technologies*, *3*(2), 1700172. https://doi.org/10.1002/admt.201700172.

Kim, J., Alspach, A., & Yamane, K. (2015). 3D printed soft skin for safe human-robot interaction. In *2015 IEEE/RSJ international conference on intelligent robots and systems (IROS)* (pp. 2419–2425).

Kong, K., & Tomizuka, M. (2009). A gait monitoring system based on air pressure sensors embedded in a shoe. *IEEE/ASME Transactions on Mechatronics*, *14*(3), 358–370. https://doi.org/10.1109/TMECH.2008.2008803.

Kumar, K., Liu, J., Christianson, C., Ali, M., Tolley, M. T., Aizenberg, J., et al. (2017). A biologically inspired, functionally graded end effector for soft robotics applications. *Soft Robotics*, *4*(4), 317–323. https://doi.org/10.1089/soro.2017.0002.

Kumbay Yildiz, S., Mutlu, R., & Alici, G. (2016). Fabrication and characterisation of highly stretchable elastomeric strain sensors for prosthetic hand applications. *Sensors and Actuators A: Physical*, *247*, 514–521. https://doi.org/10.1016/j.sna.2016.06.037.

Laschi, C., Cianchetti, M., Mazzolai, B., Margheri, L., Follador, M., & Dario, P. (2012). Soft robot arm inspired by the octopus. *Advanced Robotics*, *26*(7), 709–727. https://doi.org/10.1163/156855312X626343.

Laschi, C., Rossiter, J., Iida, F., Cianchetti, M., Margheri, L., Scharff, R. B. N., et al. (2017). Towards behavior design of a 3D-printed soft robotic hand. In *Soft robotics: Trends, applications and challenges* (pp. 23–29). Springer International Publishing.

Lee, C., Kim, M., Kim, Y. J., Hong, N., Ryu, S., Kim, H. J., et al. (2017). Soft robot review. *International Journal of Control, Automation and Systems*, *15*(1), 3–15. https://doi.org/10.1007/s12555-016-0462-3.

Lee, J., Kim, W., Choi, W., & Cho, K. (2016). Soft robotic blocks: Introducing SoBL, a fast-build modularized design block. *IEEE Robotics & Automation Magazine*, *23*(3), 30–41.

Li, S., Vogt, D. M., Rus, D., & Wood, R. J. (2017). Fluid-driven origami-inspired artificial muscles. *Proceedings of the National Academy of Sciences of the United States of America*, *114* (50), 13132–13137. https://doi.org/10.1073/pnas.1713450114.

Lin, H.-T., Leisk, G. G., & Trimmer, B. (2011). GoQBot: A caterpillar-inspired soft-bodied rolling robot. *Bioinspiration & Biomimetics*, *6*(2). https://doi.org/10.1088/1748-3182/6/2/026007, 026007.

Lina, M. C., & Alison, B. F. (2014). Smart fabric sensors and e-textile technologies: A review. *Smart Materials and Structures*, *23*(5), 053001. http://stacks.iop.org/0964-1726/23/i=5/a=053001.

Lipton, J. I., Angle, S., Banai, R. E., Peretz, E., & Lipson, H. (2016). Electrically actuated hydraulic solids. *Advanced Engineering Materials*, *18*(10), 1710–1715. https://doi.org/10.1002/adem.201600271.

Liu, Z., Zhang, X., Liu, H., Chen, Y., Huang, Y., & Chen, X. (2020). Kinematic modelling and experimental validation of a foldable pneumatic soft manipulator. *Applied Sciences*, *10*, 1447. https://doi.org/10.3390/app10041447.

Low, J. H., Lee, W. W., Khin, P. M., Thakor, N. V., Kukreja, S. L., Ren, H. L., et al. (2017). Hybrid tele-manipulation system using a sensorized 3-D-printed soft robotic gripper and a soft fabric-based haptic glove. *IEEE Robotics and Automation Letters*, *2*(2), 880–887.

Lu, Q., He, L., Nanayakkara, T., & Rojas, N. (2020). Precise in-hand manipulation of soft objects using soft fingertips with tactile sensing and active deformation. In *2020 3rd IEEE international conference on soft robotics (RoboSoft)* (pp. 52–57). https://doi.org/10.1109/RoboSoft48309.2020.9115997.

MacCurdy, R., Katzschmann, R., Youbin, K., & Rus, D. (2016). Printable hydraulics: A method for fabricating robots by 3D co-printing solids and liquids. In *2016 IEEE international conference on robotics and automation (ICRA)* (pp. 3878–3885). https://doi.org/10.1109/ICRA.2016.7487576.

Madden, J. D. W., Vandesteeg, N. A., Anquetil, P. A., Madden, P. G. A., Takshi, A., Pytel, R. Z., et al. (2004). Artificial muscle technology: Physical principles and naval prospects. *IEEE Journal of Oceanic Engineering*, *29*(3), 706–728.

Mai, H., Mutlu, R., Tawk, C., Alici, G., & Sencadas, V. (2019). Ultra-stretchable MWCNT–Ecoflex piezoresistive sensors for human motion detection applications. *Composites Science and Technology*, *173*, 118–124. https://doi.org/10.1016/j.compscitech.2019.02.001.

Majidi, C. (2013). Soft robotics: A perspective—Current trends and prospects for the future. *Soft Robotics*, *1*(1), 5–11. https://doi.org/10.1089/soro.2013.0001.

Majidi, C. (2019). Soft-matter engineering for soft robotics. *Advanced Materials Technologies*, *4*(2), 1800477. https://doi.org/10.1002/admt.201800477.

Manns, M., Morales, J., & Frohn, P. (2018). Additive manufacturing of silicon based PneuNets as soft robotic actuators. *Procedia CIRP*, *72*, 328–333. https://doi.org/10.1016/j.procir.2018.03.186.

Manti, M., Hassan, T., Passetti, G., D'Elia, N., Laschi, C., & Cianchetti, M. (2015). A bioinspired soft robotic gripper for adaptable and effective grasping. *Soft Robotics*, *2*(3), 107–116. https://doi.org/10.1089/soro.2015.0009.

Marchese, A. D., Katzschmann, R. K., & Rus, D. (2015). A recipe for soft fluidic elastomer robots. *Soft Robotics*, *2*(1), 7–25. https://doi.org/10.1089/soro.2014.0022.

Miriyev, A., Stack, K., & Lipson, H. (2017). Soft material for soft actuators. *Nature Communications*, *8*(1), 596. https://doi.org/10.1038/s41467-017-00685-3.

Miriyev, A., Xia, B., Joseph, J. C., & Lipson, H. (2019). Additive manufacturing of silicone composites for soft actuation. *3D Printing and Additive Manufacturing*, *6*(6), 309–318. https://doi.org/10.1089/3dp.2019.0116.

Mishra, A. K., Wallin, T. J., Pan, W., Xu, P., Wang, K., Giannelis, E. P., et al. (2020). Autonomic perspiration in 3D-printed hydrogel actuators. *Science. Robotics*, *5*(38), eaaz3918. https://doi.org/10.1126/scirobotics.aaz3918.

Morrow, J., Shin, H., Phillips-Grafflin, C., Jang, S., Torrey, J., Larkins, R., et al. (2016). Improving soft pneumatic actuator fingers through integration of soft sensors, position and force control, and rigid fingernails. In *2016 IEEE international conference on robotics and automation (ICRA)* (pp. 5024–5031). https://doi.org/10.1109/ICRA.2016.7487707.

Morrow, J., Hemleben, S., & Menguc, Y. (2017). Directly fabricating soft robotic actuators with an open-source 3-D printer. *IEEE Robotics and Automation Letters*, *2*(1), 277–281. https://doi.org/10.1109/LRA.2016.2598601.

Mosadegh, B., Polygerinos, P., Keplinger, C., Wennstedt, S., Shepherd, R. F., Gupta, U., et al. (2014). Pneumatic networks for soft robotics that actuate rapidly. *Advanced Functional Materials*, *24*(15), 2163–2170. https://doi.org/10.1002/adfm.201303288.

Moseley, P., Florez, J. M., Sonar, H. A., Agarwal, G., Curtin, W., & Paik, J. (2016). Modeling, design, and development of soft pneumatic actuators with finite element method. *Advanced Engineering Materials*, *18*(6), 978–988. https://doi.org/10.1002/adem.201500503.

Muth, J. T., Vogt, D. M., Truby, R. L., Mengüç, Y., Kolesky, D. B., Wood, R. J., et al. (2014). Embedded 3D printing of strain sensors within highly stretchable elastomers. *Advanced Materials*, 26(36), 6307–6312. https://doi.org/10.1002/adma.201400334.

Mutlu, R., Alici, G., in het Panhuis, M., & Spinks, G. (2016). 3D printed flexure hinges for soft monolithic prosthetic fingers. *Soft Robotics*, 3, 120–133. https://doi.org/10.1089/soro.2016.0026.

Mutlu, R., Tawk, C., Alici, G., & Sariyildiz, E. (2017). A 3D printed monolithic soft gripper with adjustable stiffness. In *IECON 2017—43rd annual conference of the IEEE industrial electronics society* (pp. 6235–6240). https://doi.org/10.1109/IECON.2017.8217084.

Park, Y.-L., Chen, B., Pérez-Arancibia, N. O., Young, D., Stirling, L., Wood, R. J., et al. (2014). Design and control of a bio-inspired soft wearable robotic device for ankle–foot rehabilitation. *Bioinspiration & Biomimetics*, 9(1). https://doi.org/10.1088/1748-3182/9/1/016007, 016007.

Park, Y., Chen, B., & Wood, R. J. (2012). Design and fabrication of soft artificial skin using embedded microchannels and liquid conductors. *IEEE Sensors Journal*, 12(8), 2711–2718. https://doi.org/10.1109/JSEN.2012.2200790.

Patel, D. K., Sakhaei, A. H., Layani, M., Zhang, B., Ge, Q., & Magdassi, S. (2017). Highly stretchable and UV curable elastomers for digital light processing based 3D printing. *Advanced Materials*, 29(15), 1606000. https://doi.org/10.1002/adma.201606000.

Payan, Y., Faure, F., Duriez, C., Delingette, H., Allard, J., Gilles, B., et al. (2012). SOFA: A multi-model framework for interactive physical simulation. In *Soft tissue biomechanical modeling for computer assisted surgery* (pp. 283–321). Berlin Heidelberg: Springer. https://doi.org/10.1007/8415_2012_125.

Plott, J., & Shih, A. (2017). The extrusion-based additive manufacturing of moisture-cured silicone elastomer with minimal void for pneumatic actuators. *Additive Manufacturing*, 17, 1–14. https://doi.org/10.1016/j.addma.2017.06.009.

Polygerinos, P., Correll, N., Morin, S. A., Mosadegh, B., Onal, C. D., Petersen, K., et al. (2017). Soft robotics: Review of fluid-driven intrinsically soft devices; manufacturing, sensing, control, and applications in human-robot interaction. *Advanced Engineering Materials*, 19(12), 1700016. https://doi.org/10.1002/adem.201700016.

Qi, X., Shi, H., Pinto, T., & Tan, X. (2020). A novel pneumatic soft Snake robot using traveling-wave locomotion in constrained environments. *IEEE Robotics and Automation Letters*, 5 (2), 1610–1617.

Renda, F., Boyer, F., Dias, J., & Seneviratne, L. (2018). Discrete Cosserat approach for multisection soft manipulator dynamics. *IEEE Transactions on Robotics*, 34(6), 1518–1533.

Rieffel, J., Saunders, F., Nadimpalli, S., Zhou, H., Hassoun, S., Rife, J., et al. (2009). In *Evolving soft robotic locomotion in PhysX Proceedings of the 11th annual conference companion on genetic and evolutionary computation conference: Late breaking papers* (pp. 2499–2504). Association for Computing Machinery. https://doi.org/10.1145/1570256.1570351.

Robertson, M. A., & Paik, J. (2017). New soft robots really suck: Vacuum-powered systems empower diverse capabilities. *Science Robotics*, 2(9). http://robotics.sciencemag.org/content/2/9/eaan6357.abstract.

Roels, E., Terryn, S., Brancart, J., Verhelle, R., Van Assche, G., & Vanderborght, B. (2020). Additive manufacturing for self-healing soft robots. *Soft Robotics*, 7(6), 711–723. https://doi.org/10.1089/soro.2019.0081.

Rosalia, L., Ang, B. W., & Yeow, R. C. (2018). Geometry-based customization of bending modalities for 3D-printed soft pneumatic actuators. *IEEE Robotics and Automation Letters*, 3(4), 3489–3496.

Rus, D., & Tolley, M. T. (2015). Design, fabrication and control of soft robots. *Nature, 521*, 467. https://doi.org/10.1038/nature14543.

Salem, M. E. M., Wang, Q., Wen, R., & Xiang, M. (2018). Design and characterization of soft pneumatic actuator for universal robot gripper. In *2018 International conference on control and robots (ICCR)* (pp. 6–10).

Schaffner, M., Faber, J. A., Pianegonda, L., Rühs, P. A., Coulter, F., & Studart, A. R. (2018). 3D printing of robotic soft actuators with programmable bioinspired architectures. *Nature Communications, 9*(1), 878. https://doi.org/10.1038/s41467-018-03216-w.

Scharff, R. B. N., Doornbusch, R. M., Doubrovski, E. L., Wu, J., Geraedts, J. M. P., & Wang, C. C. L. (2019). Color-based proprioception of soft actuators interacting with objects. *IEEE/ASME Transactions on Mechatronics, 24*(5), 1964–1973.

Scharff, R. B. N., Wu, J., Geraedts, J. M. P., & Wang, C. C. L. (2019). Reducing out-of-plane deformation of soft robotic actuators for stable grasping. In *2019 2nd IEEE international conference on soft robotics (RoboSoft)* (pp. 265–270).

Schmitt, F., Piccin, O., Barbé, L., & Bayle, B. (2018). Soft robots manufacturing: A review. *Frontiers in Robotics and AI, 5*(84). https://doi.org/10.3389/frobt.2018.00084.

Schumacher, C., Bickel, B., Rys, J., Marschner, S., Daraio, C., & Gross, M. (2015). Microstructures to control elasticity in 3D printing. *ACM Transactions on Graphics, 34*(4), 1–13. https://doi.org/10.1145/2766926.

Sencadas, V., Mutlu, R., & Alici, G. (2017). Large area and ultra-thin compliant strain sensors for prosthetic devices. *Sensors and Actuators A: Physical, 266*, 56–64. https://doi.org/10.1016/j.sna.2017.08.051.

Sencadas, V., Tawk, C., & Alici, G. (2019). Highly sensitive soft foam sensors to empower robotic systems. *Advanced Materials Technologies, 4*(10), 1900423. https://doi.org/10.1002/admt.201900423.

Sencadas, V., Tawk, C., & Alici, G. (2020). Environmentally friendly and biodegradable ultrasensitive piezoresistive sensors for wearable electronics applications. *ACS Applied Materials & Interfaces, 12*(7), 8761–8772. https://doi.org/10.1021/acsami.9b21739.

Sencadas, V., Tawk, C., Searle, T., & Alici, G. (2021). Low-hysteresis and ultrasensitive microcellular structures for wearable electronic applications. *ACS Applied Materials & Interfaces, 13*(1), 1632–1643. https://doi.org/10.1021/acsami.0c20173.

Seok, S., Onal, C. D., Cho, K., Wood, R. J., Rus, D., & Kim, S. (2012). Meshworm: A peristaltic soft robot with antagonistic nickel titanium coil actuators. *IEEE/ASME Transactions on Mechatronics, 18*(5), 1485–1497. https://doi.org/10.1109/TMECH.2012.2204070.

Shahid, T., Gouwanda, D., Nurzaman, S., & Gopalai, A. (2018). Moving toward soft robotics: A decade review of the design of hand exoskeletons. *Biomimetics, 3*. https://doi.org/10.3390/biomimetics3030017.

Shapiro, Y., Wolf, A., & Gabor, K. (2011). Bi-bellows: Pneumatic bending actuator. *Sensors and Actuators A: Physical, 167*(2), 484–494. https://doi.org/10.1016/j.sna.2011.03.008.

Shapiro, Y., Wolf, A., & Kósa, G. (2013). Piezoelectric deflection sensor for a bi-bellows actuator. *IEEE/ASME Transactions on Mechatronics, 18*(3), 1226–1230.

She, Y., Li, C., Cleary, J., & Su, H.-J. (2015). Design and fabrication of a soft robotic hand with embedded actuators and sensors. *Journal of Mechanisms and Robotics, 7*(2). https://doi.org/10.1115/1.4029497.

Shih, B., Christianson, C., Gillespie, K., Lee, S., Mayeda, J., Huo, Z., et al. (2019). Design considerations for 3D printed, soft, multimaterial resistive sensors for soft robotics. *Frontiers in Robotics and AI, 6*(30). https://doi.org/10.3389/frobt.2019.00030.

Shih, B., Shah, D., Li, J., Thuruthel, T. G., Park, Y.-L., Iida, F., et al. (2020). Electronic skins and machine learning for intelligent soft robots. *Science Robotics*, *5*(41), eaaz9239. https://doi.org/10.1126/scirobotics.aaz9239.

Shin, B., Ha, J., Lee, M., Park, K., Park, G. H., Choi, T. H., et al. (2018). Hygrobot: A self-locomotive ratcheted actuator powered by environmental humidity. *Science Robotics*, *3* (14). https://doi.org/10.1126/scirobotics.aar2629.

Shintake, J., Cacucciolo, V., Floreano, D., & Shea, H. (2018). Soft robotic grippers. *Advanced Materials*, *30*(29), 1707035. https://doi.org/10.1002/adma.201707035.

Singh, D., Tawk, C., Mutlu, R., Sariyildiz, E., Sencadas, V., & Alici, G. (2020). A 3D printed soft force sensor for soft haptics. In *2020 3rd IEEE international conference on soft robotics (RoboSoft)* (pp. 458–463). IEEE.

Singh, D., Tawk, C., Mutlu, R., Sariyildiz, E., & Alici, G. (2019). A 3D printed soft robotic monolithic unit for haptic feedback devices. In *2019 IEEE/ASME international conference on advanced intelligent mechatronics (AIM)* (pp. 388–393).

Skylar-Scott, M. A., Mueller, J., Visser, C. W., & Lewis, J. A. (2019). Voxelated soft matter via multimaterial multinozzle 3D printing. *Nature*, *575*(7782), 330–335. https://doi.org/10.1038/s41586-019-1736-8.

Slyper, R., & Hodgins, J. (2012). Prototyping robot appearance, movement, and interactions using flexible 3D printing and air pressure sensors. In *2012 IEEE RO-MAN: The 21st IEEE international symposium on robot and human interactive communication* (pp. 6–11).

Sonar, H. A., Gerratt, A. P., Lacour, S. P., & Paik, J. (2019). Closed-loop haptic feedback control using a self-sensing soft pneumatic actuator skin. *Soft Robotics*, *7*(1), 22–29. https://doi.org/10.1089/soro.2019.0013.

Sparrman, B., du Pasquier, C., Thomsen, C., Darbari, S., Rustom, R., Laucks, J., et al. (2021). Printed silicone pneumatic actuators for soft robotics. *Additive Manufacturing*, *40*. https://doi.org/10.1016/j.addma.2021.101860, 101860.

Sparrman, B., Kernizan, S., Laucks, J., Tibbits, S., & Guberan, C. (2019). Liquid printed pneumatics. In *ACM SIGGRAPH 2019 emerging technologies* Association for Computing Machinery. https://doi.org/10.1145/3305367.3340318 (p. Article 16).

Stano, G., Arleo, L., & Percoco, G. (2020). Additive manufacturing for soft robotics: Design and fabrication of airtight, monolithic bending PneuNets with embedded air connectors. *Micromachines*, *11*, 485. https://doi.org/10.3390/mi11050485.

Stoppa, M., & Chiolerio, A. (2014). Wearable electronics and smart textiles: A critical review. *Sensors*, *14*(7), 11957. http://www.mdpi.com/1424-8220/14/7/11957.

Suzumori, K., Wada, A., & Wakimoto, S. (2013). New mobile pressure control system for pneumatic actuators, using reversible chemical reactions of water. *Sensors and Actuators A: Physical*, *201*, 148–153. https://doi.org/10.1016/j.sna.2013.07.008.

Tang, Z. Q., Heung, H. L., Tong, K. Y., & Li, Z. (2020). A probabilistic model-based online learning optimal control algorithm for soft pneumatic actuators. *IEEE Robotics and Automation Letters*, *5*(2), 1437–1444. https://doi.org/10.1109/LRA.2020.2967293.

Tawk, C., & Alici, G. (2020). Finite element modeling in the design process of 3D printed pneumatic soft actuators and sensors. *Robotics*, *9*, 52. https://doi.org/10.3390/robotics9030052.

Tawk, C., & Alici, G. (2021). A review of 3D-printable soft pneumatic actuators and sensors: Research challenges and opportunities. *Advanced Intelligent Systems*. https://doi.org/10.1002/aisy.202000223, 2000223.

Tawk, C., Sariyildiz, E., Zhou, H., in het Panhuis, M., Spinks, G. M., & Alici, G. (2020). Position control of a 3D printed soft finger with integrated soft pneumatic sensing chambers. In *2020 3rd IEEE international conference on soft robotics (RoboSoft)* (pp. 446–451). IEEE.

Tawk, C., Zhou, H., Sariyildiz, E., Panhuis, M. I. H., Spinks, G., & Alici, G. (2020). Design, modeling and control of a 3D printed monolithic soft robotic finger with embedded pneumatic sensing chambers. *IEEE/ASME Transactions on Mechatronics*, 1.

Tawk, C., Gao, Y., Mutlu, R., & Alici, G. (2019). Fully 3D printed monolithic soft gripper with high conformal grasping capability. In *2019 IEEE/ASME international conference on advanced intelligent mechatronics (AIM)* (pp. 1139–1144).

Tawk, C., Gillett, A., in het Panhuis, M., Spinks, G. M., & Alici, G. (2019). A 3D-printed Omnipurpose soft gripper. *IEEE Transactions on Robotics*, *35*(5), 1268–1275.

Tawk, C., in het Panhuis, M., Spinks, G. M., & Alici, G. (2018). Bioinspired 3D printable soft vacuum actuators for locomotion robots, grippers and artificial muscles. *Soft Robotics, 5* (6), 685–694. https://doi.org/10.1089/soro.2018.0021.

Tawk, C., in het Panhuis, M., Spinks, G. M., & Alici, G. (2019). Soft pneumatic sensing chambers for generic and interactive human–machine interfaces. *Advanced Intelligent Systems, 1*(1), 1900002. https://doi.org/10.1002/aisy.201900002.

Tawk, C., in het Panhuis, M., Spinks, G. M., & Alici, G. (2020). 3D printed soft pneumatic bending sensing chambers for bilateral and remote control of soft robotic systems. In *2020 IEEE/ASME international conference on advanced intelligent mechatronics (AIM)* (pp. 922–927).

Tawk, C., Mutlu, R., & Alici, G. (2020). A 3D printed modular soft gripper for conformal grasping. In *In 2020 IEEE/ASME international conference on advanced intelligent mechatronics (AIM)* (pp. 583–588).

Tawk, C., Spinks, G. M., in het Panhuis, M., & Alici, G. (2019a). 3D printable linear soft vacuum actuators: Their modeling, performance quantification and application in soft robotic systems. *IEEE/ASME Transactions on Mechatronics*, *24*(5), 2118–2129. https://doi.org/10.1109/TMECH.2019.2933027.

Tawk, C., Spinks, G. M., in het Panhuis, M., & Alici, G. (2019b). 3D printable vacuum-powered soft linear actuators. In *2019 IEEE/ASME international conference on advanced intelligent mechatronics (AIM)* (pp. 50–55).

Terryn, S., Langenbach, J., Roels, E., Brancart, J., Bakkali-Hassani, C., Poutrel, Q.-A., et al. (2021). A review on self-healing polymers for soft robotics. *Materials Today*. https://doi.org/10.1016/j.mattod.2021.01.009.

Thalman, C., & Artemiadis, P. (2020). A review of soft wearable robots that provide active assistance: Trends, common actuation methods, fabrication, and applications. *Wearable Technologies, 1*. https://doi.org/10.1017/wtc.2020.4, e3.

Thrasher, C. J., Schwartz, J. J., & Boydston, A. J. (2017). Modular elastomer photoresins for digital light processing additive manufacturing. *ACS Applied Materials & Interfaces, 9* (45), 39708–39716. https://doi.org/10.1021/acsami.7b13909.

Thuruthel, T. G., Shih, B., Laschi, C., & Tolley, M. T. (2019). Soft robot perception using embedded soft sensors and recurrent neural networks. *Science Robotics, 4*(26), eaav1488. https://doi.org/10.1126/scirobotics.aav1488.

Tolley, M. T., Shepherd, R. F., Mosadegh, B., Galloway, K. C., Wehner, M., Karpelson, M., et al. (2014). A resilient, untethered soft robot. *Soft Robotics, 1*(3), 213–223. https://doi.org/10.1089/soro.2014.0008.

Trimmer, B. (2013). Soft robots. *Current Biology, 23*(15), R639–R641. https://doi.org/10.1016/j.cub.2013.04.070.

Truby, R. L., Wehner, M., Grosskopf, A. K., Vogt, D. M., Uzel, S. G. M., Wood, R. J., et al. (2018). Soft Somatosensitive actuators via embedded 3D printing. *Advanced Materials, 30* (15), 1706383. https://doi.org/10.1002/adma.201706383.

Truby, R. L., Katzschmann, R. K., Lewis, J. A., & Rus, D. (2019). Soft robotic fingers with embedded ionogel sensors and discrete actuation modes for somatosensitive manipulation. In *2019 2nd IEEE international conference on soft robotics (RoboSoft)* (pp. 322–329).

Vázquez, M., Brockmeyer, E., Desai, R., Harrison, C., & Hudson, S. E. (2015). In *3D printing pneumatic device controls with variable activation force capabilities Proceedings of the 33rd annual ACM conference on human factors in computing systems* (pp. 1295–1304). Association for Computing Machinery. https://doi.org/10.1145/2702123.2702569.

Viry, L., Levi, A., Totaro, M., Mondini, A., Mattoli, V., Mazzolai, B., et al. (2014). Flexible three-axial force sensor for soft and highly sensitive artificial touch. *Advanced Materials, 26*(17), 2659–2664. https://doi.org/10.1002/adma.201305064.

Walker, S., Daalkhaijav, U., Thrush, D., Branyan, C., Yirmibesoglu, O. D., Olson, G., et al. (2019). Zero-support 3D printing of thermoset silicone via simultaneous control of both reaction kinetics and transient rheology. *3D Printing and Additive Manufacturing, 6*(3), 139–147. https://doi.org/10.1089/3dp.2018.0117.

Walker, Z., Harbel, Y., Strickland, K., & Shin, H. (2020). Soft robotics: A review of recent developments of pneumatic soft actuators. *Actuators, 9*, 3. https://doi.org/10.3390/act9010003.

Wallin, T. J., Pikul, J. H., Bodkhe, S., Peele, B. N., Mac Murray, B. C., Therriault, D., et al. (2017). Click chemistry stereolithography for soft robots that self-heal. *Journal of Materials Chemistry B, 5*(31), 6249–6255. https://doi.org/10.1039/C7TB01605K.

Wallin, T. J., Simonsen, L.-E., Pan, W., Wang, K., Giannelis, E., Shepherd, R. F., et al. (2020). 3D printable tough silicone double networks. *Nature Communications, 11*(1), 4000. https://doi.org/10.1038/s41467-020-17816-y.

Wallin, T. J., Pikul, J., & Shepherd, R. F. (2018). 3D printing of soft robotic systems. *Nature Reviews Materials, 3*(6), 84–100. https://doi.org/10.1038/s41578-018-0002-2.

Wang, Z., Chathuranga, D. S., & Hirai, S. (2016). 3D printed soft gripper for automatic lunch box packing. In *2016 IEEE international conference on robotics and biomimetics (ROBIO)* (pp. 503–508).

Wang, Z., Torigoe, Y., & Hirai, S. (2017). A Prestressed soft gripper: Design, modeling, fabrication, and tests for food handling. *IEEE Robotics and Automation Letters, 2*(4), 1909–1916. https://doi.org/10.1109/LRA.2017.2714141.

Wehner, M., Truby, R. L., Fitzgerald, D. J., Mosadegh, B., Whitesides, G. M., Lewis, J. A., et al. (2016). An integrated design and fabrication strategy for entirely soft, autonomous robots. *Nature, 536*, 451. https://doi.org/10.1038/nature19100.

Wei, S., Zhang, L., Li, C., Tao, S., Ding, B., Zhu, H., et al. (2019). Preparation of soft somatosensory-detecting materials via selective laser sintering. *Journal of Materials Chemistry C, 7*(22), 6786–6794. https://doi.org/10.1039/C9TC01331H.

Wirekoh, J., Valle, L., Pol, N., & Park, Y.-L. (2019). Sensorized, flat, pneumatic artificial muscle embedded with biomimetic microfluidic sensors for proprioceptive feedback. *Soft Robotics, 6*(6), 768–777. https://doi.org/10.1089/soro.2018.0110.

Xavier, M. S., Fleming, A. J., & Yong, Y. K. (2021). Finite element Modeling of soft fluidic actuators: Overview and recent developments. *Advanced Intelligent Systems, 3*(2), 2000187. https://doi.org/10.1002/aisy.202000187.

Xi, W., Yeo, J. C., Yu, L., Zhang, S., & Lim, C. T. (2017). Ultrathin and wearable microtubular epidermal sensor for real-time physiological pulse monitoring. *Advanced Materials Technologies, 2*(5), 1700016. https://doi.org/10.1002/admt.201700016.

Xiang, C., Guo, J., & Rossiter, J. (2019). Soft-smart robotic end effectors with sensing, actuation, and gripping capabilities. *Smart Materials and Structures, 28*(5). https://doi.org/10.1088/1361-665x/ab1176, 055034.

Xie, D., Zuo, S., & Liu, J. (2020). A novel flat modular pneumatic artificial muscle. *Smart Materials and Structures*, *29*(6). https://doi.org/10.1088/1361-665x/ab84b9, 065013.

Yang, Y., & Chen, Y. (2018). Innovative design of embedded pressure and position sensors for soft actuators. *IEEE Robotics and Automation Letters*, *3*(2), 656–663.

Yang, D., Verma, M. S., So, J.-H., Mosadegh, B., Keplinger, C., Lee, B., et al. (2016). Buckling pneumatic linear actuators inspired by muscle. *Advanced Materials Technologies*, *1*(3), 1600055. https://doi.org/10.1002/admt.201600055.

Yang, H., Chen, Y., Sun, Y., & Hao, L. (2017). A novel pneumatic soft sensor for measuring contact force and curvature of a soft gripper. *Sensors and Actuators A: Physical*, *266*, 318–327. https://doi.org/10.1016/j.sna.2017.09.040.

Yang, D., Verma, M. S., Lossner, E., Stothers, D., & Whitesides, G. M. (2017). Negative-pressure soft linear actuator with a mechanical advantage. *Advanced Materials Technologies*, *2*(1), 1600164. https://doi.org/10.1002/admt.201600164.

Yap, H. K., Ng, H. Y., & Yeow, C.-H. (2016). High-force soft printable pneumatics for soft robotic applications. *Soft Robotics*, *3*(3), 144–158. https://doi.org/10.1089/soro.2016.0030.

Yeo, J. C., Yap, H. K., Xi, W., Wang, Z., Yeow, C.-H., & Lim, C. T. (2016). Flexible and stretchable strain sensing actuator for wearable soft robotic applications. *Advanced Materials Technologies*, *1*(3), 1600018. https://doi.org/10.1002/admt.201600018.

Yildiz, S. K., Mutlu, R., & Alici, G. (2018). Position control of a soft prosthetic finger with limited feedback information. In *2018 IEEE/ASME international conference on advanced intelligent mechatronics (AIM)* (pp. 700–705). https://doi.org/10.1109/AIM.2018.8452402.

Yim, M., Shen, W., Salemi, B., Rus, D., Moll, M., Lipson, H., et al. (2007). Modular self-reconfigurable robot systems [grand challenges of robotics]. *IEEE Robotics & Automation Magazine*, *14*(1), 43–52.

Yirmibeşoğlu, O. D., Oshiro, T., Olson, G., Palmer, C., & Mengüç, Y. (2019). Evaluation of 3D printed soft robots in radiation environments and comparison with molded counterparts. *Frontiers in Robotics and AI*, *6*(40). https://doi.org/10.3389/frobt.2019.00040.

Yirmibesoglu, O. D., Morrow, J., Walker, S., Gosrich, W., Cañizares, R., Kim, H., et al. (2018). Direct 3D printing of silicone elastomer soft robots and their performance comparison with molded counterparts. In *2018 IEEE international conference on soft robotics (RoboSoft)* (pp. 295–302). https://doi.org/10.1109/ROBOSOFT.2018.8404935.

Yousef, B., & Mohsen, S. (2014). A review of ionic polymeric soft actuators and sensors. *Soft Robotics*, *1*(1), 38–52. https://doi.org/10.1089/soro.2013.0006.

Yu, K., Xin, A., Du, H., Li, Y., & Wang, Q. (2019). Additive manufacturing of self-healing elastomers. *NPG Asia Materials*, *11*(1), 7. https://doi.org/10.1038/s41427-019-0109-y.

Yuk, H., Lin, S., Ma, C., Takaffoli, M., Fang, N. X., & Zhao, X. (2017). Hydraulic hydrogel actuators and robots optically and sonically camouflaged in water. *Nature Communications*, *8*, 14230. https://doi.org/10.1038/ncomms14230.

Zhang, Y.-F., Zhang, N., Hingorani, H., Ding, N., Wang, D., Yuan, C., et al. (2019). Fast-response, stiffness-tunable soft actuator by hybrid multimaterial 3D printing. *Advanced Functional Materials*, *29*(15), 1806698. https://doi.org/10.1002/adfm.201806698.

Zhang, X., Ai, J., Ma, Z., Du, Z., Chen, D., Zou, R., et al. (2019). Binary cooperative flexible magnetoelectric materials working as self-powered tactile sensors. *Journal of Materials Chemistry C*, *7*(28), 8527–8536. https://doi.org/10.1039/C9TC02453K.

Zhang, Y.-F., Ng, C. J.-X., Chen, Z., Zhang, W., Panjwani, S., Kowsari, K., et al. (2019). Miniature pneumatic actuators for soft robots by high-resolution multimaterial 3D printing. *Advanced Materials Technologies*, *4*(10), 1900427. https://doi.org/10.1002/admt.201900427.

Zhao, H., O'Brien, K., Li, S., & Shepherd, R. F. (2016). Optoelectronically innervated soft prosthetic hand via stretchable optical waveguides. *Science Robotics*, *1*(1), eaai7529. https://doi.org/10.1126/scirobotics.aai7529.

Zhou, J., Chen, Y., Chen, X., Wang, Z., Li, Y., & Liu, Y. (2020). A proprioceptive bellows (PB) actuator with position feedback and force estimation. *IEEE Robotics and Automation Letters*, *5*(2), 1867–1874.

Zhou, L.-Y., Fu, J., & He, Y. (2020). A review of 3D printing technologies for soft polymer materials. *Advanced Functional Materials*, *30*(28), 2000187. https://doi.org/10.1002/adfm.202000187.

Zhou, H., Mohammadi, A., Oetomo, D., & Alici, G. (2019). A novel monolithic soft robotic thumb for an anthropomorphic prosthetic hand. *IEEE Robotics and Automation Letters*, *4*(2), 602–609.

Zhu, M., Mori, Y., Wakayama, T., Wada, A., & Kawamura, S. (2019). A fully multi-material three-dimensional printed soft gripper with variable stiffness for robust grasping. *Soft Robotics*, *6*(4), 507–519. https://doi.org/10.1089/soro.2018.0112.

Zolfagharian, A., Kouzani, A. Z., Khoo, S. Y., Moghadam, A. A. A., Gibson, I., & Kaynak, A. (2016). Evolution of 3D printed soft actuators. *Sensors and Actuators A: Physical*, *250*, 258–272. https://doi.org/10.1016/j.sna.2016.09.028.

Zolfagharian, A., Mahmud, M. A. P., Gharaie, S., Bodaghi, M., Kouzani, A. Z., & Kaynak, A. (2020). 3D/4D-printed bending-type soft pneumatic actuators: Fabrication, modelling, and control. *Virtual and Physical Prototyping*, *15*(4), 373–402. https://doi.org/10.1080/17452759.2020.1795209.

Zolfagharian, A., Kaynak, A., & Kouzani, A. (2020). Closed-loop 4D-printed soft robots. *Materials & Design*, *188*. https://doi.org/10.1016/j.matdes.2019.108411, 108411.

4D-printed structures with tunable mechanical properties

Wael Abuzaid[a], Mohammad H. Yousuf[b], and Maen Alkhader[a]
[a]Department of Mechanical Engineering, American University of Sharjah, Sharjah, United Arab Emirates
[b]Faculty of Engineering and Information Sciences, University of Wollongong Dubai, Dubai, United Arab Emirates

Introduction

The tremendous and rapid advancements in additive manufacturing techniques and materials have already transformed traditional design and manufacturing cycles and are positioned to make a further transformative impact in virtually all industries (Bikas, Stavropoulos, & Chryssolouris, 2016; Lee, An, & Chua, 2017a, 2017b; Murr, 2016; Ngo, Kashani, Imbalzano, Nguyen, & Hui, 2018). The range of current and potential applications spans virtually all industries including aerospace (Froes & Boyer, 2019), automotive, biomedical (Miao et al., 2017), construction (Paolini, Kollmannsberger, & Rank, 2019), fashion, and even food industries (Fundamentals of 3D Food Printing and Applications, 2019). This unique level of applicability, which is not shared by almost any other manufacturing method, stems from simplifications offered by 3D printing where the synthesis of complex structures is typically accomplished with relative ease. This versatility is further enhanced by the exceptionally wide range of materials that can be utilized and adapted in continuously evolving additive manufacturing techniques. From a manufacturing and application perspective, the produced structures are typically required to have fixed properties in terms of geometry, shape, and mechanical properties. However, in four-dimensional (4D) printing, the rigidity in properties is removed, thus enabling the synthesis of adaptive structure with tunable and functional properties (Choi, Kwon, Jo, Lee, & Moon, 2015; Khoo et al., 2015; Loh, 2016; Momeni, M.Mehdi Hassani.N, Liu, & Ni, 2017; Pei, 2014; Tibbits, 2014; Zhang, Demir, & Gu, 2019). Introducing this extra dimension has tremendous advantages for the development of applications for deployable and morphing structures (Bodaghi, Noroozi, Zolfagharian, Fotouhi, & Norouzi, 2019; Neville et al., 2017; Wang et al., 2018), in structural self-assembly (Mao et al., 2015), actuator design (Mao et al., 2016), robotics (López-Valdeolivas, Liu, Broer, & Sánchez-Somolinos, 2018), artificial muscles (Peele, Wallin, Zhao, & Shepherd, 2015; Wirekoh & Park, 2017), and structures with tunable mechanical properties (Mirzaali et al., 2018; Mirzaali, Janbaz, Strano, Vergani, & Zadpoor, 2018; Rossiter, Takashima, Scarpa, Walters, & Mukai, 2014). This chapter considers one of these potential applications, namely, the tunable mechanical properties' aspect in 4D-printed periodic structures.

Relaxing the rigidity over time in 4D-printed components to achieve structural tunability (geometric and/or properties, e.g., mechanical) or to allow actuation is an active research area that continues to capture significant interest. Some of the proposed methods rely on functionally graded structural design (Bodaghi, Damanpack, & Liao, 2017; Bodaghi et al., 2019; Van Manen, Janbaz, & Zadpoor, 2017), the use of layered structures of different materials or functionally grades materials (Ge et al., 2016; Wang et al., 2018), and by using shape memory materials (Gardan, 2019; Lee, An, & Chua, 2017a; Liu, Toh, & Ng, 2015; Rastogi & Kandasubramanian, 2019; Zheng, Li, & Liu, 2018). In the case of shape memory materials (shape memory alloys (SMAs) and shape memory polymers (SMPs)), the inherent functional properties of these systems are exploited to achieve multiple configurations in a single structure while maintaining the ability to trigger recovery to the original printed shape through an external stimulus (e.g., heat or magnetic field) (Huang et al., 2010; Huang, Zheng, Liu, & Ng, 2020). The use of thermoplastic SMPs is specifically attractive due to their lightweight characteristics, large recovery strains, and relative ease of printing that relies on the widely available and heavily utilized extrusion-based 3D printers (Bodaghi et al., 2017; Bodaghi & Liao, 2019; Bodaghi et al., 2020; Hornbogen, 2006; Lendlein & Gould, 2019; Parameswaranpillai, Siengchin, George, & Jose, 2020; Yang, Chen, Wei, & Li, 2016). However, the utilization of SMP filaments in an extrusion-based 4D printing process, and in 3D printing in general, comes with significant challenges. For example, the anisotropy in properties is introduced by the processing method where the layered structure is deposited line by line (not inherent to the material). Therefore, the material deposition pattern that can be typically controlled along with the print infill density affects the structural anisotropy and final component properties. As noted earlier, this technicality relates to all deposition-based 3D-printed components. However, when thermoplastic SMP materials are considered, additional complications arise due to the need to apply heat, postprinting, to either program temporary shapes or induce original shape recovery. Such heat exposure will impact the geometry of the 4D-printed components in a manner that is time, temperature, and deposition pattern dependent (Abuzaid, Alkhader, & Omari, 2018). These dependencies are explained in detail in this chapter along with practical approaches to reduce their impact in practical applications.

In general, the utilization of the functional properties of SMP requires specific shape/structural programming followed by heat-induced recovery. Heterogeneities in the stored strains during programming (i.e., programmed strain) and the recovered strain during recovery (i.e., recovery strain) arise when complex geometric structures are considered (e.g., honeycomb periodic structures). Such complexities are further amplified following the 3D printing process (e.g., 4D printing using SMP) due to the heterogeneities introduced by the deposition process such as printing pattern and infill density. From a practical perspective, a proper understanding of the local buildup of programming, recovery, and any potential residual strains is important as these heterogeneities impact the resulting mechanical and/or shape memory properties. In addition, the heat-induced strain recovery rates will vary across the structure due to localization in strain buildup during programming. For example, 4D-printed honeycomb structures exhibited pronounced variability in the local recovery rates that consequently resulted in different regions reaching full recovery at different times. As discussed previously in Abuzaid et al. (2018) and in this chapter, such variability in strain recovery rates is dependent on stored strain levels (during programming—induced by complexity in structure) and deposition direction (induced by printing pattern).

As a technology, 4D printing can potentially be used to fabricate a wide range of complex geometries that exhibit functional triggered behaviors. However, this chapter is focused on applying 4D printing to geometries that belong to the class of materials known as periodic cellular solids or lattice structures. This class of materials is used in an extensive spectrum of industries and applications that include aerospace (Ahn, Peterson, Seferis, Nowacki, & Zachmann, 1992; Gibson & Ashby, 1997), naval (Bitzer, 1994; Daniels, Gdoutos, & Rajapakse, 2009), automotive (Tay, Lim, & Lankarani, 2014), and biomedical (Gibson, 2005; Gibson, Ashby, & Harley, 2010) applications. Periodic cellular solids were first considered as practical engineering materials in structural roles due to their exceptionally high stiffness- and strength-to-weight ratios as well as their ability to mitigate impact energy (Gibson & Ashby, 1997). These practical properties were often related to cellular solids' porosity. However, supported by advanced 3D printing and finite element modeling tools, more recent efforts showed that there exists a strong coupling between periodic cellular solids' properties and their inner structural topology (Alkhader & Vural, 2008; Deshpande, Ashby, & Fleck, 2001; Fleck, Deshpande, & Ashby, 2010). This means that cellular solids having identical porosities can exhibit starkly different behaviors. More importantly, exploiting the property–structure coupling in cellular solids showed potential for creating cellular solids with unprecedented properties. For instance, cellular solids can be designed to exhibit large negative Poisson's ratios (i.e., auxetic behavior) (Iyer, Alkhader, & Venkatesh, 2015; Lakes, 2017; Yousuf, Abuzaid, & Alkhader, 2020), controlled anisotropic elastic and plastic properties (Alkhader et al., 2019; Masters & Evans, 1996), modulated acoustic properties (Alkhader, Iyer, Shi, & Venkatesh, 2015), predesigned band gaps (Bertoldi & Boyce, 2008), and prescribed progressive failure behavior (Alkhader & Vural, 2008; Scarpa, Ciffo, & Yates, 2004). This chapter draws on the efforts aiming to realize unique cellular properties through tailoring their structure. However, it is dedicated to cellular solids' auxetic behavior.

Materials exhibiting negative Poisson's ratios, which contracts laterally when compressed axially, are referred to as Auxetic materials (Lakes, 2017). Exhibiting a negative Poisson's ratio is counterintuitive and rarely observed in naturally occurring materials. The shortlist of naturally occurring auxetic materials include α-cristobalite (Yeganeh-Haeri, Weidner, & Parise, 1992), quartz-rich sedimentary rocks comprising microcracks and micropores (Ji et al., 2018), graphene including void defects (Grima, Grech, Grima-Cornish, Gatt, & Attard, 2018), and boron arsenate (Grima-Cornish, Vella-Żarb, & Grima, 2020). The list of engineered auxetic materials, on the contrary, is long and includes composite laminates (Alderson et al., 2005; Evans, Donoghue, & Alderson, 2004), textile-based structures (Ge, Hu, & Liu, 2013), reentrant lattice materials and honeycombs (Iyer et al., 2015; Scarpa, Panayiotou, & Tomlinson,

2000; Zhang & Yang, 2016), chiral lattices (Spadoni, Ruzzene, Gonella, & Scarpa, 2009), microporous polymers (Alderson, Fitzgerald, & Evans, 2000), and rigid polyurethane foams (Brandel & Lakes, 2001; Chan & Evans, 1997). The aforementioned examples exhibit a wide range of negative Poisson's ratios that range commonly between 0 and −1. However, most of the observed large negative Poisson's ratios were delivered by engineered auxetic materials that belong to the cellular solids class of materials.

Auxetic materials, due to their negative Poisson's ratios, have advantageous practical properties. Their negative Poisson's ratio causes them to laterally contract under compressive axial loads, thus increasing their density and stiffness (Liu & Hu, 2010). Accordingly, they tend to have higher shear moduli, increased fracture toughness, increased indentation resistance, and increased resistance to impact loadings as compared to conventional materials (Wang, Zulifqar, Hu, Rana, & Fangueiro, 2016). These mechanical properties attracted interest in auxetic materials from the aerospace and marine industries, particularly to use in composite sandwich structures (Wang et al., 2016). The use of auxetic materials extends to a continuously expanding list of applications that include dilators for opening blocked arteries (Evans & Alderson, 2000), solid-state actuators, and energy harvesting applications (Farhangdoust, 2020; Ferguson, Kuang, Evans, Smith, & Zhu, 2018), and resonators (Ebrahimian, Kabirian, Younesian, & Eghbali, 2021; Eghbali, Younesian, & Farhangdoust, 2020), to name a few.

Macroscopically observed auxetic behavior, in both engineered and naturally occurring cellular and porous materials, arise from microstructurally enabled kinematic mechanisms that are contingent on the presence of concaved cavities. The latter supports the formation of internal hinges that underline the kinematic mechanisms driving auxetic behavior (Alkhader, Abuzaid, Elyoussef, & Al-Adaileh, 2019; Chan & Evans, 1997; Gibson & Ashby, 1997). Increasing the concaved porosities content and increasing their concavity result in higher negative Poisson's ratios and a more pronounced auxetic behavior.

Concaved cavities in auxetic materials can be classified as random or periodic. Random ones are introduced through deforming preexisting convex porosities. Typically, porosities in materials (e.g., foams and rocks) have convex shapes as they are caused by trapped air or gases. Subjecting porous materials to hydrostatic pressures can transform the shape of their internal porosities from convex to concave (Chan & Evans, 1997), which grants them auxetic properties. Through the aforementioned process, naturally occurring and many of the engineered auxetic materials (e.g., foams and polymers and microporous polymers) were realized. However, the compaction process has significant limitations as it causes material damage and does not allow for controlling the level of auxeticity. To have full control of the kinematic mechanisms driving the auxetic behavior in a material, the material's microstructure (e.g., shape, distribution, and location of porosities as well as solid-phase geometry) should be controlled. Auxetic cellular materials with predesigned microstructure are called periodic cellular solids or lattice structures. They are often fabricated using 3D printing techniques as they allow for building cellular structures from a predesigned geometry. More importantly, the predesigned geometry can be tailored through finite element analysis to comprise specific deformation mechanisms and exhibit a specific behavior. Most of the high-performing auxetic materials (e.g., chiral and reentrant honeycombs) and lattice structures are fabricated using 3D printing. This chapter takes such high-performing auxetic materials one step forward by rendering them functional and adaptive (Abuzaid et al., 2018; Iyer & Venkatesh, 2010) through using a smart constituent material (SMP) in the 3D printing process (i.e., 4D printing). The latter is demonstrated through multiple examples in this chapter. In these examples, 4D-printed periodic cellular solid structures are programmed to exhibit unprecedented functional behavior. For instance, in one example, a 4D-printed lattice is programmed to exhibit a wide range of Poisson's ratios, extending from negative to even positive magnitudes, depending on the programming level. This unique functional behavior was achieved by exploiting the unique properties of cellular solids using 4D printing.

Periodic cellular solids with functional, adaptive, and tunable properties, which are enabled by 4D printing, are sought after in many applications. For example, sensors with tunable resonance frequencies or acoustic impedances (Iyer & Venkatesh, 2010) that better match their dynamic environments and provide optimal mechanical–electrical energy conversion (Iyer, Alkhader, & Venkatesh, 2016; Iyer & Venkatesh, 2010), tunable phononic crystals, and metamaterials that are designed to exhibit unique band-gap frequency ranges and can be controlled through external triggers (e.g., force, electric field, or temperature) (Bertoldi & Boyce, 2008; Bertoldi, Boyce, Deschanel, Prange, & Mullin, 2008; Bertoldi, Vitelli, Christensen, & van Hecke, 2017), and tunable biomedical implants and stents that better match patients' needs and overcome incompatibilities resulting from the natural variability in humans (Yakacki et al., 2007). The aforementioned examples rely on adaptive components with tunable behaviors that respond to external triggers. 4D printing is particularly qualified to create such components. This is shown through multiple examples in this chapter.

In summary, this chapter presents a comprehensive review of recent work by the authors on 4D printing thermoplastic SMP structures. Detailed analysis and discussion focused on the effect of heat exposure and material deposition pattern on the shape stability and shape recovery properties are provided. An extensive assessment of the mechanical properties of the synthesized 4D-printed specimens, simple as well as complex periodic structures, is also provided. In particular, the work explores the tunable mechanical properties of a 4D-printed cellular periodic structure (Poisson's ratio and structural stiffness). Systematic analysis of the global and local strain measurements during programming (tuning properties), mechanical loading, and heat-induced recovery was conducted to provide a deep insight into the degradation of mechanical and shape memory properties following multiple tuning cycles (i.e., functional fatigue). Full-field deformation measurements are extensively utilized to provide a quantitative assessment of the average global response in various configurations as well as the localization developing at different stages of the SMP programming and recovery cycles. By evaluating the localization in strain accumulation, a better understanding of the different aspects affecting shape memory functionality degradation is provided. The presented work explores these aspects and highlights their major contributions to shape recovery rates and functionality degradation in

4D-printed structures. Overall, the presented work demonstrates that 4D-printed structures can be designed and fabricated to exhibit adjustable mechanical properties. Many aspects, including material deposition pattern, heat exposure temperature and duration, deformation level (magnitude of stored strain), and the number of applied programming/recovery cycles affect the tunability of mechanical properties. These effects are carefully studied and quantified to provide a better insight into the challenges, and future research and optimization need to further improve the mechanical properties of thermoplastic 4D-printed structures.

Shape memory polymer material

All the specimens reported in this chapter were 3D printed using a thermoplastic-based SMP obtained from SMP technologies, Japan. The glass transition temperature (T_g) of the 1.75-mm-diameter filament was 55°C, as reported by the manufacturer. Specimens of different geometries, simple dog bone, and complex shape periodic structures, were 3D printed using a standard fused deposition 3D printer (Flashforge Finder). A deposition temperature of 200°C and a layer height of 0.1 mm was utilized across all specimens. The deposition direction relative to specimen geometry and infill density were altered, however, in order to elucidate the effect of these aspects on the stability and functionality of the 4D-printed specimens. The full details of the exact specimen geometries, 3D printing conditions, and applied loading are provided along with the corresponding results.

Stability and functional properties of 4D-printed specimens

In this section, different aspects related to 4D printing simple and complex structures using SMP filaments are discussed. Specimens of different geometries were printed and analyzed in terms of geometric stability following heat exposure, recovery properties, mechanical properties, and functional properties under cyclic loading.

Geometric stability following heat exposure

As discussed previously, the challenge associated with the dimensional stability of 4D-printed SMP structures is induced by postprinting temperature exposure. Heating above the glass transition temperature (T_g) is an integral part of the structural programming and recovery processes in SMP materials. As shown in this section, and in previous works (Abuzaid et al., 2018), the heating affects the geometry of the printed components and can complicate the measurement and interpretation of the programming and recovery processes and the associated strains. This issue, although shared by most SMP materials, in particular, thermoplastic-based, is further complicated in 3D-printed structures that are inherently heterogeneous due to strong influence from deposition-related aspects (e.g., deposition speed, temperature, direction, infill density, and layer height). In this section, the effects of deposition direction, infill density, and exposure temperature postprinting are considered. The aim is to assess the dimensional stability of 4D-printed specimens having simple structures as shown in Fig. 6.1. The dog bone specimens were 3D printed with different settings within the slicer software to generate structures with different deposition directions: (1) aligned with the loading direction (designated as 0 degree lines in Fig. 6.1); (2) deposition direction perpendicular to the loading direction (0 degree lines); (3) a 90-degree pattern (+grid); and finally, (4) a 45-degree pattern or × grid. For the + and × grid patterns, 50% infill density specimens were also printed and analyzed to shed insight into the effect of infill density. The dimensional stability and any accumulation of strain were quantified using Digital Image Correlation (DIC). A high-temperature black paint was used to create a stable speckle pattern that does not experience degradation during temperature exposure. An environmental chamber was utilized to heat the specimens. The chamber was preheated to the desired exposure temperature before inserting the specimen. Optical images of the sample's surface were captured every 30 s throughout the experiment and utilized to calculate the accumulation of strain (i.e., shrinkage strain). Different exposure temperatures were considered ranging between 45 and 65°C (i.e., slightly below, at, and slightly above T_g).

The evolution of the average strains with temperature exposure time was evaluated within the sample's gauge section. The full-field and optical strain measurements using DIC allow for the calculation of the in-plane strain components; the normal strain ε_{yy} (i.e., along the loading direction of the dog bone specimen), horizontal strain ε_{xx}, and shear strain ε_{xy}. The results from a representative specimen are shown in Fig. 6.2. For the shown case, the deposition direction was aligned with the loading direction of the samples (0 degree lines as shown in Fig. 6.1) and the exposure

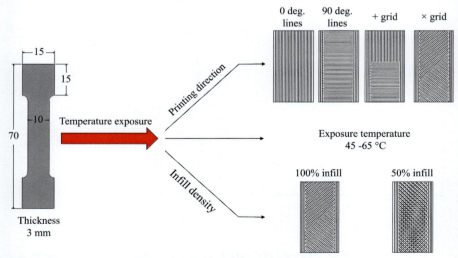

Fig. 6.1 The analyzed 3D printing aspects affecting the geometric stability of simple 4D-printed specimens.
No permission required

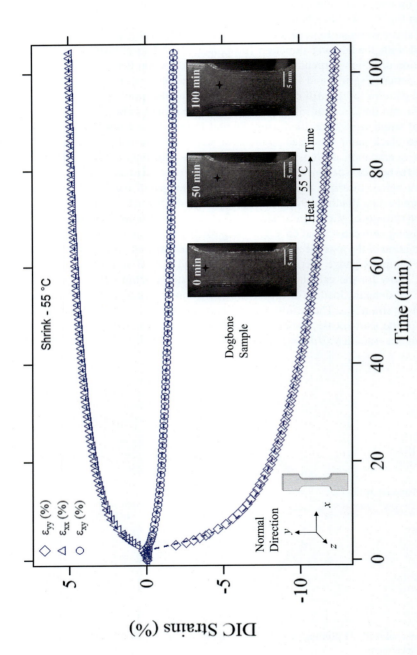

Fig. 6.2 Average DIC strains following temperature exposure at 55°C, 0-degree lines, and 100% infill density. The inset shows selected deformed images of the actual sample at different times.
From Abuzaid, W., Alkhader, M., & Omari, M. (2018a). Experimental analysis of heterogeneous shape recovery in 4d printed honeycomb structures. Polymer Testing. https://doi.org/10.1016/j.polymertesting.2018.03.050.

temperature was 55°C. A clear and significant accumulation of strains (shrinkage strains) can be observed with the highest magnitude being along the deposition/loading direction of the sample (ε_{yy}). The results clearly highlight the effect of temperature exposure, postprinting, on the dimensions of the 3D-printed component. It should be pointed out that this issue is not unique to the utilized SMP material. In fact, most thermoplastic-based polymers will exhibit such changes once heated, in particular to temperatures around their glass transition. However, in most cases, heating is not typically necessary or required for such materials under normal operating conditions. The issue is therefore more relevant to the 4D-printed components considered here as heating around T_g is required for the shape programming and recovery processes that provide material and structural functionality.

At different exposure temperatures, the relative changes in the specimen dimension differed significantly. Fig. 6.3 shows the evolution of the normal strain (ε_{yy}) for three different specimens following heating to 45°C, 55°C, and 65°C. As all of these specimens were 3D printed with 0 degree lines (i.e., deposition along the loading direction similar to the case shown in Fig. 6.2), the ε_{yy} represents the largest strain magnitude and was therefore used for direct comparison across the shown specimens in Fig. 6.3. The exposure temperatures affected both, the total amount of accumulated strain (total shrinkage strain) and the rate at which these strains evolved. However, the accumulation rate, which again alters the dimensions of the as-printed specimens, was not linear exhibiting a transition between a relatively high strain accumulation rate (strain rate 1 (SR1) as marked in Fig. 6.3) to a much lower rate (strain rate 2 (SR2)). A summary of the average results for the different considered temperatures is shown in Table 6.1. The results clearly highlight that the measured dimensional changes are both time and temperature dependent.

All the results shown just now were obtained using a single deposition pattern (i.e., 0-degree lines). Similar experiments were also conducted on different specimens with the other three printing patterns considered here. Fig. 6.4 shows the strain evolution for the average DIC strain components (ε_{xx}, ε_{yy}, and ε_{yy}) and the resulting effective strain (ε_{eff}) which was calculated using the following equation and the known strain components: $\varepsilon_{eff} = \sqrt{\frac{2}{3} \varepsilon_{ij} \varepsilon_{ij}}$.

The effective strain was calculated to provide simple means for comparison across the various considered printing patterns. The results presented in Fig. 6.4 were collected following exposure to the same temperature of 55°C. A strong influence from the printing pattern can be observed. For example, the 0-degree line specimen exhibits a pronounced accumulation of strain along the deposition/normal direction (ε_{yy}). However, the highest accumulation of strain for the 90-degree line sample took place in the x-direction (ε_{xx}) which happened to be the deposition direction as well. This suggests a strong correlation between the material deposition direction and the amount of strain accumulation. This aspect is discussed in detail in "Stress-free shape recovery" section of this chapter.

Fig. 6.5 shows the accumulation of shrinkage strains for the 50% infill density specimens. Both the + and × patterns were considered here as these two conditions produced the best results among the considered printing patterns (i.e., least amount of

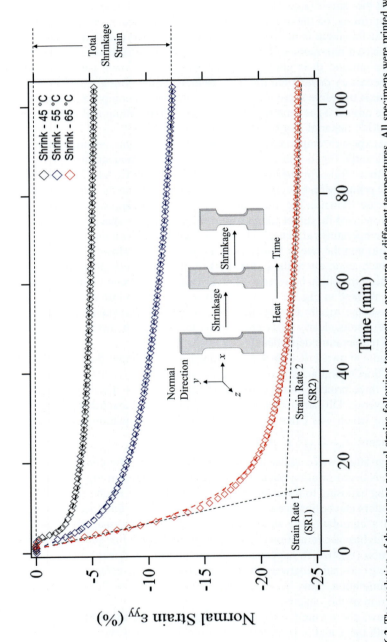

Fig. 6.3 The evolution of the average normal strains following temperature exposure at different temperatures. All specimens were printed with 0-degree lines and 100% infill density.
From Abuzaid, W., Alkhader, M., & Omari, M. (2018b). Experimental analysis of heterogeneous shape recovery in 4d printed honeycomb structures. Polymer Testing. https://doi.org/10.1016/j.polymertesting.2018.03.050.

Table 6.1 Summary of total shrinkage and strain accumulation rates at different exposure temperatures.

Heat exposure temperature (°C)	Strain rate 1—SR1 (%/min)	Strain rate 2—SR2 (%/min)	Total shrinkage strain (%)
45	−0.56	−0.01	−5.57
55	−0.58	−0.02	−12.75
65	−1.88	−0.01	−23.50

From Abuzaid, W., Alkhader, M., & Omari, M. (2018a). Experimental analysis of heterogeneous shape recovery in 4d printed honeycomb structures. Polymer Testing. https://doi.org/10.1016/j.polymertesting.2018.03.050.

strain accumulation as shown in Fig. 6.4) and are widely utilized in 3D printing, in general.

A comparison plot of the shrinkage response, focusing only on the calculated effective strain (ε_{eff}), is shown in Fig. 6.6. The total effective strain was calculated at the end of the exposure time and represents the total accumulation of strain (i.e., shrinkage). Obviously, as this strain accumulation is indicative of dimensional changes, or lack of dimensional stability, lower magnitudes highlight a better response. A summary of the total effective strain for the considered specimens is shown in Fig. 6.7. The 0-degree line case is obviously the worst-case scenario from a shrinkage or stability perspective. The × grid pattern has shown the least amount of strain accumulation compared to all other conditions. Decreasing the infill density to 50% had a negative impact on the dimensional stability as the level of total strains increased by ≈28% and 42% for the + grid and × grid patterns, respectively.

So far, the average shrinkage or heat-induced strain accumulation in simple structures was considered. However, in almost all potential practical applications, 4D-printed components will have a rather complex structure that will induce heterogeneity in response, including in the case of heat-induced shrinkage strain accumulation. Single-cell honeycomb specimens (shown in Fig. 6.8A) were utilized to shed further insight into the dimensional stability issue following heat exposure, however, in this case for a relatively complex structure. To enable unambiguous assessment of strain buildup relative to the deposition direction, the deposition direction was aligned with the cell walls of the structure (i.e., deposition parallel to wall ligaments). An optical image showing a representative specimen before heating is shown in Fig. 6.8B. Changes in the dimensions of the sample following 15 min and 60 min of temperature exposure at 55°C can be visually observed in Fig. 6.8C and D. Note that the horizontal white lines and vertical red lines, shown in Fig. 6.8B and C, are fixed in space, thus providing quantitative insight into the level of dimensional changes in the specimen. All-optical images of the specimen were captured with the sample placed inside the environmental chamber. Thus, the specimen was in all cases at the desired 55°C temperature.

Full-field strain assessment was conducted by correlating the collected optical images during heat exposure to the reference images captured inside the chamber (shown in Fig. 6.9A). The contour plots of the accumulated strains, ε_{xx}, ε_{yy}, and

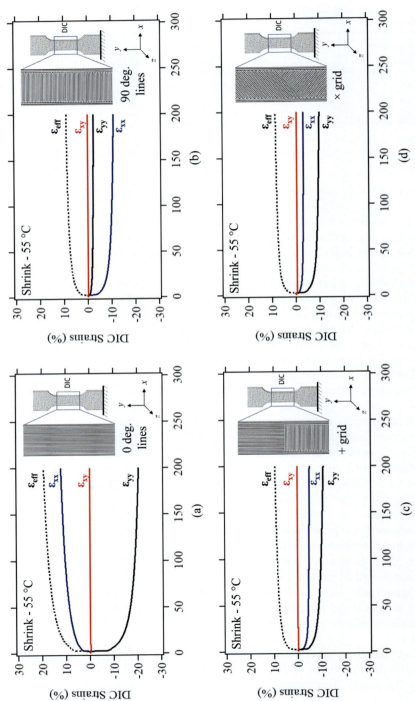

Fig. 6.4 Average DIC strains following temperature exposure at 55 °C for (a) 0-degree line deposition direction. (b) 90-degree lines. (c) +grid printing pattern. (d) ×grid printing pattern. All specimens were 3D printed using 100% infill density. No permission required.

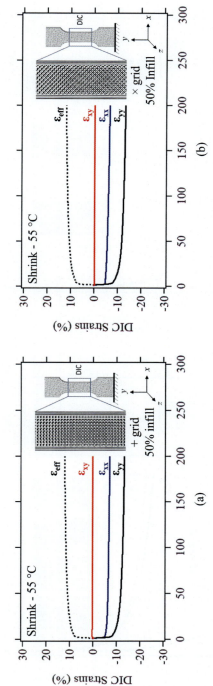

Fig. 6.5 Average DIC strains following temperature exposure at 55 °C for (A) + grid printing pattern. (B) × grid printing pattern. All specimens were 3D printed using 50% infill density.
No permission required.

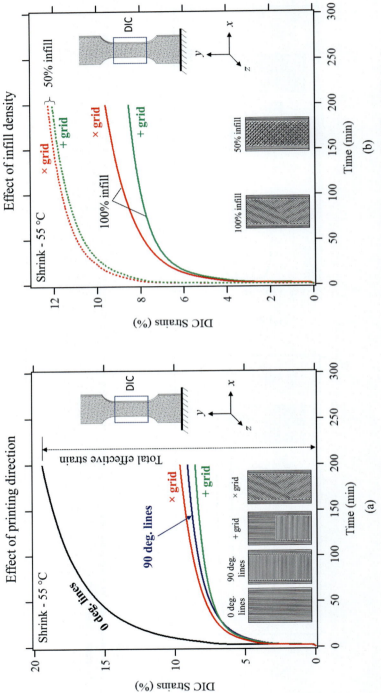

Fig. 6.6 The evolution of the effective strain following temperature exposure. (A) The effect of printing pattern. (B) Affect infill density. No permission required.

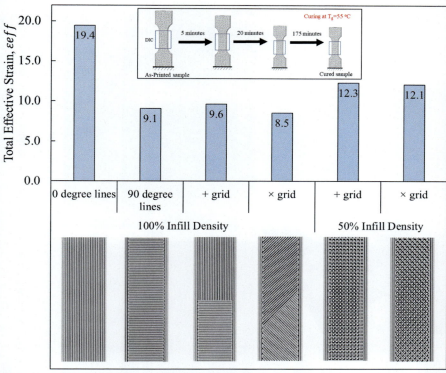

Fig. 6.7 A summary of the total shrinkage strains for the different considered printing patterns and infill densities. See Fig. 6.6 for strain evolution with time.
No permission required.

ε_{xy}, following 60 min of exposure are shown in Fig. 6.9 as a representative case. As expected, the buildup of shrinkage strains (i.e., heat-induced permanent strain accumulation) was highly nonuniform. For example, the horizontal strain contour plot (ε_{xx}) points to a high level of heterogeneity with the horizontal ligament exhibiting negative strains while the inclined cell walls show positive strains.

As discussed earlier for the simple dog bone specimens, the buildup of strains following heat exposure exhibits not only temperature and time dependence but is also influenced by the deposition direction. For the honeycomb sample shown in Figs. 6.8 and 6.9, the deposition direction was aligned with cell walls as clearly seen in the actual specimen shown in Fig. 6.10A (captured prior to adding the DIC speckle pattern). This deposition pattern enables a more in-depth analysis of the source of heterogeneity in strain accumulation (Fig. 6.9). The three marked regions (R1, R2, and R3) in Figs. 6.9 and 6.10 exhibit different strain levels and material deposition directions. By monitoring the strains in these local regions, the effect of deposition direction can be clarified further. Note that all the calculated strains are in the sample coordinate frame (x-y frame, see Fig. 6.10A). However, to enable a direct comparison of strain buildup relative to material deposition direction, a transformed coordinate

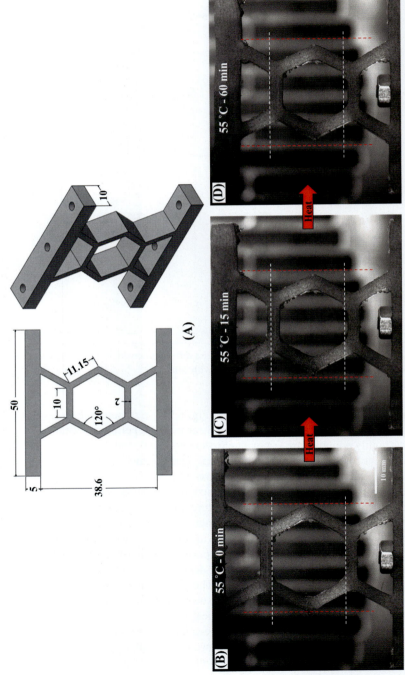

Fig. 6.8 (A) Schematic of the single-cell honeycomb specimen. (B) Optical image of an actual honeycomb specimen before temperature exposure. (C) After 15 min exposure time at 55°C. (D) After 60 min exposure time. From Abuzaid, W., Alkhader, M., & Omari, M. (2018a). Experimental analysis of heterogeneous shape recovery in 4d printed honeycomb structures. Polymer Testing. https://doi.org/10.1016/j.polymertesting.2018.03.050.

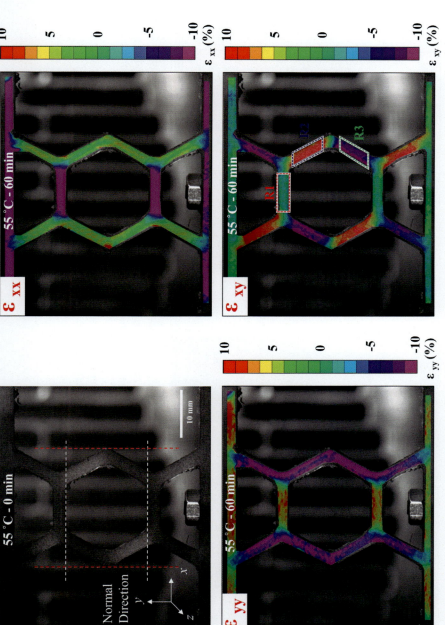

Fig. 6.9 Full-field DIC strain contour plots following 60min of heat exposure at 55°C. Localized response with a heterogeneous accumulation of strain can be observed in the full-field measurements.

From Abuzaid, W., Alkhader, M., & Omari, M. (2018a). Experimental analysis of heterogeneous shape recovery in 4d printed honeycomb structures. Polymer Testing. https://doi.org/10.1016/j.polymertesting.2018.03.050.

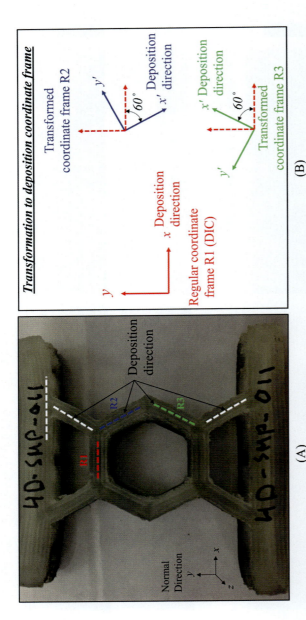

Fig. 6.10 (A) Actual specimen showing the deposition direction relative to the structure of the honeycomb sample. Regions R1, R2, and R3 (marked in Fig. 6.9) have different deposition directions. (B) The transformation from the regular coordinate frame to the deposition coordinate frame in regions R2 and R3.

From Abuzaid, W., Alkhader, M., & Omari, M. (2018a). Experimental analysis of heterogeneous shape recovery in 4d printed honeycomb structures. Polymer Testing. https://doi.org/10.1016/j.polymertesting.2018.03.050.

4D-printed structures with tunable mechanical properties

frame was defined with the x' direction aligned with the corresponding deposition direction. Both frames, original sample and transformed, are shown for the three considered regions (R1-R3) in Fig. 6.10B. The transformed strains in the x'-y' coordinated frame (aligned with deposition direction) can easily be calculated from the measured strain in the sample x-y frame using the following equation: $[\varepsilon_{x'-y'}] = [Q^T][\varepsilon_{x-y}][Q]$

where $[\varepsilon_{x-y}]$ is the strain tensor in the sample coordinate frame as measured using DIC, $[\varepsilon_{x'-y'}]$ is the transformed strain tensor in the deposition coordinate frame (different for R1 and R2), and $[Q]$ is the transformation matrix between the coordinate frames (see Fig. 6.10B).

The average strain in each of the three regions shown in Figs. 6.9 and 6.10 was evaluated for all of the collected images during the temperature exposure. The local strain evolution with time for all three regions (R1-R3) is presented in Fig. 6.11A. The figure shows only the ε_{xx} component in the sample coordinate frame. Consistent with contour plots in Fig. 6.9, the results highlight a significant difference in the localized response for the different regions. Note that for region R1, the x-direction is aligned with the deposition direction in the ligament and the reported strains were the highest compared to the other two regions. However, once the strains in regions R2 and R3 are transformed to the deposition direction frame (x'-y'), a similar shrinkage response in all regions was observed as shown in Fig. 6.11B. This observation was consistent for the other strain components, ε_{yy} as shown in Fig. 6.11C and ε_{xy} (Fig. 6.11D). The results clearly highlight the impact of material deposition direction on the nature and magnitude of heat-induced shape changes in 4D-printed structures.

Stress-free shape recovery

In this section, the shape recovery properties of simple and complex 4D-printed specimens are evaluated. However, as noted in "Geometric stability following heat exposure" section, the heat exposure around T_g induces shape changes and permanent strain accumulation. As the programming and heat-induced recovery of the SMP material require heating, the induced changes and strain buildup in as-printed specimens will hinder the ability to unambiguously distinguish the shrinkage strains from programming, recovery, and any residual strain accumulation. In other words, during SMP programming and recovery phases, shrinkage strains will coexist along with programming and/or recovery strains and that will induce significant error in measuring these important SMP properties. This is particularly significant in the first few minutes of temperature exposure as the shrinkage strain accumulation rate is the highest (see Fig. 6.3 and Table 6.1). However, and as explained earlier, this rate of strain accumulation drops to much lower values with time, and an almost saturated response is achieved. Once that is reached, the rate at which dimensional changes take place becomes relatively insignificant compared to the level of applied programming and recovery strains. Therefore, and motivated by the measured reduction in shrinkage rate with continued exposure time, a preheating step around the glass transition temperature T_g for an extended period of time was introduced before any programming or recovery strain assessment (henceforward referred to as sample curing). It should be

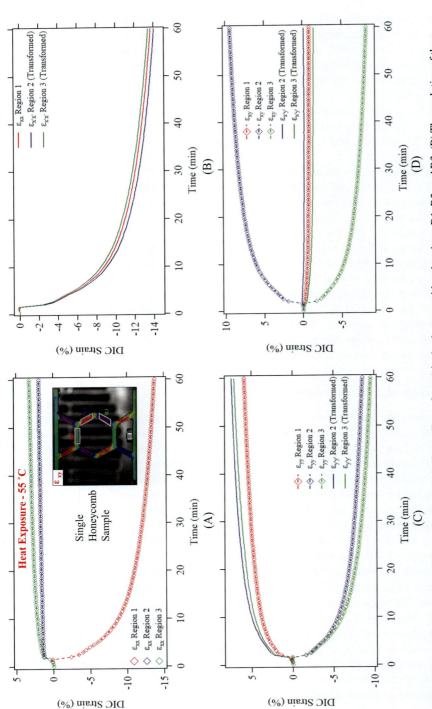

Fig. 6.11 (A) The evolution of the ε_{xx} strains (original coordinate frame) in the three considered regions, R1, R2, and R3. (B) The evolution of the $\varepsilon_{x'x'}$ strains (transformed coordinate frame) in the same three regions. Note that x′ is aligned with the deposition direction as shown in Fig. 6.10. (C) The evolution of the ε_{yy} (original) and $\varepsilon_{y'y'}$ (transformed) strains. (D) The evolution of the ε_{xy} (original) and $\varepsilon_{x'y'}$ (transformed) strains. From Abuzaid, W., Alkhader, M., & Omari, M. (2018a). Experimental analysis of heterogeneous shape recovery in 4d printed honeycomb structures. Polymer Testing. https://doi.org/10.1016/j.polymertesting.2018.03.050.

emphasized that although this process does provide 100% stability beyond curing, the additional changes during programming and recovery are relatively insignificant and do not alter the observations and measurements related to programming and recovery strains. Here, it is worth mentioning that the heat exposure time during either the programming or recovery phases does not exceed 6 min which would result in insignificant shrinkage strains once the curing procedure described previously is followed.

Consistent with the previous aspects considered for shrinkage strain accmulation, the effect of deposition direction and infill density were also analyzed for their recovery properties under stress-free conditions. The results from representative samples are shown in Fig. 6.12 for the 0-degree line pattern and in Fig. 6.13 for the × grid deposition pattern, both at 100% infill density. In the programming phase of the as-cured sample, each specimen was heated to 55°C. Once the desired temperature was reached, a tensile deformation was applied to induce a change in length of $\Delta L = 2$ mm. Following deformation, the displacement was held constant while allowing the sample to cool down back to room temperature. Once cooled, the specimen would be in the programmed state and the retained strains are programmed strains. To trigger recovery, the specimen was reheated to 55°C which initiated shape recovery under stress-free conditions. The amount of strain reduction from the programmed state level is the recovery strain. Any remaining strains at the end of recovery are residual (permanent) strains in the sample. Ideally, the recovery strain would be equal to the programmed strain in the case of full recovery with zero residual strain. A finite

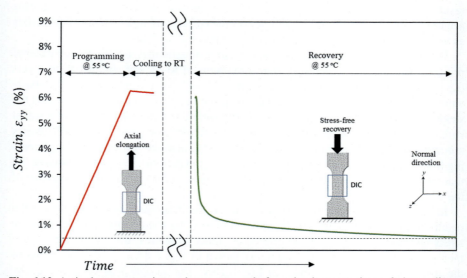

Fig. 6.12 A single programming and recovery cycle for a dog bone specimen, 0-degree lines, and 100% infill density. Stress-free recovery was conducted for 200 min.
From Yousuf, M. H., Abuzaid, W., & Alkhader, M. (2020). 4D Printed Auxetic Structures with Tunable Mechanical Properties. Additive Manufacturing, 101,364–101,364. https://doi.org/10.1016/j.addma.2020.101364.

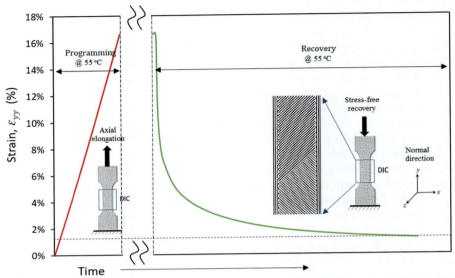

Fig. 6.13 A single programming and recovery cycle for a dog bone specimen, × grid pattern, and 100% infill density. Stress-free recovery was conducted for 200 min.
No permission required.

residual strain of about 0.5% was measured for the case presented in Fig. 6.12. This accumulation of residual strains will impact the corresponding mechanical properties as will be shown in "Tunable mechanical properties" section.

Fig. 6.14 shows the resulting residual strain magnitudes for all the considered printing patterns. All specimens were programmed to almost similar strain levels (i.e., programmed strains) as summarized in Table 6.2. Again, and similar to the observation made from shrinkage assessment, the printing pattern had a clear impact on the shape memory properties. The × grid printing pattern had the least amount of residual strain accumulation which is indicative of better recovery properties. The infill density reduction from 100% to 50% resulted in a degradation in the recovery levels (increasing from 1.2% to 1.9%, a 58% increase for the × grid pattern). It should be pointed out that the + grid pattern exhibited also good properties with a similar response to that of the × grid pattern.

In the dog bone specimens discussed earlier, the structure is rather simple and the resulting global stresses/strain are rather uniform. To shed further insight into the stress-recovery properties of 4D-printed complex structures, the honeycomb specimen geometry was considered (same as shown in Fig. 6.8A). An optical image of the *as-cured* specimen is shown in Fig. 6.15A. The sample was programmed by heating to 55°C inside the environmental chamber followed by applying a tensile displacement and finally holding the applied displacement while allowing the sample to cool back to room temperature. Fig. 6.15B shows an optical image of the programmed specimen after removing the applied load during programming and cooling the sample. To initiate recovery, the sample was again re-heated to 55°C which induced

4D-printed structures with tunable mechanical properties

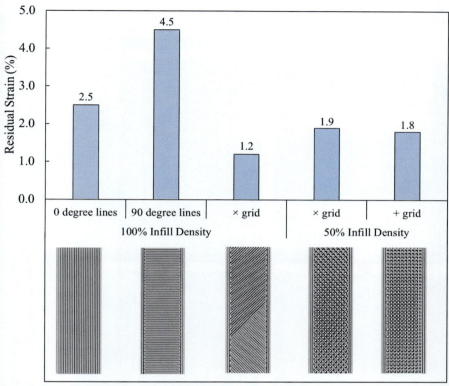

Fig. 6.14 Effect of printing pattern on the residual strain accumulation under stress-free conditions.
No permission required.

stress-free recovery (see Fig. 6.15C for the final recovered shape). The strain contour plots of the programmed strains (Fig. 6.15D) and residual strains (Fig. 6.15E) were obtained by correlating the corresponding optical images at these states to the original reference image shown in Fig. 6.15A. The normal strain fields ε_{yy} are shown in Fig. 6.15 (i.e., along the deformation direction); however, the other two strain

Table 6.2 Summary of the stress-free recovery properties for different deposition patterns.

Infill density	Printing pattern	Programmed strain (%)	Recovery strain (%)	Residual strain (%)
100%	0 deg	16.8	14.3	2.5
	90 deg	17.4	12.9	4.5
	× grid	16.6	15.4	2.9
50%	+ grid	17.0	15.2	1.8
	× grid	14.0	12.1	1.9

No permission required.

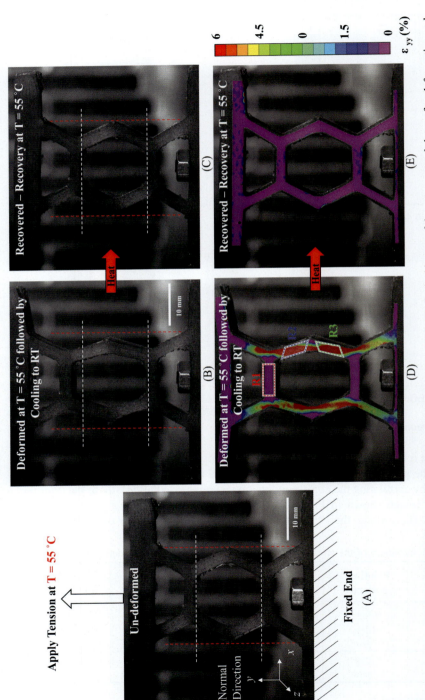

Fig. 6.15 (A) Single honeycomb sample following heating for 60 min at $T = 55\,°C$. (B) Optical image of the programmed shape after deformation and cooling back to room temperature. (C) Optical image of the recovered shape following heating to $55\,°C$ for 3 min. (D and E) Full-field contour plots of the normal strain field (along the y-direction—ε_{yy}) for the programmed and recovered shapes, respectively. From Abuzaid, W., Alkhader, M., & Omari, M. (2018a). Experimental analysis of heterogeneous shape recovery in 4d printed honeycomb structures. Polymer Testing. https://doi.org/10.1016/j.polymertesting.2018.03.050.

components were also measured (ε_{xx} and ε_{xy}). The full-field strain measurements reveal the rather complex and heterogeneous strain accumulation during programming, and consequently in the recovery phase, for the considered honeycomb structure.

To evaluate the localization in programming and recovery properties, three different regions were defined (R1-R3) as shown in Fig. 6.15D. The evolution of the local recovery strain components (ε_{xx}, ε_{yy}, and ε_{xy}), with time, are shown in Fig. 6.16 for the three considered regions. The results clearly highlight the dissimilar levels of strains in the different regions (some tensile while others are compressive) and point to significant differences in the local recovery rates (i.e., rate of local strain recovery). A summary of the local strain recovery rates is shown in Table 6.3 for the normal strain component (ε_{yy}). The results clearly show that the programming level, which is manifested as local programming strain accumulation, affects the local recovery rates in the corresponding regions.

In addition to the three regions considered in Fig. 6.16 and Table 6.3, which exhibit a rather homogeneous response, the localization in the honeycomb cell corner was also analyzed. The heterogeneity in this location results in both, negative (region R4 as shown in Fig. 6.16D) and positive strain magnitudes (region R5). The evolution of the recovery strains in these two corner regions is shown in Fig. 6.16D. Note that although both regions exhibited comparable high levels of recovery, the recovery rates differed significantly. For example, the compressive R4 region experienced a delayed recovery response as highlighted in the figure. Such delay in the triggering or recovery was not observed in the tensile region. Similar conclusions were drawn from programming the honeycomb specimen under compressive loading as shown with further details in Abuzaid et al. (2018).

The local recovery properties differed in horizontal ligaments (R1 in Fig. 6.15) compared to inclined cell walls (R2 and R3). As noted previously, the main difference was in the recovery rate and not in the final state of recovery as in all cases, high levels of recovery were achieved. The amount of stored or programmed strains has a major influence on the recovery rates (see Table 6.3). However, the material deposition direction also plays a role in affecting the recovery rates. To clarify this point further, the local recovery along the deposition directions, which as noted already is always aligned with cell walls, was evaluated using DIC virtual extensometers as shown in Fig. 6.17A. The first extensometer (VE1 as shown in Fig. 6.17A) was aligned with the deposition direction of the horizontal cell wall while the second (VE2) was aligned with one of the inclined ligaments. The local recovery along the deposition directions for the two considered cell walls/deposition directions is shown in Fig. 6.17B. As expected, the strain recovery rates differed as the levels of programmed strains were also different (i.e., about 2.5% for VE1 compared to 5.2% for VE2 as shown in Fig. 6.17B). The table shown in Fig. 6.17C provides a summary of the programmed strain levels and recovery rates for both extensometers. Even when the strains were analyzed along with the deposition directions, the results point to differences in the recovery rates. However, once the rates were normalized by the corresponding programmed strain, both extensometers (VE1 and VE2) displayed similar normalized recovery rates. This analysis supports the conclusion that the shape memory strain

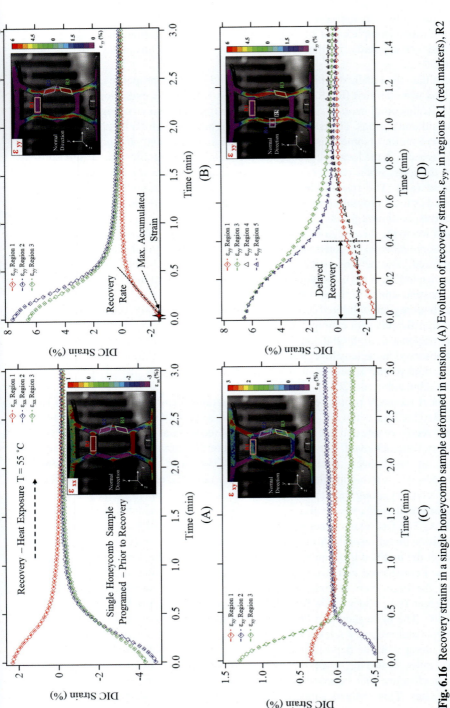

Fig. 6.16 Recovery strains in a single honeycomb sample deformed in tension. (A) Evolution of recovery strains, ε_{xx} in regions R1 (red markers), R2 (blue markers), and R3 (green markers). (B) Evolution of the ε_{yy} recovery strains. (C) Evolution of the ε_{xx} recovery strains. (D) Recovery strains at honeycomb cell corner. Region R4 exhibits compressive (negative) strain accumulation while the other side of the corner (R5) exhibits tensile (positive) strains.

From Abuzaid, W., Alkhader, M., & Omari, M. (2018a). Experimental analysis of heterogeneous shape recovery in 4d printed honeycomb structures. Polymer Testing. https://doi.org/10.1016/j.polymertesting.2018.03.050.

Table 6.3 Strain recovery rates for the εyy strains in regions R1-R3 marked in Fig. 6.16.

Region (Fig. 6.16)	Max. accumulated strain (%)	Strain recovery rate (%/min)	Strain recovery rate/max. Accumulated strain
R1	−2.54	5.79	−2.29
R2	7.67	−13.61	−1.77
R3	6.58	−11.4	−1.73

From Abuzaid, W., Alkhader, M., & Omari, M. (2018a). Experimental analysis of heterogeneous shape recovery in 4d printed honeycomb structures. Polymer Testing. https://doi.org/10.1016/j.polymertesting.2018.03.050.

recovery in 4D-printed structures exhibits dependence on both, the programmed strain level and the deposition direction. This is clearly shown in the reported results for the strain recovery rates.

Tunable mechanical properties

The inherent shape memory properties of SMP can be exploited to alter the geometry and consequently the structural and mechanical properties of 4D-printed components. In this section, the different practical aspects affecting the tunability of properties are investigated. Simple dog bone specimens, as well as complex periodic structures, are considered to shed insight into both; the inherent material response changes and structure/geometric-induced changes following material programming and recovery. In addition, further insight into the functional behavior of 4D-printed specimens is provided under cyclic loading conditions (i.e., multiple programming and recovery cycles). A typical evaluation cycle in this context is explained schematically in Fig. 6.18. The experiment is divided into five main stages, starting with mechanical loading on the original sample, followed by programming, then testing of the programmed shape, followed by recovery, ending with testing the final recovered shape. The exact same steps are followed for all experiments in this section, regardless of sample structure.

Tunability in simple structures

By considering dog bone tensile specimens, the inherent tunable properties of the SMP material were investigated. In particular, two main questions are raised: 1—does material programming alter the stiffness (i.e., elastic modulus) and 2—what sort of impact does the programming level (i.e., magnitude of programmed strains) have on the elastic modulus of the tensile specimens. In both experiments, the properties were evaluated through nondestructive uniaxial tensile tests conducted in the elastic regions of original, programmed (to different strain levels), and recovered specimens. Representative test results for one cycle of constant programming level experiment ($\varepsilon_{prog} \approx 6.25\%$) are shown in Fig. 6.19. Prior to the programming stage, the elastic modulus of the original shape was measured at 2025 MPa. The stiffness drops to

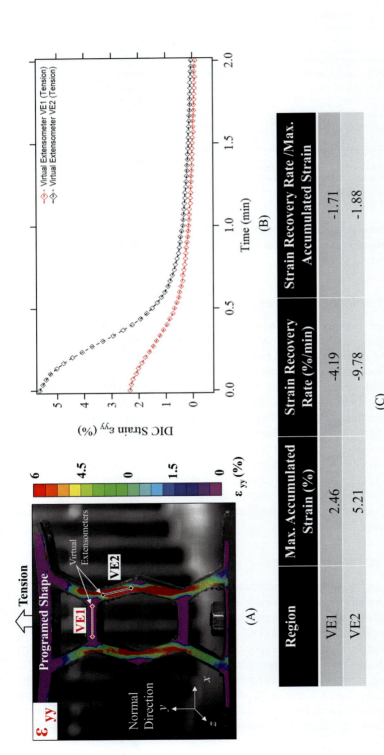

Fig. 6.17 (A) Full-field contour plot of the normal strains ε_{yy} for a sample deformed (programmed) in tension. The recovery strains were evaluated along the deposition directions using virtual extensometers VE1 and VE2. (B) The evolution of recovery strains with time. (C) Summary of the strain recovery rates along the deposition direction.
From Abuzaid, W., Alkhader, M., & Omari, M. (2018a). Experimental analysis of heterogeneous shape recovery in 4d printed honeycomb structures. Polymer Testing. https://doi.org/10.1016/j.polymertesting.2018.03.050.

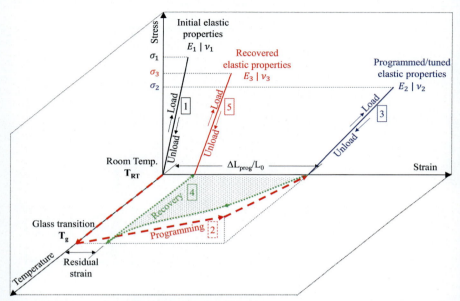

Fig. 6.18 A schematic explaining a typical evaluation cycle. 1—Test start with elastic tension test to evaluate elastic properties, followed by 2—programming, which consists of heating the sample to Tg temperature and deforming it to a new desired shape, followed by cooling to lock the deformed shape. 3—Elastic property evaluation of the programmed shape. 4—Stress-free recovery by heating the sample back to Tg. 5—Elastic property evaluation of the recovered shape.
From Yousuf, M. H., Abuzaid, W., & Alkhader, M. (2020). 4D Printed Auxetic Structures with Tunable Mechanical Properties. Additive Manufacturing, 101,364–101,364. https://doi.org/10.1016/j.addma.2020.101364.

1605 MPa after the programming stage (about 20% change). Following heat-induced recovery, the stiffness was measured at 1940 MPa. The results highlight two important conclusions: first, the programming process alters the stiffness of the specimens; second, the mechanical properties are partially recovered following heat-induced shape recovery. As will be discussed later in this section, this lack of mechanical property recovery is attributed to the accumulation of residual strains (i.e., lack of complete shape recovery).

To evaluate the functional properties, multiple SMP evaluation cycles were applied on a 4D-printed tensile specimen. In the three considered cycles, the changes in the programmed and recovered elastic moduli were evaluated as a function of cycle number N. The measured moduli along with the magnitude of programmed and residual strains are shown in Table 6.4. With an average residual strain accumulation of $\varepsilon_{Relative} \approx 0.65\%$ per cycle, the results point to a correlation between the buildup of residual strains and the systematic decrease in the elastic moduli of the recovered shapes. However, it should be pointed out that the programmed shape showed minimal variation as a function of repeated cycles.

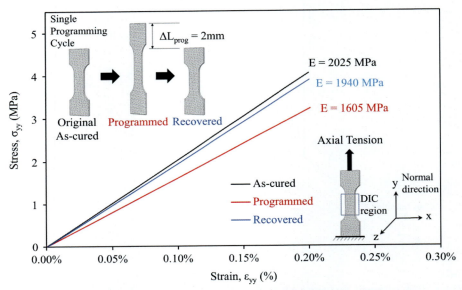

Fig. 6.19 Stress–strain behavior for a single programming/recovery cycle done using $\Delta L_{prog} = 2$ mm.
From Yousuf, M. H., Abuzaid, W., & Alkhader, M. (2020). 4D Printed Auxetic Structures with Tunable Mechanical Properties. Additive Manufacturing, 101,364–101,364. https://doi.org/10.1016/j.addma.2020.101364.

In addition to the cyclic experiment shown earlier, which was conducted to constant programming levels in all cycles, another cyclic test was conducted with varying levels of programming ranging from 0 to 10 mm, as shown in Fig. 6.20. The results show a nonlinear trend, especially at low values of programming levels where tunability of elastic modulus was most significant. However, as the programming level

Table 6.4 Summary of strain measurements following three programming/recovery cycles using $\Delta L\_prog = 2$ mm.

Cycle	Shape	Accumulated programming and recovery strain	Relative residual strain	Absolute residual strain	E (MPa)
As-cured	Original	–	0%	0%	2025
1	Programmed	6.25%	0.76%	0.76%	1605
	Recovered	−5.49%			1940
2	Programmed	6.19%	0.51%	1.27%	1648
	Recovered	−5.68%			1842
3	Programmed	6.27%	0.68%	1.95%	1637
	Recovered	−5.59%			1818

From Yousuf, M. H., Abuzaid, W., & Alkhader, M. (2020). 4D Printed Auxetic Structures with Tunable Mechanical Properties. Additive Manufacturing, 101,364–101,364. https://doi.org/10.1016/j.addma.2020.101364.

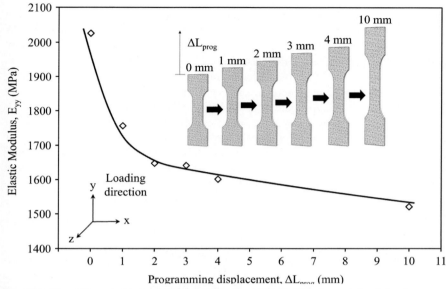

Fig. 6.20 Tunability of elastic modulus in 3D-printed SMP. Elastic modulus behavior as a function of programming level.
From Yousuf, M. H., Abuzaid, W., & Alkhader, M. (2020). 4D Printed Auxetic Structures with Tunable Mechanical Properties. Additive Manufacturing, 101,364–101,364. https://doi.org/10.1016/j.addma.2020.101364.

increased beyond $\Delta L_{prog} = 2$ mm, the change in elastic modulus was less pronounced. The incremental programming level experiment proves that the tunability in elastic modulus is achievable, as evident by the 25% change from 2025 MPa to 1467 MPa. In summary, the results presented in this section indicate that in 4D-printed materials, the mechanical properties are affected systematically by the number of programming/recovery cycles in addition to the programming levels.

Tunability in complex periodic structures

In contrast to the previously explored dog bone simple structure, applications in the real world require rather complicated configurations. Local strain heterogeneities will develop in such components along with more complex interactions from the deposition pattern. The localization in strain accumulation during programming, recovery, and any buildup of residual strains affects the resulting mechanical properties. The focus of this section is to investigate the utilization of SMPs in complex periodic structures to achieve tunable mechanical properties. In particular, negative Poisson's ratio (NPR) reentrant variant of the hexagonal honeycomb structure (auxetic honeycomb structure) will be 4D-printed and its mechanical properties will be explored subject to constant programming cycles, as well as incrementally increasing programming levels. The mechanical properties of cellular solids, and auxetic structures, in particular, depend on two main variables: the constituent material and the dimensions of the structure. The ability of SMP to change shape allows exploration of the dimensional aspect of the auxetic structure. By controlling the shape of the auxetic structure, new properties are obtained on the structural level. Generally, the properties of interest in auxetic cellular structures are stiffness and Poisson's ratio. This section aims to: first, provide insight into the effect of the bulk strain accumulation on the elastic properties in 4D-printed auxetic structures subjected to constant programming/recovery cycles and secondly, investigate the tunability of mechanical properties in 4D-printed auxetic structures under different programming levels. Two types of specimens were 4D-printed: 3×4 multicell auxetic honeycomb and a simpler 2×2 auxetic honeycomb with a single central cell. Both specimens are shown schematically in Fig. 6.21. The same printing pattern was selected for all specimens with a deposition direction parallel to cell walls as shown in Fig. 6.21C.

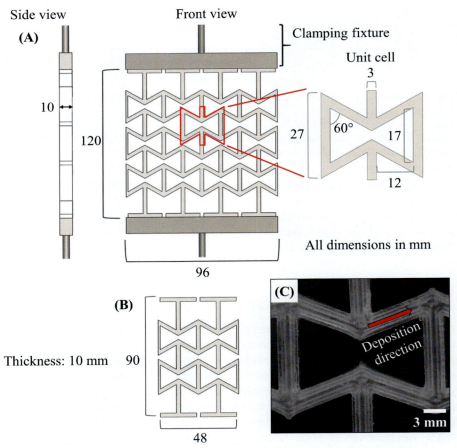

Fig. 6.21 Schematic of the (A) 3×4 and (B) 2×2 auxetic structures. (C) Optical image showing the deposition direction that was aligned with cell walls.
From Yousuf, M. H., Abuzaid, W., & Alkhader, M. (2020). 4D Printed Auxetic Structures with Tunable Mechanical Properties. Additive Manufacturing, 101,364–101,364. https://doi.org/10.1016/j.addma.2020.101364.

The mechanical properties of 4D-printed specimens were evaluated by applying an elastic tensile loading. The response was measured before programming, in the programmed state, and finally following heat-induced recovery (all conducted at room temperature). The programming and recovery phases were achieved using a load-frame equipped with an environmental chamber to allow deformation under a controlled temperature atmosphere. A representative case showing optical images of an actual specimen in its original, programmed, and recovered shapes are shown in Fig. 6.22.

During the elastic mechanical loading, the applied load was recorded along with capturing optical images of the specimen's surface. Digital image correlation (DIC) was used to quantitatively measure the full-field strains during deformation and following the programming and recovery phases. As shown in Fig. 6.23, the representative full-field strains show the locations of the local strain accumulation that will be further investigated in the coming sections. In addition, as shown in Fig. 6.23, virtual horizontal and vertical extensometers were used in the measurement of Poisson's ratio of the auxetic cell.

As mentioned previously, 4D-printed parts can accumulate permanent residual strains in the programming and recovery stages of the shape memory cycle. The buildup of residual strains is of significant importance due to its evident effect on the geometry and properties of 4D-printed components. Therefore, the evaluation of these strains in the auxetic structure is essential as it could affect the tunability of the elastic properties on the structural level. The 2×2 specimen shown in Fig. 6.23 was subjected to repeated programming and recovery cycles with a constant programming level of 10 mm. The accumulated strains as a function of repeated functional programming are shown in Fig. 6.24. The reported global strains were calculated using virtual extensometers as explained in Fig. 6.23. It is noted that the programming strain levels for each cycle N ($N = 1$–4) were almost constant (around 13%). However, the buildup of residual strains after each programming/recovery cycle resulted in an incremental increase in the magnitude of the total measured strains. The magnitude of residual strain accumulation per cycle was almost constant and equal to 2.4% ($\varepsilon_{R,relative} = 2.4\%$). The total or absolute residual strain ($\varepsilon_{R,absolute}$) in the sample increases after each cycle by $\varepsilon_{R,relative}$. The resulting residual strain accumulation ($\varepsilon_{R,absolute}$) reached 8.7% after 4 cycles as marked in Fig. 6.24.

The programming process induces a significant change in the geometry of the 4D-printed structure (intentionally) and consequently alters the corresponding mechanical properties. Such changes give the ability to tune the properties of 4D-printed structures. For example, Poisson's ratio of the 2×2 auxetic structure increased from -0.57 to -0.38 following a 10-mm structural programming (about 33% tunability). As shown in Fig. 6.25, this level of tunability remained unchanged through the four applied programming/recovery cycles. However, the incremental accumulation of residual strains following repeated cycling resulted in an incremental change in Poisson's ratio that increased from -0.57 in the first cycle ($\varepsilon_{R,absolute} = 0\%$) to -0.5 at the end of the fourth cycle where the residual strain $\varepsilon_{R,absolute}$ was equal to 8.7%.

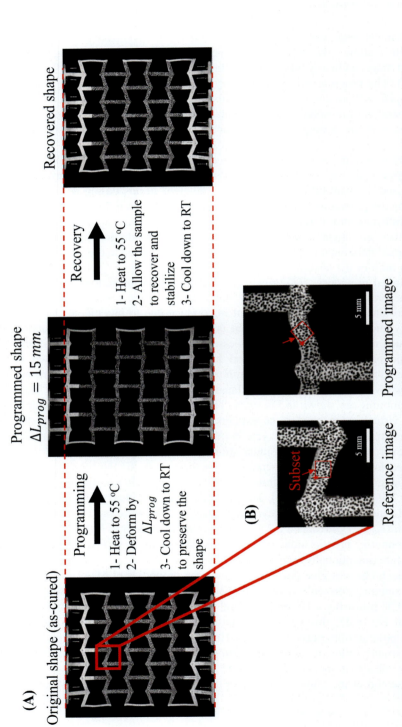

Fig. 6.22 (A) Programming and recovery cycle of 4D-printed 3 × 4 auxetic structure. Programming is achieved by heating the sample to Tg and applying 15-mm bulk deformation using a uniaxial tensile force followed by cooling down to room temperature. Stress-free recovery is achieved by heating the sample back to Tg. (B) Enlarged local region. The red box represents the subset used for DIC measurements. From Yousuf, M. H., Abuzaid, W., & Alkhader, M. (2020). 4D Printed Auxetic Structures with Tunable Mechanical Properties. Additive Manufacturing, 101364–101364. https://doi.org/10.1016/j.addma.2020.101364.

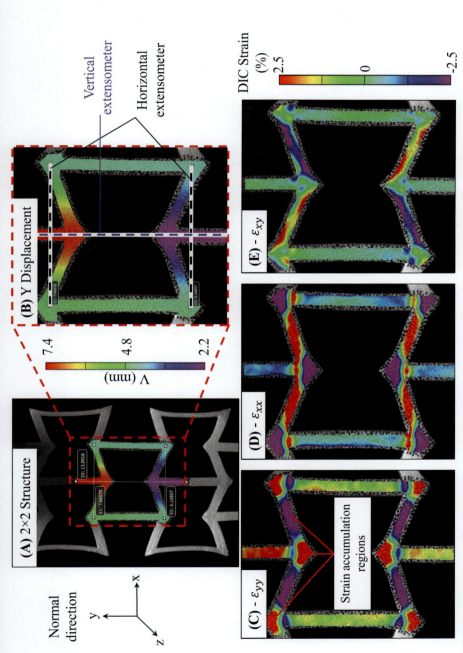

Fig. 6.23 Representative strain contour plots on the central cell of the 2 × 2 auxetic structure post 10mm. (A) Entire view. (B) y-direction displacement. (C) y-direction strains ε_{yy}. (D) x-direction strains ε_{xx}. (E) shear strains ε_{xy}. From Yousuf, M. H., Abuzaid, W., & Alkhader, M. (2020). 4D Printed Auxetic Structures with Tunable Mechanical Properties. Additive Manufacturing, 101,364–101,364. https://doi.org/10.1016/j.addma.2020.101364.

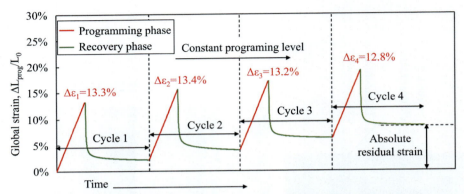

Fig. 6.24 Programming and recovery constant cycles on a 2 × 2 auxetic structure using a programming level of 10 mm for four repeated cycles. The programming phase was done using a displacement rate of 1 mm/min, while the recovery phase was conducted for 200 min.
From Yousuf, M. H., Abuzaid, W., & Alkhader, M. (2020). 4D Printed Auxetic Structures with Tunable Mechanical Properties. Additive Manufacturing, 101,364–101,364. https://doi.org/10.1016/j.addma.2020.101364.

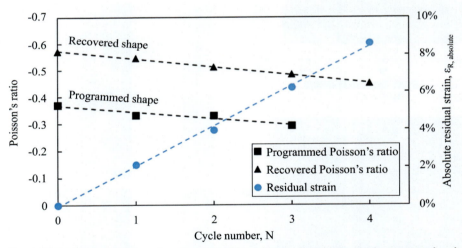

Fig. 6.25 Summary of Poisson's ratio results for 2 × 2 auxetic structure at the recovered and programmed shapes following multiple constant programming cycles. A degradation in Poisson's ratio is caused by the accumulation of residual strains.
From Yousuf, M. H., Abuzaid, W., & Alkhader, M. (2020). 4D Printed Auxetic Structures with Tunable Mechanical Properties. Additive Manufacturing, 101,364–101,364. https://doi.org/10.1016/j.addma.2020.101364.

In addition to Poisson's ratio, the structural stiffness κ was evaluated for the 2 × 2 auxetic specimens at each programmed and recovered states. Similar to Poisson's ratio, the stiffness was measured by subjecting the specimen to an elastic tensile loading. The collected load and displacement levels were used to calculate κ. In the first programming cycle, the tunability in Poisson's ratio was about 33%. However, the

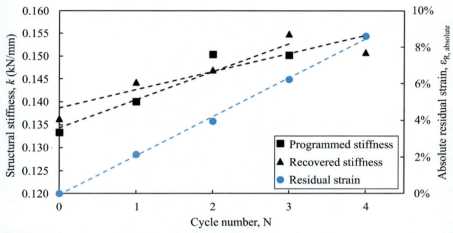

Fig. 6.26 Summary of structural stiffness results for 2 × 2 auxetic structure at the recovered and programmed shapes following multiple constant programming cycles. The accumulation of residual strain induces variation in the structural stiffness.
From Yousuf, M. H., Abuzaid, W., & Alkhader, M. (2020). 4D Printed Auxetic Structures with Tunable Mechanical Properties. Additive Manufacturing, 101,364–101,364. https://doi.org/10.1016/j.addma.2020.101364.

change in structural stiffness was only ≈2% for the exact conditions (see Fig. 6.26). From a pure geometric perspective, the programming process and the associated changes in cell angle should have a larger impact on structural stiffness. However, as discussed earlier, the programming process induces changes in the material properties as shown in Fig. 6.20. The reduction in material modulus following programming negates the stiffening effect of tensile programming for the auxetic structure (Yousuf et al., 2020). Such complex interactions limit the ability to tune the structural stiffness in the considered structure. It should be pointed out that such counteracting effects (i.e., reduction in stiffness due to programming and stiffening due to geometry change), do not limit the ability to achieve tunable properties through 4D printing. However, these interactions have to be well quantified and understood, as attempted in this section, to address their influence during the design phase and realize the desired tunability in properties. Also, as highlighted for the tunability in Poisson's ratio earlier (33% tunability), the reduction in SMP's stiffness following programming does not have an identical counteracting effect on all related mechanical properties, as in the case for structural stiffness.

The accumulation of residual strains upon cycling distorts the geometry of the auxetic structure and can impact the resulting mechanical properties. Such an effect was clearly observed for Poisson's ratio as shown in Fig. 6.25. The structural stiffness was also evaluated to better understand the impact of residual strain accumulation. Although very limited, the cycling process had a more pronounced impact on κ compared to structural programming as shown in Fig. 6.26. The stiffness increased incrementally, which is expected given the changes in the structure as it becomes less concave with residual strain accumulation (Scarpa, Panayiotou, & Tomlinson, 2000; Scarpa, Smith, Chambers, & Burriesci, 2003; Yang, Harrysson, West, & Cormier, 2015; Zhang & Yang, 2016).

The results presented already for the tunability of Poisson's ratio and structural stiffness were obtained at a constant programming level of 10 mm. A clear change in mechanical properties was demonstrated and the effect of repeated cycling was highlighted (i.e., the impact of incremental residual strain buildup). To further quantify the range of properties that can be obtained through structural programming of the 4D-printed auxetic component, a similar complex periodic structure was investigated with a wide range of programming levels. Incremental programming and recovery cycles were applied on a 3 × 4 auxetic structure and the stiffness and Poisson's ratio were investigated. As shown in Fig. 6.27, the 3 × 4 structure was subjected to five varied programming levels, ranging from 5 to 40 mm. The induced strains during programming (programmed strains ε_{prog}) were 4.4%, 8.4%, 12.7%, 17.2%, and 34.7%. Note that as the programming level increased, the relative accumulation of residual strain also increased (see Fig. 6.28). Also, as multiple cycles were applied, and as shown in Fig. 6.24, the total or absolute level of residual strain reported in Figs. 6.27 and 6.28 is influenced by both, the cycling process that induces incremental accumulation of residual strains and the amplification in residual strain magnitude induced by the increasing programming level (i.e., the relative accumulation of residual strain). However, and based on the results in Fig. 6.28, the magnitude of programming has a more pronounced influence on the total level of residual strain in the sample compared to the cycling process.

As explained earlier, the programming process aims to alter the geometry and consequently the mechanical properties of the 4D-printed auxetic structure. The 3 × 4 auxetic specimen discussed in Fig. 6.27 was subjected to an elastic loading at the

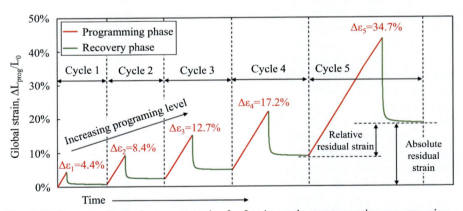

Fig. 6.27 Programming and recovery cycles for 3 × 4 auxetic structure using programming levels of 5, 10, 15, 20, and 40 mm. The programming phase was done using a displacement rate of 1 mm/min, while the recovery was conducted stress-free for 200 min.
From Yousuf, M. H., Abuzaid, W., & Alkhader, M. (2020). 4D Printed Auxetic Structures with Tunable Mechanical Properties. Additive Manufacturing, 101364–101364. https://doi.org/10.1016/j.addma.2020.101364.

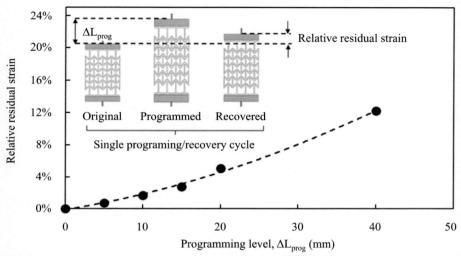

Fig. 6.28 Effect of programming level on the relative residual strain accumulation. From Yousuf, M. H., Abuzaid, W., & Alkhader, M. (2020). 4D Printed Auxetic Structures with Tunable Mechanical Properties. Additive Manufacturing, 101,364–101,364. https://doi.org/10.1016/j.addma.2020.101364.

end of each programming level and subsequent recovery (2 mm extension). DIC was utilized to measure the *global* lateral strains ε_{xx} and the axial strains ε_{yy} (along the loading direction) during the elastic tensile loading of the structure. The measured strains were used to calculate the in-plane Poisson's ratio for the periodic structure before programming (original shape), after programming (programmed shape), and after heat-induced recovery (recovered shape). Representative cases are shown in Fig. 6.29 for the 5- and 40-mm programming levels. The results of 5-mm programming show a noticeable reduction in Poisson's ratio from −0.42 to −0.33, indicating a certain tunability of Poisson's ratio. At such a magnitude of programming where the accumulation of residual strain is insignificant (see Fig. 6.28), the recovered shape exhibited a very similar Poisson's ratio (−0.41) compared to the original shape (−0.42). At a significantly higher level of programming (40 mm), the Poisson's ratio exhibited a remarkable change from $\nu_{xy} = -0.35$ to a positive 0.69 as the structure was converted to a regular honeycomb during programming (see Fig. 6.29B). Despite the extreme change in properties from a negative to positive Poisson's ratio in this case, high recovery was still attainable with Poisson's ratio going back to its original negative magnitude ($\nu_{xy} = -0.31$). The lack of full recovery in properties is induced by the incremental accumulation of residual strains.

A summary of the measured Poisson's ratio for all the considered programming levels (5–40 mm) is presented in Fig. 6.30. The inset images highlight the changes in cell geometry (i.e., inclined member angle) due to programming. Poisson's ratio increased incrementally with higher programming levels. However, it should be pointed out that the measured response includes the effects of both, the dimensional

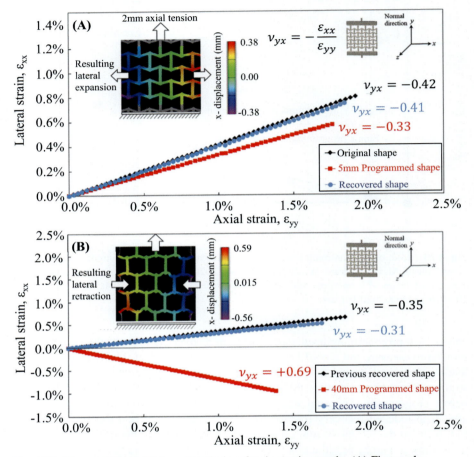

Fig. 6.29 The tunability of Poisson's ratio in a 3×4 auxetic sample. (A) First cycle ($\Delta L_{prog} = 5$ mm) and (B) cycle with most deformation ($\Delta L_{prog} = 40$ mm). The ratio between the lateral and axial strains represents the Poisson's ratio of the structure.
From Yousuf, M. H., Abuzaid, W., & Alkhader, M. (2020). 4D Printed Auxetic Structures with Tunable Mechanical Properties. Additive Manufacturing, 101,364–101,364. https://doi.org/10.1016/j.addma.2020.101364.

changes introduced during programming and the impact of incremental residual strain accumulation due to incomplete SMP recovery. Nevertheless, and by observing the relatively minimal impact of residual strain accumulation on Poisson's ratio in the recovered states (see the black point in Fig. 6.30), the reported tunability (red points in Fig. 6.30) is affected significantly more by the programming process rather than the changes induced by residual strain accumulation.

In addition to Poisson's ratio, the structural stiffness was investigated as a function of incremental programming levels. Fig. 6.31 shows a summary of structural stiffness as a function of the programming level. At low programming levels, minimal variation

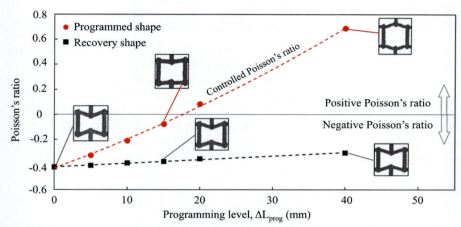

Fig. 6.30 Poisson's ratio of the 3 × 4 auxetic structure at different programming levels. The insets show representative cells of the programmed shapes for selected programming levels.
From Yousuf, M. H., Abuzaid, W., & Alkhader, M. (2020). 4D Printed Auxetic Structures with Tunable Mechanical Properties. Additive Manufacturing, 101364–101364. https://doi.org/10.1016/j.addma.2020.101364.

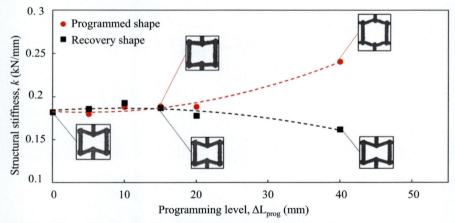

Fig. 6.31 The structural stiffness of the 3 × 4 auxetic structure at different programming levels. The insets show representative cells of the programmed shapes for selected programming levels.
From Yousuf, M. H., Abuzaid, W., & Alkhader, M. (2020). 4D Printed Auxetic Structures with Tunable Mechanical Properties. Additive Manufacturing, 101364–101364. https://doi.org/10.1016/j.addma.2020.101364.

in stiffness was observed, with an apparent increase as the programming level increased. The increase in the stiffness resulted from the conversion from the auxetic honeycomb to a regular honeycomb. As discussed earlier, the limitation in achieving a wider window of tunability in structural stiffness stems from the reduction in material modulus following programming.

The development of heterogeneities—Local response

In the previous section, the major aspects affecting the tunability of mechanical properties in 4D-printed auxetic structures were analyzed. Strain accumulation during programming and the buildup of residual strains following repeated cycling both affect the resulting mechanical properties. The strain analysis was based on the global structural response (i.e., average cell response). However, the complex geometry and the heterogeneities introduced by the deposition process induce localizations in the measured strains (elastic, programmed, recovered, and residual). The development of such strain localizations affect the buildup of residual strains in the structure. Such heterogeneities are therefore important to understand and quantify as they directly impact the development of residual strains in the structure and consequently the mechanical properties of the 4D-printed component. Fig. 6.32A presents the local member axial

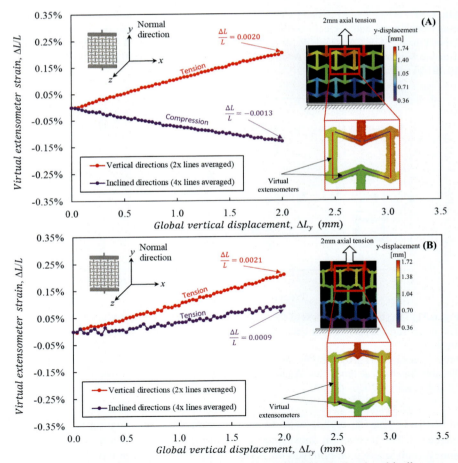

Fig. 6.32 Local member axial strain of a representative cell within the 3 × 4 multicell structure at (A) 0-mm programming level (B) 40-mm programming level.
No permission required.

strain of a representative cell within the 3 × 4 multicell structure at 0-mm programming level (subjected to 2-mm elastic tensile loading). Despite the bulk tensile strain, the auxetic structure exhibits a complex combination of local tensile and local compressive strains. The results show tensile strains in the vertical directions reaching up to 0.20% and 0.13% compressive strain in the inclined directions. The compressive strains in the inclined directions are expected due to negative Poisson's ratio characteristic of the considered cell structure. This behavior was significantly altered during the programming/tuning process. For example, Fig. 6.32B shows the local member axial strain during elastic loading of a representative cell within the 3 × 4 multicell structure at the 40-mm programming level. At such a high programming level, Poisson's ratio was tuned from negative (original structure) to a positive magnitude as shown in Fig. 6.30 based on global and average measurements. On the local level, the previously compressive members experienced tensile strains reaching 0.09%. The vertical members, tensile in both cases, exhibited no noticeable differences between the considered extreme cases; a local tensile strain of 0.20% in the original shape compared to 0.21% for positive Poisson's ratio programmed shape. A summary of the local member axial strain for the vertical and inclined directions as a function of programming level is presented in Fig. 6.33. The reported strains represent the maximum strains at the highest applied load. It can be seen that the vertical struts showed minimal variation as the programming level increased. A transition in the inclined strut strains from compressive to tensile is evident at ($\Delta L_{prog} \approx 20$ mm) which is a result of the inclined strut angle reaching zero or less, therefore, altering the structure from an auxetic honeycomb with negative Poisson's ratio to regular honeycomb with positive Poisson's ratio. This indicates that the elastic property change (Poisson's ratio and stiffness) due to programming is primarily influenced by the deformation and structure of the inclined members.

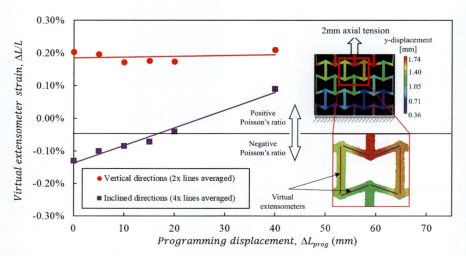

Fig. 6.33 Summary of local member axial strain of vertical and inclined directions as a function of programming level.
No permission required.

As noted previously, the buildup of residual strains is affected by the heterogeneities at the local level that develops during programming and recovery. For example, Fig. 6.34 shows the localized strain accumulation after programming and heat-induced recovery for the 10- and 40-mm programming levels. The full-field strain contour plots highlight the localizations that develop following programming and locations where residual strains accumulate after recovery. This buildup of localized strains eventually results in finite magnitudes of geometric distortion (i.e., global residual strain) that affects the mechanical properties of the 4D-printed structure.

As shown in Fig. 6.34, the programming and recovery cycles induce localized residual strains with magnitudes that depend on the severity of deformation during structural programming. To further clarify this correlation, the strain evolution in the vertical and inclined struts was quantified during the programming and recovery phases for all the considered bulk programming levels (i.e., 5, 10, 15, 20, and 40 mm).

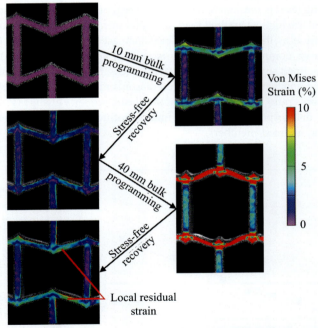

Fig. 6.34 Local residual strain accumulation following programming (10 and 40 mm) and recovery. Higher programming levels induce additional local residual strain accumulation. From Yousuf, M. H., Abuzaid, W., & Alkhader, M. (2020). 4D Printed Auxetic Structures with Tunable Mechanical Properties. Additive Manufacturing, 101,364–101,364. https://doi.org/10.1016/j.addma.2020.101364.

The evolution of programming and recovery strains for the inclined struts is presented in Fig. 6.35 and summarized in Table 6.5. Similarly, the programming and recovery strains of the vertical struts are presented in Fig. 6.36 and summarized in Table 6.6. By comparing the local response in the inclined struts to that of the vertical struts, it is noted that the magnitude of the compressive strains in the inclined struts was larger than the tensile strains in the vertical struts. This indicates that the major contributor to the bulk programming strains is the deformation and rotation of the inclined struts. A notable difference in the recovery behavior between the inclined and vertical struts is also seen. For the inclined struts, the recovery starts in a typical manner (i.e., reduction in the programmed strains); however, relaxation was observed at the later stages of the recovery as indicated by the nonsaturated strains during recovery (note the green lines showing the recovery in Fig. 6.35). This behavior indicates additional shrinkage caused by heating during the heat-induced recovery. The deposition pattern, being aligned with the strut axis, contributes to the additional accumulation of shrinkage strains. Recall that this configuration is the most deposition pattern conducive to the accumulation of shrinkage strains as shown in Fig. 6.7. However, the additional shrinkage tends to stabilize as the number of cycles increases, as per results shown in Tables 6.5 and 6.6.

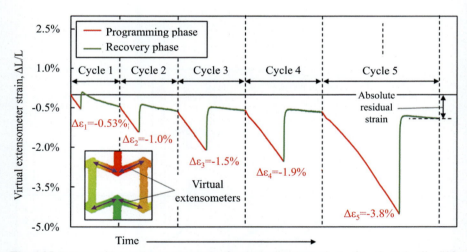

Fig. 6.35 Programming and recovery cycles for the inclined members of a selected cell within the 3 × 4 auxetic structure for programming levels of 5, 10, 15, 20, and 40 mm. The small cell indicates the location of extensometers used to extract the local strains in the inclined struts. From Yousuf, M. H., Abuzaid, W., & Alkhader, M. (2020). 4D Printed Auxetic Structures with Tunable Mechanical Properties. Additive Manufacturing, 101,364–101,364. https://doi.org/10.1016/j.addma.2020.101364.

Table 6.5 Summary of local residual strain accumulation in the inclined members for the case of a multicell auxetic structure with incremental programming levels.

Programming level (mm)	Phase	Accumulated programming strain and accumulated recovery strain	Relative residual strain	Absolute residual strain
5	Programming	−0.53%	−0.45%	−0.45%
	Recovery	0.08%		
10	Programming	−1.0%	−0.2%	−0.65%
	Recovery	0.8%		
15	Programming	−1.5%	0.0%	−0.65%
	Recovery	1.5%		
20	Programming	−1.9%	0.0%	−0.65%
	Recovery	1.9%		
40	Programming	−3.8%	−0.2%	−0.85%
	Recovery	3.6%		

From Yousuf, M. H., Abuzaid, W., & Alkhader, M. (2020). 4D Printed Auxetic Structures with Tunable Mechanical Properties. Additive Manufacturing, 101,364–101,364. https://doi.org/10.1016/j.addma.2020.101364.

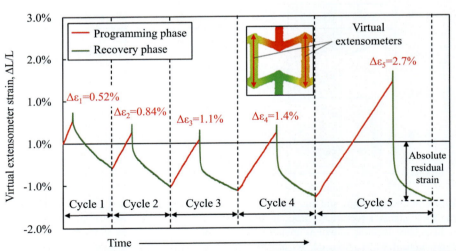

Fig. 6.36 Programming and recovery cycles for the vertical members of a selected cell within the 3 × 4 auxetic structure for programming levels of 5, 10, 15, 20, and 40 mm. The small cell indicates the location of extensometers used to extract the local strains in the vertical struts. From Yousuf, M. H., Abuzaid, W., & Alkhader, M. (2020). 4D Printed Auxetic Structures with Tunable Mechanical Properties. Additive Manufacturing, 101,364–101,364. https://doi.org/10.1016/j.addma.2020.101364.

4D-printed structures with tunable mechanical properties

Table 6.6 Summary of local residual strain accumulation in the vertical members for the case of a multicell auxetic structure with incremental programming levels.

Programming level (mm)	Phase	Accumulated programming strain and accumulated recovery strain	Relative residual strain	Absolute residual strain
5	Programming	0.52%	−0.58%	−0.58%
	Recovery	−1.1%		
10	Programming	0.86%	−0.44%	−1.02%
	Recovery	−1.3%		
15	Programming	1.1%	−0.10%	−1.12%
	Recovery	−1.2%		
20	Programming	1.6%	0.10%	−1.02%
	Recovery	−1.5%		
40	Programming	2.7%	−0.10%	−1.12%
	Recovery	−2.8%		

From Yousuf, M. H., Abuzaid, W., & Alkhader, M. (2020). 4D Printed Auxetic Structures with Tunable Mechanical Properties. Additive Manufacturing, 101,364–101,364. https://doi.org/10.1016/j.addma.2020.101364.

As shown previously, the bulk axial deformation on the 3×4 auxetic structure results in tensile strains in the vertical struts. The recovery strains plotted in Fig. 6.36 for the vertical struts clearly surpassed the initial tensile strains for all programming levels. Similar to the case of inclined struts, the vertical struts suffered additional shrinkage that caused a negative value of absolute residual strains. It is important to note here that the trends observed from local measurements (i.e., the inclined and vertical struts in Figs. 6.35 and 6.36) differ from the results obtained based on bulk strain measurements (see Fig. 6.27). The study of local strains showed evidence of additional shrinkage in the inclined and vertical struts, which, as a result, will contribute to the reduction in the overall size of the sample. This explains one of the main reasons for the accmulation of residual strains that was observed in the entire strcuture through bulk or global deformation measurements. The observation that strut elements experience full recovery followed by the potential for additional accumulation of shrinkage strains has been revealed by local measurements within cell walls with no indication of such behavior from bulk/average structure response. This highlights the importance and the need to conduct response assessments on both, global as well as local levels to better understand the source of residual strain accumulation and strain localization in general.

Summary and concluding remarks

The work discussed in this chapter highlights the great potential of 4D printing to induce tunable properties in complex 3D-printed components. Various specimen types, printing patterns, and loading conditions have been considered to clarify the applicability, challenges, and limitations of utilizing a thermoplastic-based SMP in a 4D printing process. The provided quantitative and qualitative analyses exemplify the ability to tune the mechanical properties of solid structures to various levels while maintaining the ability to recover the original properties through a heat-induced recovery process. The results also point to two main challenges associated with the practical side of the process: (1) dimensional instability during first-time heat exposure (shrinkage), and its strong dependence on the printing pattern and infill density, and (2) local and heterogeneous residual strain accumulation during programming and recovery cycles within the structural members and its negative impact on the tunability of Poisson's ratio and the structural stiffness in auxetic structures. Potential solutions to minimize the effect of these challenges through the careful selection of printing parameters and by implementing a postprinting treatment step were also thoroughly discussed. In summary, the chapter supports the following conclusions:

4D-printed structures, manufactured using thermoplastic SMP, are susceptible to dimensional changes following initial heat exposure above the glass transition temperature, T_g. These changes alter the dimensions of the printed component (i.e., shrinkage) in a manner, that is, time, temperature, and deposition pattern dependent. For example, using a deposition direction aligned with tensile specimens loading axis or cell walls in the case of periodic honeycomb structures results in relatively large magnitudes of shrinkage or dimensional changes along the deposition direction. By utilizing higher infill densities along with alternating deposition patterns, such as + and × grids, the amount of shrinkage and the associated dimensional changes can be significantly reduced. From a practical perspective, and to reduce the impact of the unavoidable dimensional changes following initial heat exposure, a curing process can be implemented that reduces further shrinkage during heat-induced programming and recovery of the 4D-printed structures. This curing process relies on the observation that the rate of shrinkage strain accumulation is significantly reduced to almost a saturated response after about 10 min of temperature exposure. By implementing this step, further changes during programming and recovery are significantly reduced. However, the finite amount of dimensional distortion endured during the curing step has to be accounted for in the design phase of the 4D-printed component to achieve the desired response.

The assessment of the shape memory properties of 4D-printed components, quantitatively, is rather complex and requires strain measurements at various stages including programming and recovery, which both require heating above T_g. With the presence of shrinkage strains at such temperatures, distinguishing between all these different strain magnitudes becomes complex if not impossible. The curing process adopted here limits the magnitude of additional shrinkage strains during programming and recovery to levels where their contribution becomes insignificant relative to the programmed and recovery strains, thus allowing for proper measurements of strain accumulation during programming, recovery strains during heat-induced recovery, and any potential residual or unrecoverable strains. The importance of such an approach is realized by noting that the distortion induced by residual strains affects the resulting mechanical properties. Therefore, accurate quantification of the residual strains is crucial.

4D-printed components are in general subjected to large levels of deformation during programming. Significant heterogeneities in the accumulated programmed and recovered strains were revealed in the relatively complex periodic structures considered here. Although such localizations are anticipated from a purely structural perspective, the reported results also highlight a significant contribution to the development of strain heterogeneities from the deposition pattern used during 3D printing.

The ability to alter the geometry of the 4D-printed periodic structure through programming has been utilized to achieve tunability in the mechanical properties. For example, in the auxetic honeycomb specimen discussed in this chapter, Poisson's ratio was altered between an original magnitude of -0.35 to an impressive $+0.69$ (about 300% change) following bulk tensile programming of about 35%. The corresponding change in structural stiffness was, however, limited to about 37% (0.177–0.242 kN/mm). The limitation in achieving a wider window of tunability in structural stiffness stems from the reduction in material modulus following programming. This was revealed by investigating the material modulus at different programming levels where the results displayed a nonlinear correlation between the magnitude of stored/programmed strains and material stiffness.

In practical applications, generating a predictable programmable response, in terms of geometry and/or mechanical properties, that is, both tunable and recoverable, is extremely important. Following multiple programming and recovery cycles, the considered specimens exhibited high levels of recovery in terms of geometry and mechanical properties, thus pointing to a repeatable process within acceptable tolerances. In addition, the results highlight not only the applicability of 4D printing using SMP to achieve tunable properties but also that all the associated aspects, such as localization and recovery rates, are predictable and can be accounted for with proper process design and understanding of the source of strain localization.

The buildup of residual strains following the programming and recovery cycles has a notable impact on the resulting mechanical properties. These undesirable strains were found to negatively influence the tunability of mechanical properties by degrading Poisson's ratio and structural stiffness subsequent to repeated programming/recovery cycling (i.e., functional fatigue). The magnitude of the residual strains is also affected by the programming level in addition to the number of programming and recovery cycles. As the buildup of residual strains induces distortions to the structure and impacts the mechanical properties, it is important to properly quantify their magnitude and the different factors influencing their accumulation.

References

Abuzaid, W., Alkhader, M., & Omari, M. (2018). Experimental analysis of heterogeneous shape recovery in 4d printed honeycomb structures. *Polymer Testing*. https://doi.org/10.1016/j.polymertesting.2018.03.050.

Ahn, K. J., Peterson, L., Seferis, J. C., Nowacki, D., & Zachmann, H. G. (1992). Prepreg aging in relation to tack. *Journal of Applied Polymer Science*, *45*(3), 399–406.

Alderson, K. L., Fitzgerald, A., & Evans, K. E. (2000). The strain dependent indentation resilience of auxetic microporous polyethylene. *Journal of Materials Science*, *35*(16), 4039–4047. https://doi.org/10.1023/A:1004830103411.

Alderson, K. L., Simkins, V. R., Coenen, V. L., Davies, P. J., Alderson, A., & Evans, K. E. (2005). How to make auxetic fibre reinforced composites. *Physica Status Solidi B*, *242* (3), 509–518. https://doi.org/10.1002/pssb.200460371.

Alkhader, M., Abuzaid, W., Elyoussef, M., & Al-Adaileh, S. (2019). Localized strain fields in honeycomb materials with convex and concaved cells. *European Journal of Mechanics. A, Solids*. https://doi.org/10.1016/j.euromechsol.2019.103890.

Alkhader, M., Iyer, S., Shi, W., & Venkatesh, T. A. (2015). Low frequency acoustic characteristics of periodic honeycomb cellular cores: The effect of relative density and strain fields. *Composite Structures*, *133*, 77–84. https://doi.org/10.1016/j.compstruct.2015.07.102.

Alkhader, M., & Vural, M. (2008). Mechanical response of cellular solids: Role of cellular topology and microstructural irregularity. *International Journal of Engineering Science*, *46*(10), 1035–1051. https://doi.org/10.1016/j.ijengsci.2008.03.012.

Bertoldi, K., & Boyce, M. C. (2008). Mechanically triggered transformations of phononic band gaps in periodic elastomeric structures. *Physical Review B*, *77*(5). https://doi.org/10.1103/PhysRevB.77.052105.

Bertoldi, K., Boyce, M. C., Deschanel, S., Prange, S. M., & Mullin, T. (2008). Mechanics of deformation-triggered pattern transformations and superelastic behavior in periodic elastomeric structures. *Journal of the Mechanics and Physics of Solids*, *56*(8), 2642–2668. https://doi.org/10.1016/j.jmps.2008.03.006.

Bertoldi, K., Vitelli, V., Christensen, J., & van Hecke, M. (2017). Flexible mechanical metamaterials. *Nature Reviews Materials*, *2*(11), 17066.

Bikas, H., Stavropoulos, P., & Chryssolouris, G. (2016). Additive manufacturing methods and modeling approaches: A critical review. *International Journal of Advanced Manufacturing Technology*, *83*(1–4), 389–405. https://doi.org/10.1007/s00170-015-7576-2.

Bitzer, T. (1994). Honeycomb marine applications. *Journal of Reinforced Plastics and Composites*, *13*(4), 355–360. https://doi.org/10.1177/073168449401300406.

Bodaghi, M., Damanpack, A. R., & Liao, W. H. (2017). Adaptive metamaterials by functionally graded 4D printing. *Materials and Design*, *135*, 26–36. https://doi.org/10.1016/j.matdes.2017.08.069.

Bodaghi, M., & Liao, W. H. (2019). 4D printed tunable mechanical metamaterials with shape memory operations. *Smart Materials and Structures*, *28*(4), 045019. https://doi.org/10.1088/1361-665X/AB0B6B.

Bodaghi, M., Noroozi, R., Zolfagharian, A., Fotouhi, M., & Norouzi, S. (2019). 4D printing self-morphing structures. *Materials (Basel, Switzerland)*, *12*(8). https://doi.org/10.3390/ma12081353.

Bodaghi, M., Serjouei, A., Zolfagharian, A., Fotouhi, M., Rahman, H., & Durand, D. (2020). Reversible energy absorbing meta-sandwiches by FDM 4D printing. *International Journal of Mechanical Sciences*, *173*, 105451. https://doi.org/10.1016/j.ijmecsci.2020.105451.

Brandel, B., & Lakes, R. S. (2001). Negative Poisson's ratio polyethylene foams. *Journal of Materials Science*, *36*(24), 5885–5893. https://doi.org/10.1023/A:1012928726952.

Chan, N., & Evans, K. E. (1997). Fabrication methods for auxetic foams. *Journal of Materials Science*, *32*(22), 5945–5953. https://doi.org/10.1023/A:1018606926094.

Choi, J., Kwon, O. C., Jo, W., Lee, H. J., & Moon, M. W. (2015). 4D printing technology: A review. In *Vol. 2. 3D printing and additive manufacturing* (pp. 159–167). Mary Ann Liebert Inc. https://doi.org/10.1089/3dp.2015.0039. Issue 4.

Daniels, I. M., Gdoutos, E. E., & Rajapakse, Y. D. S. (2009). *Major accomplishments in composite materials and sandwich structures: An anthology of ONR sponsored research* (1st ed.). Springer.

Deshpande, V. S., Ashby, M. F., & Fleck, N. A. (2001). Foam topology: Bending versus stretching dominated architectures. *Acta Materialia, 49*(6), 1035–1040. https://doi.org/10.1016/S1359-6454(00)00379-7.

Ebrahimian, F., Kabirian, Z., Younesian, D., & Eghbali, P. (2021). Auxetic clamped-clamped resonators for high-efficiency vibration energy harvesting at low-frequency excitation. *Applied Energy, 295*, 117010. https://doi.org/10.1016/j.apenergy.2021.117010.

Eghbali, P., Younesian, D., & Farhangdoust, S. (2020). Enhancement of the low-frequency acoustic energy harvesting with auxetic resonators. *Applied Energy, 270*, 115217. https://doi.org/10.1016/j.apenergy.2020.115217.

Evans, K. E., & Alderson, K. L. (2000). Auxetic materials: The positive side of being negative. *Engineering Science and Education Journal, 9*(4), 148–154. https://doi.org/10.1049/esej:20000402.

Evans, K. E., Donoghue, J. P., & Alderson, K. L. (2004). The design, matching and manufacture of auxetic carbon fibre laminates. *Journal of Composite Materials, 38*(2), 95–106. https://doi.org/10.1177/0021998304038645.

Farhangdoust, S. (2020). Auxetic cantilever beam energy harvester. In K. Gath, & N. G. Meyendorf (Eds.), *Vol. 113820. Smart structures and NDE for industry 4.0, smart cities, and energy systems* SPIE. https://doi.org/10.1117/12.2559327.

Ferguson, W. J. G., Kuang, Y., Evans, K. E., Smith, C. W., & Zhu, M. (2018). Auxetic structure for increased power output of strain vibration energy harvester. *Sensors and Actuators A: Physical, 282*, 90–96. https://doi.org/10.1016/j.sna.2018.09.019.

Fleck, N. A., Deshpande, V. S., & Ashby, M. F. (2010). Micro-architectured materials: Past, present and future. In *Vol. 466. Proceedings of the royal society A: Mathematical, physical and engineering sciences* (pp. 2495–2516). Royal Society. https://doi.org/10.1098/rspa.2010.0215. Issue 2121.

Froes, F., & Boyer, R. (2019). *Additive manufacturing for the aerospace industry*. Elsevier.

Fundamentals of 3D Food Printing and Applications. (2019). *Fundamentals of 3D food printing and applications*. Elsevier. https://doi.org/10.1016/c2017-0-01591-4.

Gardan, J. (2019). Smart materials in additive manufacturing: State of the art and trends. In *Vol. 14. Virtual and physical prototyping* (pp. 1–18). Taylor and Francis Ltd. https://doi.org/10.1080/17452759.2018.1518016. Issue 1.

Ge, Q., Sakhaei, A. H., Lee, H., Dunn, C. K., Fang, N. X., & Dunn, M. L. (2016). Multimaterial 4D printing with tailorable shape memory polymers. *Scientific Reports, 6*(April), 1–11. https://doi.org/10.1038/srep31110.

Gibson, L. J. (2005). Biomechanics of cellular solids. *Journal of Biomechanics, 38*(3), 377–399. https://doi.org/10.1016/j.jbiomech.2004.09.027.

Ge, Z., Hu, H., & Liu, Y. (2013). A finite element analysis of a 3D auxetic textile structure for composite reinforcement. *Smart Materials and Structures, 22*(8). https://doi.org/10.1088/0964-1726/22/8/084005, 084005.

Gibson, J. L., & Ashby, F. M. (1997). *Cellular solids: Structure and properties*. Cambridge University Press.

Gibson, L. J., Ashby, M. F., & Harley, B. A. (2010). *Cellular materials in nature and medicine*. Cambridge University Press.

Grima, J. N., Grech, M. C., Grima-Cornish, J. N., Gatt, R., & Attard, D. (2018). Giant auxetic behaviour in engineered graphene. *Annalen der Physik, 530*(6), 1700330. https://doi.org/10.1002/andp.201700330.

Grima-Cornish, J. N., Vella-Żarb, L., & Grima, J. N. (2020). Mechanical metamaterials: Negative linear compressibility and auxeticity in boron arsenate (Ann. Phys. 5/2020). *Annalen der Physik, 532*(5), 2070023. https://doi.org/10.1002/andp.202070023.

Hornbogen, E. (2006). Comparison of shape memory metals and polymers. *Advanced Engineering Materials, 8*(1–2), 101–106. https://doi.org/10.1002/adem.200500193.

Huang, W. M., Ding, Z., Wang, C. C., Wei, J., Zhao, Y., & Purnawali, H. (2010). Shape memory materials. *Materials Today, 13*(7–8), 54–61. https://doi.org/10.1016/S1369-7021(10)70128-0.

Huang, R., Zheng, S., Liu, Z., & Ng, T. Y. (2020). Recent advances of the constitutive models of smart materials-hydrogels and shape memory polymers. *International Journal of Applied Mechanics*. https://doi.org/10.1142/s1758825120500143.

Iyer, S., Alkhader, M., & Venkatesh, T. A. (2015). Electromechanical behavior of auxetic piezoelectric cellular solids. *Scripta Materialia, 99*, 65–68. https://doi.org/10.1016/j.scriptamat.2014.11.030.

Iyer, S., Alkhader, M., & Venkatesh, T. A. (2016). On the relationships between cellular structure, deformation modes and electromechanical properties of piezoelectric cellular solids. *International Journal of Solids and Structures, 80*, 73–83. https://doi.org/10.1016/j.ijsolstr.2015.10.024.

Iyer, S., & Venkatesh, T. A. (2010). Electromechanical response of porous piezoelectric materials: Effects of porosity connectivity. *Applied Physics Letters, 97*(7), 72903–72904.

Ji, S., Li, L., Motra, H. B., Wuttke, F., Sun, S., Michibayashi, K., et al. (2018). Poisson's ratio and auxetic properties of natural rocks. *Journal of Geophysical Research - Solid Earth, 123*(2), 1161–1185. https://doi.org/10.1002/2017JB014606.

Khoo, Z. X., Teoh, J. E. M., Liu, Y., Chua, C. K., Yang, S., An, J., et al. (2015). 3D printing of smart materials: A review on recent progresses in 4D printing. *Virtual and Physical Prototyping, 10*(3), 103–122. https://doi.org/10.1080/17452759.2015.1097054.

Lakes, R. S. (2017). Negative-Poisson's-ratio materials: Auxetic solids. *Annual Review of Materials Research, 47*.

Lee, A. Y., An, J., & Chua, C. K. (2017a). Two-way 4D printing: A review on the reversibility of 3D-printed shape memory materials. *Engineering, 3*(5), 663–674. https://doi.org/10.1016/J.ENG.2017.05.014.

Lee, J. Y., An, J., & Chua, C. K. (2017b). Fundamentals and applications of 3D printing for novel materials. In *Vol. 7. Applied materials today* (pp. 120–133). Elsevier Ltd. https://doi.org/10.1016/j.apmt.2017.02.004.

Lendlein, A., & Gould, O. E. C. (2019). Reprogrammable recovery and actuation behaviour of shape-memory polymers. In *Vol. 4. Nature reviews materials* (pp. 116–133). Nature Publishing Group. https://doi.org/10.1038/s41578-018-0078-8. Issue 2.

Liu, Y., & Hu, H. (2010). A review on auxetic structures and polymeric materials. *Scientific Research and Essays, 5*(10), 1052–1063. https://doi.org/10.5897/SRE.9000104.

Liu, Z., Toh, W., & Ng, T. Y. (2015). Advances in mechanics of soft materials: A review of large deformation behavior of hydrogels. *International Journal of Applied Mechanics, 7*(5). https://doi.org/10.1142/S1758825115300011.

Loh, X. J. (2016). *Four-dimensional (4D) printing in consumer applications* (pp. 108–116). Royal Society of Chemistry.

López-Valdeolivas, M., Liu, D., Broer, D. J., & Sánchez-Somolinos, C. (2018). 4D printed actuators with soft-robotic functions. *Macromolecular Rapid Communications, 3*. https://doi.org/10.1002/marc.201700710.

Mao, Y., Ding, Z., Yuan, C., Ai, S., Isakov, M., Wu, J., et al. (2016). 3D printed reversible shape changing components with stimuli responsive materials. *Scientific Reports, 6*(April), 1–13. https://doi.org/10.1038/srep24761.

Mao, Y., Yu, K., Isakov, M. S., Wu, J., Dunn, M. L., & Jerry Qi, H. (2015). Sequential self-folding structures by 3D printed digital shape memory polymers. *Scientific Reports, 5*, 1–12. https://doi.org/10.1038/srep13616.

Masters, I. G., & Evans, K. E. (1996). Models for the elastic deformation of honeycombs. *Composite Structures*, *35*(4), 403–422. https://doi.org/10.1016/S0263-8223(96)00054-2.

Miao, S., Castro, N., Nowicki, M., Xia, L., Cui, H., Zhou, X., et al. (2017). 4D printing of polymeric materials for tissue and organ regeneration. *Materials Today*, *20*(10), 577–591. https://doi.org/10.1016/j.mattod.2017.06.005.

Mirzaali, M. J., Caracciolo, A., Pahlavani, H., Janbaz, S., Vergani, L., & Zadpoor, A. A. (2018). Multi-material 3D printed mechanical metamaterials: Rational design of elastic properties through spatial distribution of hard and soft phases. *Applied Physics Letters*, *113*(24), 241903. https://doi.org/10.1063/1.5064864.

Mirzaali, M. J., Janbaz, S., Strano, M., Vergani, L., & Zadpoor, A. A. (2018). Shape-matching soft mechanical metamaterials. *Scientific Reports*, *8*(1), 1–7. https://doi.org/10.1038/s41598-018-19381-3.

Momeni, F., M.Mehdi Hassani.N, S., Liu, X., & Ni, J. (2017). A review of 4D printing. *Materials and Design*, *122*, 42–79. https://doi.org/10.1016/j.matdes.2017.02.068.

Murr, L. E. (2016). Frontiers of 3D printing/additive manufacturing: From human organs to aircraft fabrication. *Journal of Materials Science and Technology*, *32*(10), 987–995. Chinese Society of Metals https://doi.org/10.1016/j.jmst.2016.08.011.

Neville, R. M., Chen, J., Guo, X., Zhang, F., Wang, W., Dobah, Y., et al. (2017). A Kirigami shape memory polymer honeycomb concept for deployment. *Smart Materials and Structures*, *26*(5). https://doi.org/10.1088/1361-665X/aa6b6d.

Ngo, T. D., Kashani, A., Imbalzano, G., Nguyen, K. T. Q., & Hui, D. (2018). Additive manufacturing (3D printing): A review of materials, methods, applications and challenges. In *Vol. 143. Composites part B: Engineering* (pp. 172–196). Elsevier Ltd. https://doi.org/10.1016/j.compositesb.2018.02.012.

Paolini, A., Kollmannsberger, S., & Rank, E. (2019). Additive manufacturing in construction: A review on processes, applications, and digital planning methods. In *Vol. 30. Additive manufacturing* Elsevier B.V. https://doi.org/10.1016/j.addma.2019.100894.

Parameswaranpillai, J., Siengchin, S., George, J. J., & Jose, S. (Eds.). (2020). *Shape memory polymers, blends and composites: Advances and applications* Springer. https://www.springer.com/gp/book/9789811385735.

Peele, B. N., Wallin, T. J., Zhao, H., & Shepherd, R. F. (2015). 3D printing antagonistic systems of artificial muscle using projection stereolithography. *Bioinspiration & Biomimetics*, *10*(5). https://doi.org/10.1088/1748-3190/10/5/055003.

Pei, E. (2014). 4D printing: Dawn of an emerging technology cycle. In *Vol. 34. Assembly automation* (pp. 310–314). Emerald Group Publishing Ltd. https://doi.org/10.1108/AA-07-2014-062. Issue 4.

Rastogi, P., & Kandasubramanian, B. (2019). Breakthrough in the printing tactics for stimuli-responsive materials: 4D printing. *Chemical Engineering Journal*, *366*, 264–304. https://doi.org/10.1016/J.CEJ.2019.02.085.

Rossiter, J., Takashima, K., Scarpa, F., Walters, P., & Mukai, T. (2014). Shape memory polymer hexachiral auxetic structures with tunable stiffness. *Smart Materials and Structures*, *23*(4), 45007.

Scarpa, F., Ciffo, L. G., & Yates, J. R. (2004). Dynamic properties of high structural integrity auxetic open cell foam. *Smart Materials and Structures*, *13*(1), 49.

Scarpa, F., Panayiotou, P., & Tomlinson, G. (2000). Numerical and experimental uniaxial loading on in-plane auxetic honeycombs. *Journal of Strain Analysis for Engineering Design*, *35*(5), 383–388. https://doi.org/10.1243/0309324001514152.

Scarpa, F., Smith, F. C., Chambers, B., & Burriesci, G. (2003). Mechanical and electromagnetic behaviour of auxetic honeycomb structures. *The Aeronautical Journal*, *107*(1069), 175–183.

Spadoni, A., Ruzzene, M., Gonella, S., & Scarpa, F. (2009). Phononic properties of hexagonal chiral lattices. *Wave Motion, 46*(7), 435–450. https://doi.org/10.1016/j.wavemoti. 2009.04.002.

Tay, Y. Y., Lim, C. S., & Lankarani, H. M. (2014). A finite element analysis of high-energy absorption cellular materials in enhancing passive safety of road vehicles in side-impact accidents. *International Journal of Crashworthiness, 19*(3), 288–300. https://doi.org/ 10.1080/13588265.2014.893789.

Tibbits, S. (2014). 4D printing: Multi-material shape change. *Architectural Design, 84*(1), 116–121. https://doi.org/10.1002/ad.1710.

Van Manen, T., Janbaz, S., & Zadpoor, A. A. (2017). Programming 2D/3D shape-shifting with hobbyist 3D printers. *Materials Horizons, 4*(6), 1064–1069. https://doi.org/10.1039/ c7mh00269f.

Wang, Q., Tian, X., Huang, L., Li, D., Malakhov, A. V., & Polilov, A. N. (2018). Programmable morphing composites with embedded continuous fibers by 4D printing. *Materials & Design, 155*, 404–413. https://doi.org/10.1016/J.MATDES.2018.06.027.

Wang, Z., Zulifqar, A., Hu, H., Rana, S., & Fangueiro, R. (2016). *7—Auxetic composites in aerospace engineering* (pp. 213–240). Woodhead Publishing. https://doi.org/10.1016/ B978-0-08-100037-3.00007-9.

Wirekoh, J., & Park, Y. L. (2017). Design of flat pneumatic artificial muscles. *Smart Materials and Structures, 26*(3). https://doi.org/10.1088/1361-665X/aa5496.

Yakacki, C. M., Shandas, R., Lanning, C., Rech, B., Eckstein, A., & Gall, K. (2007). Unconstrained recovery characterization of shape-memory polymer networks for cardio-vascular applications. *Biomaterials, 28*(14), 2255–2263.

Yang, L., Harrysson, O., West, H., & Cormier, D. (2015). Mechanical properties of 3D re-entrant honeycomb auxetic structures realized via additive manufacturing. *International Journal of Solids and Structures, 69*, 475–490.

Yang, Y., Chen, Y., Wei, Y., & Li, Y. (2016). 3D printing of shape memory polymer for functional part fabrication. *International Journal of Advanced Manufacturing Technology, 84*(9–12), 2079–2095. https://doi.org/10.1007/s00170-015-7843-2.

Yeganeh-Haeri, A., Weidner, D. J., & Parise, J. B. (1992). Elasticity of α-cristobalite: A silicon dioxide with a negative Poisson's ratio. *Science, 257*(5070), 650. https://doi.org/10.1126/ science.257.5070.650.

Yousuf, M. H., Abuzaid, W., & Alkhader, M. (2020). 4D printed auxetic structures with tunable mechanical properties. *Additive Manufacturing*, 101364. https://doi.org/10.1016/j.addma. 2020.101364.

Zhang, X., & Yang, D. (2016). Mechanical properties of auxetic cellular material consisting of re-entrant hexagonal honeycombs. *Materials (Basel, Switzerland), 9*(11), 900. https://doi. org/10.3390/ma9110900.

Zhang, Z., Demir, K. G., & Gu, G. X. (2019). Developments in 4D-printing: A review on current smart materials, technologies, and applications. *International Journal of Smart and Nano Materials, 10*(3), 205–224. https://doi.org/10.1080/19475411.2019.1591541.

Zheng, S., Li, Z., & Liu, Z. (2018). The fast homogeneous diffusion of hydrogel under different stimuli. *International Journal of Mechanical Sciences, 137*, 263–270. https://doi.org/ 10.1016/j.ijmecsci.2018.01.029.

4D-printed shape memory polymer: Modeling and fabrication

Reza Noroozi[a,b], Ali Zolfagharian[c], Mohammad Fotouhi[d], and Mahdi Bodaghi[a]
[a]Department of Engineering, School of Science and Technology, Nottingham Trent University, Nottingham, United Kingdom
[b]Faculty of Engineering, School of Mechanical Engineering, University of Tehran, Tehran, Iran
[c]School of Engineering, Deakin University, Geelong, VIC, Australia
[d]Department of Materials, Mechanics, Management & Design (3MD), Delft University of Technology, Delft, Netherlands

Introduction

Additive manufacturing (AM), also known as three-dimensional (3D) printing, as one of the most potent manufacturing methods, is fabricating a 3D structure using computer-aided design (CAD) data or computed tomography (CT) scan under computer control by adding material layer by layer (Askari et al., 2021; Attaran, 2017; Shirzad, Zolfagharian, Matbouei, & Bodaghi, 2021; Soltani, Noroozi, Bodaghi, Zolfagharian, & Hedayati, 2020; Wong & Hernandez, 2012; Yan et al., 2018; Zolfagharian, Kaynak, et al., 2020; Zolfagharian, Denk, et al., 2020). In contrast to traditional manufacturing processes that start from raw block material and remove waste materials, the AM process can demonstrate better aspects, such as freedom of design, reduced postprocessing, waste reduction, faster construction, and lower cost, and provide the opportunity for the fabrication of a complex structure (Levy et al., 2018; Ngo, Kashani, Imbalzano, Nguyen, & Hui, 2018; Soltani et al., 2020; Sun, Ma, Huang, Zhang, & Qian, 2020; Zadpoor & Malda, 2017). Researchers have been developed various methods for 3D printing, such as fused deposition modeling (FDM), selective laser melting (SLM), and stereolithography (SLA), due to the importance and application of 3D printing technology in different industries (Dudek, 2013; Ngo et al., 2018; Shahrubudin, Lee, & Ramlan, 2019). The SLM is one of the AM-metal-base processes in which a 3D part is created by applying laser energy to powder beds. In this method, by using a laser in the powder area, each layer is melted to the required cross section, and then, the melted powder particles bond and solidify. This process continues until the 3D object is created (Olakanmi, Cochrane, & Dalgarno, 2015). The FDM process is one of the most popular methods among printing methods due to its simplicity, accessibility, low cost, and ability to use different materials (Rayegani & Onwubolu, 2014). In this method, in order to create a specific object, a moving nozzle fused a thermoplastic continuous filament on a bed (Penumakala, Santo, & Thomas, 2020; Rayegani & Onwubolu, 2014). Although 3D printing technology has evolved over time, there are still some limitations and challenges, such as scalability, directional dependency, and structure rigidity, which could not change their shape and property over time (Ahmed et al., 2020; El-Sayegh, Romdhane, & Manjikian, 2020). For example, recent researchers have found that 3D-printed sample characterizations are dependent on various input parameters, such as printing speed, orientation, print temperature, which make the predictability of the sample features challenging.

With the advent of 3D printing technology, the concept of 4D printing technology was introduced by Tibbits in 2013, and some of the challenges and limitations of 3D printing technology were overcome (Tibbits, 2014). 4D printing technology has added a fourth dimension to 3D printing technology, which displays the time. In other words, the fourth dimension is that printed structures are a function of time (Askari et al., 2021; Attaran, 2017; Bodaghi et al., 2020; Bodaghi & Liao, 2019; Damanpack, Bodaghi, & Liao, 2020; Shirzad et al., 2021; Soltani et al., 2020; Wong & Hernandez, 2012; Yan et al., 2018; Zolfagharian, Denk, et al., 2020; Zolfagharian et al., 2021). In 4D-printed structures, in contrast to 3D printing technology, which constructs 3D-static objects, dynamic structures with the ability to change shape and color are built. Static structures, which are printed with active materials, under external stimuli such as thermal, humidity, light, and electric or magnetic field, exhibit a dynamic behavior such as self-sensing, self-actuating, and shape-changing (Khoo et al., 2015; Shie et al., 2019; Zhang, Demir, & Gu, 2019). Active materials, such as shape memory polymers, should be used in the printing process in order to print structures with dynamic properties. Shape memory polymers (SMPs) have received more attention due to their lower cost, lower density, high recoverable strain, simple shape programming process, and excellent controllability over the recovery process (Bodaghi, Noroozi, Zolfagharian, Fotouhi, & Norouzi, 2019; Jebellat, Baniassadi, Moshki, Wang, & Baghani, 2020; Leng, Lu, Liu, Huang, & Du, 2009). Significant research has been conducted on 4D printing technology due to its vast potential. For example, in a study, the commercial polymer, SU-8, was employed in 4D printing self-morphing structures in which the shape transformation is driven via a swellable guest medium (Su et al., 2018). Using the self-folding mechanism by other researchers, a biomedical application that can be used as a cardiac stent was introduced. To model the thermomechanical behavior of structures, they used the phase transformation approach, drove equilibrium equations, and solved them using an in-house code (Bodaghi, Damanpack, & Liao, 2017). Since the printing parameters have a significant effect on sample performance, the effect of filament size, orientation, air pump pressure, and the filling pattern was studied on the shape transformation. To model the shape transformation mechanism, a hyperelastic model was employed for the SU-8 polymer (Su et al., 2018). Another research investigated programming SMP during the FDM process. In this regard, the dynamic structures shape their configuration from 1D to 2D and 1D to 2D-3D under the thermal stimulus. After shape transformation, self-folding structures made of SMP could not be folded or unfolded at temperatures lower than glass transition. To settle this issue, in a study, by using the FDM process, a novel self-folding mechanism was designed. They fabricated a hybrid hinge that was made of a soft elastomer and a 4D printing element. They also validated their experimental result with COMSOL-based finite element simulation (Yamamura & Iwase, 2021).

The literature review reveals that the fabrication and modeling of 4D printing components are extremely important and many researchers have done various studies on 4D printing and its applications. Therefore, in this chapter, first, the concept of the shape memory effect is described, and then, based on the SMPs cycle, the 4D printing programming during the FDM process is explained. Due to the complexity and difficulties of employing shape memory constitutive equations in finite element software to model 4D-printed elements, a simple method based on the 4D printing concept is proposed. The proposed model has been calibrated based on experiments and, as case studies, some of the smart structures are designed and modeled in finite element software.

4D printing programming

In this section, by considering the shape memory effect (SME) and FDM process, the concept of 4D printing, for 3D-printed samples during manufacturing, is described. The temperature-sensitive SMPs are a class of active material in which temperature plays a triggering role. As shown in Fig. 7.1, shape memory cycling consists of four stages. At the first step, the sample is heated up to the T_h, which is higher than the glassy transition temperature (T_g). At the constant temperature (T_h), the sample is loaded and transformed to the desired shape, and then, the sample is cooled down until T_g, which is lower than glassy transition temperature. In the next step, the mechanical constraints are removed and an inelastic strain, called prestrain, remains in the sample, which causes programmed shape (after a small elastic deformation). The programmed shape is a free-state configuration that is stable in lower temperatures than T_g. When it is needed, the sample could be reheated and recovered to its original (permanent) shape; this step is called the shape recovery process.

In the FDM technology, a thermomechanical cycle similar to shape memory programming is applied to the sample during the printing process. So, it may have the potential to print a 4D SMP architecture with shape programming simultaneously. Fig. 7.2 shows a schematic of 3D printing technology. The FDM technology is based on melting raw materials. Therefore, in the first step, the filament heated up until T_{in}

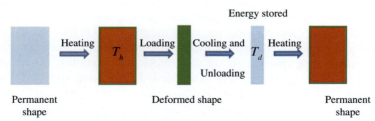

Fig. 7.1 Cycle of SMPs. Thermomechanical cycle of SMPs.
No permission required.

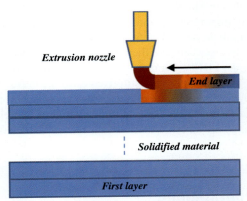

Fig. 7.2 3D Printing technology. Schematic of 3D printing technology. No permission required.

in the liquefier, which is higher than the glassy temperature. So, the nozzle, by applying the force into the filament, extrudes the filament onto the base (called platform or table) and simultaneously prints it on the platform with printing speed S_p. In this step, the filament is heated and stretched, so this step of the printing process is similar to the heating-loading process in the SMP programming, in which the prestrain is created in the sample. By increasing the printing speed, the filament stretches with a greater force; hence, the amount of prestrain induced in the sample is greater. After deposition, the printed layer cools and solidifies in the same way as the cooling step in the programming process. When one layer is printed, the platform moves down, and the printing nozzle moves to build the next layer. Finally, the thermomechanical programming process is completed when the 3D-printed sample is detached from the platform.

It is noteworthy that during the printing process, layers experience several heating–cooling processes. For example, during the printing of the second layer, the first layer heats up somewhat, so some of its prestrain release; hence, its shape memory effect decreases. In such a way, other layers undergo partial heating and some of their prestrain release too. Since the first layer experiences more heating processes, it has lower prestrain than other layers. In the same way, the latest layer does not experience any heating and has the maximum prestrain. The rate of the heating–cooling process plays a significant role in the release of the prestrain. On the contrary, by controlling the printing speed, the rate of the heating–cooling process can be changed. Therefore, the printing speed is one of the critical parameters that can be affected by shape memory programming. For example, at low printing speeds, the sample has more time to cool, so the heating–cooling process slows down, and less amount of prestrain is released. As previously stated, each layer undergoes a different heating–cooling process, resulting in a different amount of induced prestrain in each layer; thus, shape memory programming can be thought of as functionally graded (FG) 4D printing.

Constitutive equations

SMPs are a class of programmable material whose shape changes under external stimuli such as temperature, magnetic, or electric field. Due to the significant development of SMPs and the importance of understanding their complex thermomechanical mechanism, the development of a constitutive equation that is able to predict SMPs behavior is taken into consideration. In this regard, for modeling the SMP mechanism, there are various viewpoints that mainly can be classified as thermoviscoelastic modeling and phase transformation approaches. In this section, these two approaches are described.

Thermoviscoelastic approach

Due to the viscoelastic nature of polymers, thermoviscoelastic modeling can be employed to describe and predict SMP behavior. In this viewpoint, the constitutive equations are based on a rheological model in which the parameters are time and temperature property-dependent. Dashpot, spring, and frictional elements are common components of thermoviscoelastic models. The viscosity is one of the most important features of the SMPs that is due to the mobility between polymer chains, which in such modeling has been considered. Also, in this viewpoint, the modeling predicts SME and predicts the viscoelastic behavior of polymers. Tobushi, Hara, Yamada, and Hayashi (1996) proposed a simple viscoelastic model for investigating SMPs' behavior. In this model, they used a standard linear viscoelastic model (SLV) for modeling the SMP mechanism, in which a Maxwell model is parallel with a spring element. Fig. 7.3 shows the SLV model that consists of three elements.

For the SLV model, the stress–strain relationship can be expressed as follows:

$$\dot{\varepsilon} = \frac{\dot{\sigma}}{E} + \frac{\dot{\sigma}}{\mu} - \frac{\varepsilon}{\lambda} \qquad (7.1)$$

where σ and ε show stress and strain, respectively, and E, λ, and μ denote elastic modules, time retardation, and viscosity, respectively.

The SLV model, however, has some limitations, such as a disability in predicting mechanical properties in the glassy transition region and the difference between creep recovery above and below the glass transition region. To settle these issues, they expressed model parameters in the form of an exponential function to solve the drastic

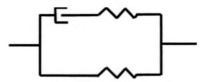

Fig. 7.3 SLV model. The standard linear viscoelastic (SLV) model.
No permission required.

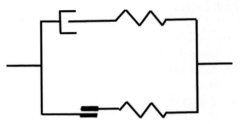

Fig. 7.4 Modified SLV model.
No permission required.

changes in mechanical properties in the glassy region. Also, they added a slip mechanism to solve the difference between creep recovery above and below the glass transition region. Fig. 7.4 shows the modified SLV model.

The stress–strain relationship for the modified SLV model could be expressed as the following equation:

$$\dot{\varepsilon} = \frac{\dot{\sigma}}{E} + \frac{\sigma}{\mu} - \frac{\varepsilon - \varepsilon_s}{\lambda} \tag{7.2}$$

The significant change in mechanical properties of SMPs in the glass transition region could be described as the following form:

$$E = E_g \exp\left(a_E\left(\frac{T_g}{T} - 1\right)\right) \quad \mu = \mu_g \exp\left(a_\mu\left(\frac{T_g}{T} - 1\right)\right) \tag{7.3}$$

where E_g and μ_g are the values of E and μ at $T = T_g$.

where E_g and μ_g are the values of E and μ at $T = T_g$, and a_E and a_μ show the slope of straight line.

Phase transformation approach

Polymers are a combination of many chains with different lengths and angles, which each has different entropies and different temperature transitions. During the cooling process, the phase transition first occurs in the chain with lower entropy, so the SMPs have a phase heterogeneity consisting of glassy and rubbery phases. These phases could change into each other under thermal conditions, and during the thermal process, the amount of the changes constantly. By considering the fact that SMPs are a combination of two phases, a new class of constitutive equations was proposed by researchers (Bodaghi et al., 2017). In contrast to the thermoviscoelastic model, or a rheological model, the phase transition model is a phonological model with a physical meaning. These models not only cover the viscosity of the polymers but also predict the phase transformation of SMPs. In this section, one of the phase transition modeling approaches is described.

The volume fraction of glassy and rubbery phases during the thermal process can be expressed by scalar variables ξ_g and ξ_r that can be defined as follows:

$$\xi_g = \frac{V_g}{V} \quad \xi_r = \frac{V_r}{V} \tag{7.4}$$

in which V_g shows the volume of the glassy phase and V_r denotes the rubbery-phase volume. Since the SMPs only consist of two phases, rubbery and glassy, the summation of the two phases' volume fractions must be equal to unity ($\xi_g + \xi_r = 1$). The change of the rubbery phase to the glassy phase or vice versa is considered only a function of temperature. Therefore, ξ_g and ξ_r only depend on the temperature, and the volume fraction of the rubbery phase can be rewritten as a function of the glassy volume fraction, as the following form:

$$\xi_r = 1 - \xi_g \tag{7.5}$$

So, the glassy volume and the volume fraction of the glassy phase are independent variables and could be expressed as follows:

$$\xi_g = \xi_g(T) \quad V_g = V_g(T) \tag{7.6}$$

There are different methods to determine mechanical properties. Dynamic mechanical thermal analysis (DMTA) is one of the important tests that determine material properties by considering temperature dependency. In this test, sinusoidal stress is applied to the sample, and its strain is measured. Also, the temperature of the sample changes during the test; therefore, this test can be used to determine temperature-dependent mechanical properties. So, using the DMTA test, the glassy volume fraction can be expressed explicitly as follows:

$$\eta_g = \frac{\tan h\left(a_1 T_g - a_2 T\right) - \tan h\left(a_1 T_g - a_2 T_h\right)}{\tan h\left(a_1 T_g - a_2 T_h\right) - \tan h\left(a_1 T_g - a_2 T_l\right)} \tag{7.7}$$

in which a_i ($i = 1,2$) are specified by fitting the explicit function on the DMTA curve.

It is assumed that the rubbery and glassy phases in SMPs are linked to each other in series form. Since printed structures are under small deformation and large rotation, the total strain can be written as the sum of four parts; glassy, rubbery, thermal, and inelastic, which is due to phase transformation; Eq. (7.8) shows the total strain:

$$\varepsilon = \xi_g \, \varepsilon_g + \left(1 - \xi_g\right) \varepsilon_r + \varepsilon_{in} + \varepsilon_{th} \tag{7.8}$$

where ε shows total strain, and ε_g and ε_r denote strain of the glassy and rubbery phases, respectively; ε_{in} is inelastic strain and ε_{th} represents the thermal strain, which can be defined as an integral part such as the following:

$$\varepsilon_{th} = \int_T^{T_0} \alpha_e \, dT \tag{7.9}$$

in which T_0 expresses the reference temperature and α_e is the equivalent thermal expansion that is dependent on the temperature and defined as follows:

$$\alpha_e = \alpha_r + (\alpha_g - \alpha_r)\,\xi_g\,(T) \tag{7.10}$$

where α_r and α_g express thermal expansion for rubbery and glassy phases, respectively.

During phase change from rubbery phase to glassy phase, cooling process, the inelastic strain ε_{in}, is stored in the SMP that can be formulated as the following:

$$\dot{\varepsilon}_{in} = \xi_g \dot{\varepsilon}_r \tag{7.11}$$

in which the dot denotes the rating function.

In contrast to the cooling process in which the inelastic strain is stored in the SMPs, in the heating process, the stored strain releases gradually in proportion to the volume fraction of the glassy phase with respect to the preceding glassy phase. The strain storage is expressed as follows:

$$\dot{\varepsilon}_{in} = \frac{\dot{\xi}_g}{\xi_g}\,\varepsilon_{in} \tag{7.12}$$

In order to derive the stress state, the second law of thermodynamic, the Clausius-Duhem inequality, must be satisfied. The ε and T variables are selected as external control variables, also, ε_g, ε_r, ε_{in}, and ξ_g are chosen as internal variables. By considering Helmholtz free energy density functions, stress, σ, can be derived as follows:

$$\sigma = \sigma_g = \sigma_r \tag{7.13}$$

in which the stress of glassy and rubbery phases can be formulated as follows:

$$\sigma_g = C_g\,\varepsilon_g \quad \sigma_r = C_r\,\varepsilon_r \tag{7.14}$$

where C is the elasticity tensor which for an isotropic material has two independent constants and defined as follows:

$$C = \frac{E}{(1+\nu)(1-2\nu)}
\begin{bmatrix}
1-\nu & \nu & \nu & 0 & 0 & 0 \\
\nu & 1-\nu & \nu & 0 & 0 & 0 \\
\nu & \nu & 1-\nu & 0 & 0 & 0 \\
0 & 0 & 0 & \dfrac{1-2\nu}{2} & 0 & 0 \\
0 & 0 & 0 & 0 & \dfrac{1-2\nu}{2} & 0 \\
0 & 0 & 0 & 0 & 0 & 0
\end{bmatrix} \tag{7.15}$$

in which E and ν denote Young's modules and Passion's ratio, respectively. Now by substituting Eq. (7.14) in Eq. (7.8), the stress obtains as follows:

$$\sigma = C_e \left(\varepsilon - \varepsilon_{in} - \varepsilon_{th} \right) \tag{7.16}$$

where C_e is the equivalent elasticity tensor and expressed with $C_e = (S_r + \xi_g (S_g - S_r))^{-1}$, in which S shows the inverse matrix of C ($C = S^{-1}$), the so-called compliance matrix. The stress–strain relationship can be described as follows:

$$\sigma = \left(S_r + \xi_g \left(S_g - S_r \right) \right)^{-1} \left(\varepsilon - \varepsilon_{in} - \varepsilon_{th} \right) \tag{7.17}$$

In order to solve the nonlinear equation of SMP behavior, it can be considered as an explicit time-discrete stress/strain-temperature-driven problem. The total $[0, t]$ time is broken down into subdomains, and the constitutive equation of SMP is solved in the subdomains $[t^k, t^{k+1}]$. For all variables, the superscript k expresses the value of the variable in the previous step, while the superscript $k+1$ indicates the current step. The inelastic strain expressed with Eqs. (7.18) and (7.19), now by applying the linearized implicit backward Euler integration method to the flow rule, can be discretized as (7.18) and (7.19):

$$\varepsilon_{in}^{k+1} = \varepsilon_{in}^{k} + \Delta \zeta_g^{k+1} \varepsilon_r^{k+1} \tag{7.18}$$

$$\varepsilon_{in}^{k+1} = \varepsilon_{in}^{k} + \frac{\Delta \xi_g^{k+1}}{\xi_g^{k+1}} \varepsilon_{in}^{k+1} \tag{7.19}$$

where

$$\Delta \xi_g^{k+1} = \xi_g^{k+1} - \xi_g^{k} \tag{7.20}$$

By using Eqs. (7.14), (7.17), (7.19), and (7.20) along with a mathematical simplification, we could explicitly update the inelastic strain for the cooling and heating processes. For the cooling process, for stress control, it could be written as follows:

$$\varepsilon_{in}^{k+1} = \varepsilon_{in}^{k} + \Delta \xi_g^{k+1} S_r^{k+1} \sigma^{k+1} \tag{7.21}$$

and for strain control, it could be written as follows:

$$\varepsilon_{in}^{k+1} = \left(I + \Delta \xi_g^{k+1} S_r^{k+1} C_e^{k+1} \right)^{-1} \left(\varepsilon_{in}^{k} \Delta \xi_g^{k} S_r^{k+1} C_e^{k+1} \left(\varepsilon^{k+1} - \varepsilon_{th}^{k+1} \right) \right) \tag{7.22}$$

For the heating process, Eq. (7.22) can be simplified as the following:

$$\varepsilon_{in} = \frac{\xi_g^{k+1}}{\xi_g^k} \, \varepsilon_{in}^k \tag{7.23}$$

Now, by substituting the updated inelastic strain into Eq. (7.17), the stress–strain relationship for heating and cooling processes can be obtained as follows:

$$\sigma^{k+1} = C_D^{k+1} \left(\varepsilon^{k+1} - \delta \, \varepsilon_{in}^{k+1} - \varepsilon_{th}^{k+1} \right) \tag{7.24}$$

where elasticity tensor C_D, and the parameter for the heating and cooling processes δ, are defined as follows:

$$C_D^{k+1} = \left(I + \Delta \xi_g^{k+1} \, S_r^{k+1} \, C_e^{k+1} \right)^{-1} C_e^{k+1}, \delta = 1 \, \dot{T} < 0 \, C_D^{k+1} = C_e, \delta$$

$$= \frac{\xi_g^{k+1}}{\xi_g^k} \, \dot{T} > 0 \tag{7.25}$$

Fabrication and modeling 4D-printed elements

Materials

In order to fabricate active elements, the polylactic acid (PLA) has been used as a SMP filament with 1.75 mm diameters and the glass transition temperature of 65°C. In this regard, a 3DGence DOUBLE printer developed by 3DGence has been used, which extrudes a filament 1.75 mm with a $0.4 - \text{mm}$ nozzle. The 3DGence Slicer software was used to convert the STL file to G-code and control printing parameters. The printing parameters such as the liquefier, chamber, and platform temperatures are set at 210, 24,and 24°C, respectively. The active elements (beam-like elements) were 4D-printed in the shape of a straight beam with the dimension $(30 \times 1.6 \times 1)$ mm for length, width, and thickness, respectively.

One of the most important properties that must be employed in thermomechanical analysis is the dependence of the properties on temperature. Therefore, using a dynamic-mechanical analyzer (DMA, Q800, TA Instruments, New Castle, DE, USA), the temperature-dependent properties for 3D-printed beam-like samples were determined. In this regard, the DMTA test was carried out in the axial tensile mode, in which the frequency of forced oscillation and heating rate were considered 1 Hz and 5°C/min, respectively. Therefore, the results of DMTA 4D-printed beam-like were reported as storage modules, E_s, and phase lag, δ,which is shown in Fig. 7.5. According to the phase lag result, the sample shows its peak curve at 65°C, which indicates the glass transition temperature. It can also be deduced from the diagram of storage modules that at 60°C,the sample experiences a drastic change in its storage modules.

4D-printed shape memory polymer: Modeling and fabrication

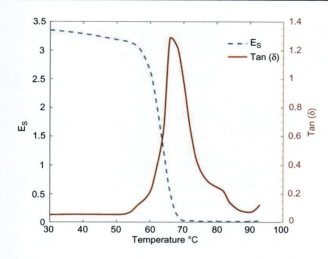

Fig. 7.5 DMA results. The DMA results for 3D-printed PLA. No permission required.

4D printing elements

As mentioned in the 4D printing programming section, the print speed plays a significant role in inducing the SME on 3D-printed samples. Therefore, five beam-like elements were printed at five different speeds $S_p = 5, 10, 20, 40, 70\ \frac{mm}{s}$. Fig. 7.6 indicates the straight configuration of the 3D-printed beam-like element. After preparation of the 3D-printed sample, samples were heated by immersion in hot water at a specified temperature of 85°C, which is higher than the glass temperature transition, and then, the samples were placed at room temperature and allowed to cool and reach room temperature (24°C). Fig. 7.7 shows the configuration of the 3D-printed sample after the heating–cooling process. As shown in Fig. 7.7, the straight beam-like element configuration changes to a curved beam under the thermal stimulus. Fig. 7.7A shows the 3D-printed sample, which has been printed at low speed $S_p = 5$ mm/s. In this printing speed, the beam configuration remains unchanged under temperature stimulation,

Fig. 7.6 Straight beam-like configuration. The straight beam-like configuration after 4D printing.
From Noroozi, R., Bodaghi, M., Jafari, H., Zolfagharian, A., & Fotouhi, M. (2020). Shape-adaptive metastructures with variable bandgap regions by 4D printing. *Polymers, 12*(3), 519. https://doi.org/10.3390/polym12030519.

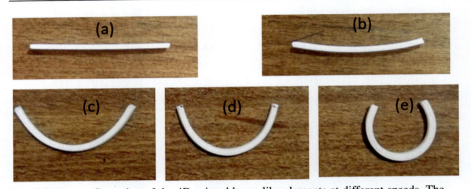

Fig. 7.7 The configuration of the 4D-printed beam-like elements at different speeds. The configuration of the 4D-printed beam-like elements at different speeds of (A) 5, (B) 10, (C) 20, (D) 40, and (E) 70 mm/s after the heating–cooling process.
From Noroozi, R., Bodaghi, M., Jafari, H., Zolfagharian, A., & Fotouhi, M. (2020). Shape-adaptive metastructures with variable bandgap regions by 4D printing. *Polymers*, *12*(3), 519. https://doi.org/10.3390/polym12030519.

and this is because no prestrain has not been induced in the sample. Therefore, the low printing speeds could not be employed in the 4D printing programming. Fig. 7.7B shows the configuration of the sample with print speed $S_p = 10$ mm/s after heating–cooling. In contrast to the previous speed, the beam-like element after the heating–cooling process transforms to a slight curve element at this printing speed, demonstrating that increasing the print speed increases the amount of the induced prestrain. Because of the slight curve at the $S_p = 10$ mm/s, this print speed could be thought of as a transition speed which printed samples with lower than S_p are static elements and higher than are dynamic elements. The self-bending feature is due to the unbalanced prestrain through the sample thickness, in which the lower layer has less prestrain, and the upper layers have more prestrain. As shown in the experiment, by increasing the 4D printing speed, the self-bending/folding feature increases; therefore, by controlling 4D printing speed during the manufacturing process, the shape programming process can be controlled. Nevertheless, the FDM 4D printing technology has high potential in the fabricating and programming active elements with self-bending/folding features.

Finite element modeling

The implementation of constitutive equations in commercial finite element software such as ABAQUS, COMSOL, and ANSYS has been gaining attention due to its extensive application. The constitutive equation of SMPs, due to their complexity, has a lot of difficulties using in finite element software. In this section, to settle this issue by considering the concept of SME and FG 4D printing, a simple method for modeling the thermomechanical behavior of the 4D-printed self-folding element has been proposed. In this regard, the temperature-dependent Young's modulus of the 3D-printed

Table 7.1 Temperature-dependent Young's modulus for 3D-printed PLA.

T (°C)	30	40	50	60	70	80	90
E (Mpa)	3350	3280	3166	2554	48	18	14

component has been imported into COMSOL finite element software according to Table 7.1:

As previously stated, the 4D-printed beam-like element can be regarded as an FG material, with the induced prestrain changing gradually along with its thickness. Therefore, to model 4D-printed elements with COMSOL Multiphysics, the sample is divided into multiple sections whose thermal expansion is different. For this purpose, the 4D-printed sample is discretized into six sections. Although considering a large number of sections can lead to a more accurate answer, considering six sections is acceptable in terms of computational cost and accuracy. Fig. 7.8 shows the divided form of a 3D-printed sample in which black and white color shows the first and last printed layers, respectively.

The coefficients of thermal expansion of layers are selected so that the simulated configurations replicate the experimental configurations after the heating–cooling process. Table 7.2 shows the thermal expansion coefficient for different printing speeds.

Considering the obtained coefficients of thermal expansion and applying them into COMSOL finite element software, the beam-like configuration after the heating–cooling process is shown in Fig. 7.9. To describe the geometric features of 4D-printed samples, three parameters are selected, as shown in Fig. 7.9C. The parameters R_1, R_2, and R_3 represent the outer length, opening, and depth length, respectively.

Fig. 7.8 Discretized printed sample. Discretized 4D-printed element.
No permission required.

Table 7.2 The thermal expansion coefficients for different printing speeds.

α_i (1/°C)	S_p (mm/s) 10	20	40	70
α_1	−0.0006	−0.0016	−0.0018	−0.00252
α_2	−0.0004	−0.0011	−0.0016	−0.00222
α_3	−0.0002	−0.0011	−0.0013	−0.0022
α_4	−0.00009	−0.0008	−0.0011	−0.00172
α_5	−0.00007	−0.0006	−0.0008	−0.00152
α_6	−0.00005	−0.0004	−0.0005	−0.00122

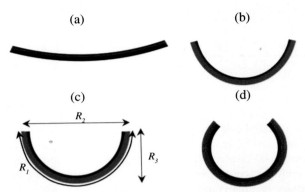

Fig. 7.9 Finite element simulation of the 4D-printed samples. Finite element simulation of the 4D-printed samples for different printing speeds (A) 10, (B) 20, (C) 40, and (D) 70 mm/s after the heating–cooling process.
From Noroozi, R., Bodaghi, M., Jafari, H., Zolfagharian, A., & Fotouhi, M. (2020). Shape-adaptive metastructures with variable bandgap regions by 4D printing. *Polymers*, *12*(3), 519. https://doi.org/10.3390/polym12030519.

As shown in Fig. 7.9, the 4D-printed sample with printing speed, $S_p = 70$ mm/s has much more deformation than other printing speeds, which selected parameters clearly indicate this fact. In addition to the experimental and finite element analysis (FEA), an in-house code for solving the constitutive equation of SMPs is employed. Hence, to investigate the accuracy of the simple method used in the finite element results, the geometric features obtained from experimental, finite element analyses, and in-house code are compared and listed in Table 7.3. As shown in Table 7.3, the in-house code results are consistent with the simple method and experimental results. Therefore, the simple method accuracy and efficiency for the 4D-printed sample with the self-folding feature are validated.

Table 7.3 The geometric parameters of the 4D-printed active element after the heating–cooling process.

Method	S_p (mm/s)	R_1 (mm)	R_2 (mm)	R_3 (mm)
Experiment	10	29.8	28.2	3.1
	20	29.3	19	8.3
	40	29.1	16.3	9.2
	70	29.0	7.1	10.5
In-house FE method	10	29.9	28.3	3.0
	20	29.1	19.2	8.4
	40	29.4	16.3	9.2
	70	29.2	7.2	10.3
FE COMSOL Multiphysics	10	29.7	28.3	3.2
	20	29.4	19.1	8.4
	40	29.2	16.2	9.1
	70	28.9	7.0	10.4

Case studies

In this section, using 4D-printed beam-like elements, which in the previous sections were fabricated and modeled in the finite element software, some of the 4D printing applications have been shown. First, a soft actuator under thermal stimulus is fabricated and modeled. Second, the 4D-printed composites with self-assembly and self-folding features are investigated. Finally, by using passive and active elements, adaptive metastructures with variable bandgap regions are proposed.

Self-folding structures

Self-folding is one of the most important mechanisms that provide the range of motion and changing configuration in nature. The self-folding structure can be used for a wide range of applications such as deployable biomedical devices and self-assembling robots due to their excellent features, including reducing time and effort to assemble complex structures and remote and spontaneous performances. The SMPs are good candidates for being employed in self-folding structures due to their unique properties, such as sensitivity to environmental changes (Felton et al., 2013; Mao et al., 2015). In this section, by employing the 4D-printed beam-like elements, different patterns of self-folding/twisting structures are designed, and their behavior under thermal stimulus has been investigated.

Gripper actuator

The design and fabrication of a practical gripper play an important role in robot performance which leads to overcome robot inaccuracy and enhance overall system performance (Honarpardaz, Tarkian, Ölvander, & Feng, 2017). Therefore, the design and fabrication of various grippers with different patterns and applications have been considered by researchers. In this subsection, by employing the fabricated active beam-like elements, a smart gripper is designed in which four active elements are connected to a passive pyramid. Fig. 7.10 shows the configuration of the gripper before the heating–cooling process. After modeling the gripper in finite element software, it

Fig. 7.10 Gripper configuration. Gripper configuration before the heating–cooling process. No permission required.

Fig. 7.11 The gripper deformed configuration. The FE Abaqus simulation of the gripper at printing speed: (A) 20 and (B) 40 mm/s.
No permission required.

is exposed to external heat, and its temperature reaches 65 °C and then cooled down to the temperature room. Fig. 7.11 represents the deformed configuration of the gripper. As shown in Fig. 7.11, due to the self-folding feature of active elements, the gripper after thermal stimulus exhibits a self-folding behavior so that it can be used as a soft robotic gripper. The amount of bending angle can be tuned by changing printing speed. For instance, beam-like components with higher 4D printing speeds have more bending angles, and as a result, the various deformed configurations can be achieved by controlling the printing speed, which makes it possible to work in different situations.

Self-folding smart composites

There are different strategies to generate a self-folding mechanism. One of them that during the last decades has been considered by researchers is the embedded rigid particles in the soft matrix and controlling their orientation (Schmied, Le Ferrand, Ermanni, Studart, & Arrieta, 2017; Sydney Gladman, Matsumoto, Nuzzo, Mahadevan, & Lewis, 2016; Wang et al., 2018). One of the critical parameters that significantly affect the controllability of composite deformation is rigid particle orientation. To address this issue, rigid particles can be replaced with continuous fiber to gain better controllability. In this subsection, using the active elements, three patterns of self-folding smart composites are designed and modeled.

In the first pattern, a self-assemble structure is created using 4D-printed elements. In this regard, a flat paper sheet is reinforced with eight 4D-printed elements. Fig. 7.12 shows the undeformed configuration of the self-assemble structure. In order to predict the thermomechanical behavior of the self-assemble structure under the thermal stimulus, the simple method is employed in the finite element software. According to the DMTA test, PLA has a low Young's modulus at 85 °C. Therefore, if the active elements are heated up to 85 °C, the stiffness of the paper prevents elements from deformation and backs them to their straight configuration. To solve this issue, in the FEM, the structure is heated up to 65 °C. The interaction between 4D-printed elements and

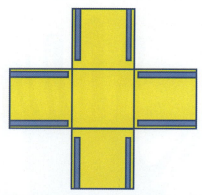

Fig. 7.12 Configuration of the smart composite. Configuration of the smart composite before the heating–cooling process.
No permission required.

the paper sheet is considered perfect bound, in which there is no relative displacement among them. The heating–cooling process is applied to the structure so that at first, the structure is heated up to 65 °C and then cooled to room temperature. Fig. 7.13 indicates the deformed self-assemble configuration after the heating–cooling process.

As shown in Fig. 7.13, the self-assembly structure's deformation, with the initial flat state form, can be considered a 2D-to-3D shape changing, which can be called a self-assembly mechanism. By increasing the printing speed, the structure undergoes more bending so that the printing speed of $S_p = 70$ mm/s has more bending, and on the contrary, the speed of $S_p = 20$ mm/s has the least bending. It should be noted that in to have the best self-assemble structure, the optimum printing speed should be used, in

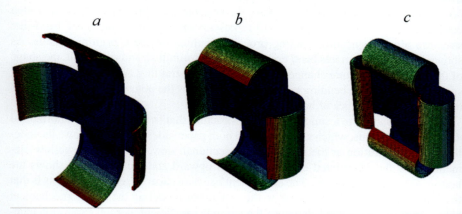

Fig. 7.13 Simulated configuration of the self-assembly composite. The simulated configuration of the self-assembly composite after the heating–cooling process for (A) 20, (B) 40, and (C) 70 mm/s.
No permission required.

Fig. 7.14 The hand-like gripper. Undeformed configuration of the hand-like gripper. No permission required.

which in this structure the printing speed of $S_p = 40$ mm/s is optimum printing speed. Also, by changing the arrangement of 4D-printed elements along the flat paper sheet, different self-assembly patterns can be obtained.

As a second pattern, a smart hand-like structure is proposed in which a hand-like flat paper sheet is reinforced with 4D-printed elements. Fig. 7.14 shows the undeformed configuration of the hand-like structure. To model the thermomechanical behavior of the hand-like structure, the proper thermal and mechanical boundary conditions are applied to the FEM. For this purpose, it is assumed that the interaction between 4D-printed elements and the paper sheet is tie-type, which means there is no relative motion between them. Also, the structure is heated up to 65 °C and then cooled to room temperature. Fig. 7.15 represents the configuration of the hand-like structure after the heating–cooling process for three different printing speeds. As shown, the active elements bend under the thermal stimulus, and as a result, the shape of the hand-like structure changes under the thermal stimulus. Also, different printing speeds have different bending angles, which indicates the potential of printing speeds to generate wide ranges of hand motion patterns.

The third pattern is a smart corrugated structure that is reinforced with horizontal 4D-printed elements, which are attached to both paper sides. Fig. 7.16 shows its undeformed configuration at the top and cross-sectional view, in which the yellow elements show the 4D-printed element attached upward and the black line shows the bottom 4D-printed elements. The attachment between paper sheer and elements is that the first printed layer is connected to the paper. For modeling the thermomechanical behavior of the 4D-printed elements, the variable coefficient thermal expansion technique has been employed in finite element software. The thermal and mechanical boundary conditions are tie-type and 65 °C, respectively. After the heating up structure to 65 °C, it is placed at a specific temperature (room temperature) to cool down. Fig. 7.17 represents the deformed configuration of the rectangular structure after the

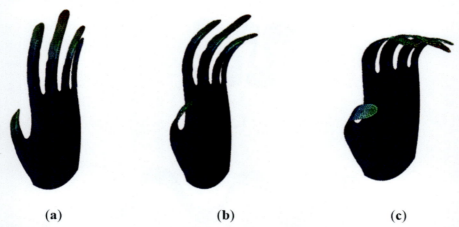

Fig. 7.15 The deformed configuration of the hand-like gripper. The deformed configuration of the hand-like gripper after heating–cooling process for (A) 20, (B) 40, and (C) 70 mm/s.
No permission required.

Fig. 7.16 Configuration of the smart corrugated structure. Configuration of the smart corrugated structure with horizontal 4D-printed beams.
No permission required.

heating–cooling process for three different speeds. After applying an external stimulus, the 2D reinforced rectangular paper wrinkles to form a 3D corrugated structure. As shown, different printing speeds have different pitches, so the higher the printing speed, the lower the pitch value. Moreover, one of the parameters that significantly affect the corrugated feature is the distance between 4D-printed elements.

Adaptive dynamic structures

One of the most important issues in our modern daily life is the analysis of acoustic waves. Acoustic metamaterials are a type of architect material that is designed to control and manipulate waves passing through it. The periodic structures as a kind of metamaterials have the ability to damp some specific frequency ranges of elastic waves, called bandgap or stopband (Bertoldi & Boyce, 2008; D'Alessandro, Zega, Ardito, & Corigliano, 2018; Li, Wang, & Yan, 2021; Meaud & Che, 2017). In other words, while elastic wave propagation occurs in all directions in the propagating frequency ranges, bandgap areas provide a frequency range where elastic wave propagation is stopped, and this ability can be used in a variety of applications, including controlling and managing wave propagation and acoustic mirrors. The geometry and material type of periodic structure determine the bandgap width, frequency level, and modal location of the structure (D'Alessandro et al., 2018).

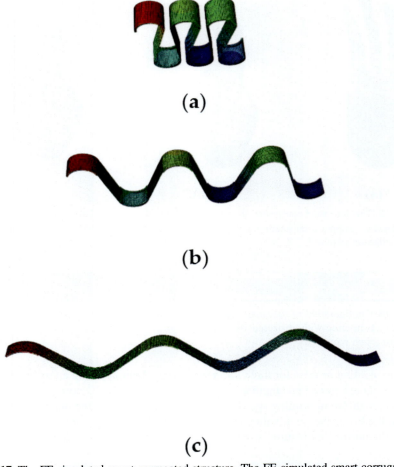

Fig. 7.17 The FE simulated smart corrugated structure. The FE-simulated smart corrugated structure after the heating–cooling process: (A) 20, (B) 40, and (C) 70 mm/s. No permission required.

Traditional materials and structures do not change their configuration and properties, so they can only be used in a constant range of frequency. The researchers have designed adaptive structures with variable and uncommon behavior in different applications. Therefore, by employing different methods, researchers have tried to design and fabricate adaptive and tunable structures (Bertoldi & Boyce, 2008; Bertoldi, Boyce, Deschanel, Prange, & Mullin, 2008; D'Alessandro et al., 2018; Li et al., 2021; Li, Wang, Chen, Wang, & Bao, 2019; Meaud & Che, 2017; Shim, Wang, & Bertoldi, 2015; Singh & Pandey, 2018).

Wave propagation formulation

To study wave propagation in architected period structures, the Bloch theorem has been used. Based on the Bloch theorem, the displacement of any point in a periodic architected structure can be defined as a function of the displacement of its corresponding point in the reference unit cell ($\vec{U}_{Ref}(\vec{r})$). As a result, wave propagation analyses for the entire periodic structure were simplified to solve equations for a unit cell.

Where $\vec{U}(\vec{r},t)$ expresses displacement for an arbitrary point in the periodic structure and $\vec{U}_{Ref}(\vec{r})$ indicates its corresponding point in the reference unit cell. Also, $\vec{R}$ and $\vec{r}$ show the lattice and position vector, respectively. Besides, $\vec{k} = (k_1, k_2)$ is the wave vector and k_1 and k_2 are its components along the x and y directions, respectively. Also, ω and t indicate frequency and time, respectively. In order to define periodically in the period structures, the direct vector must be specified. Fig. 7.18 shows a periodic structure in which a_1 and a_2 are direct vectors that according to them, the lattice vector defines as follows:

$$\vec{R} = n_1 \vec{a}_1 + n_2 \vec{a}_2 \quad n_1, n_2 = 0, \pm 1, \pm 2, \ldots \tag{7.26}$$

Bloch wave vector varies in the Brillouin zone. Therefore, to analyze the wave propagation, the reciprocal space corresponding to the direct space should be defined, which is shown in the following equation.

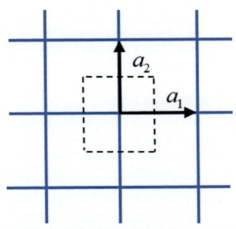

Fig. 7.18 Direct vectors of periodic structure. Periodic structure with its direct vectors. No permission required.

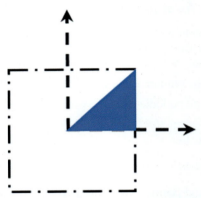

Fig. 7.19 First Brillouin zone and irreducible Brillouin zone. First Brillouin zone (FBZ, dashed square) and irreducible Brillouin zone (IBZ, colored triangle).
No permission required.

$$\vec{a}_i \cdot \vec{b}_j = 2\pi\, \delta_{ij} = \begin{cases} \vec{b}_1 = 2\pi \dfrac{a_2 \times a_3}{a_1.(a_2 \times a_3)} \\ \vec{b}_2 = 2\pi \dfrac{a_3 \times a_1}{a_1.(a_2 \times a_3)} \end{cases} \quad a_3 = (0, 0, 1) \tag{7.27}$$

where δ_{ij} is the Kronecker delta.

Fig. 7.19 shows the first Brillouin zone (FBZ) and irreducible Brillouin zone (IBZ). To analyze the wave propagation, the IBZ has been employed which is a small part of the FBZ (Fig. 7.19).

According to the equation just now mentioned, the wave propagation is a linear eigenvalue problem with three unknown parameters ω, k_1, and k_2 which the wave vector investigates in the IBZ. To solve the wave propagation equations, COMSOL Multiphysics software is employed. Also, convergence studies on mesh size have been performed to achieve reliable and accurate results.

Design adaptive periodic structures

In this section, using the concept of active and passive elements, two periodic structures are conceptually proposed. The term passive elements imply elements that under thermal stimulus do not change their configuration and dynamic behavior. As mentioned in the previous sections, 4D-printed beam-like elements during the printing process can be programmed to show a dynamic behavior under the stimulus. No induced prestrain is generated in the 4D-printed sample at low printing speeds. Therefore, low printing speeds, such as 5 mm/s, can be used to print passive elements. On the contrary, active elements are printed at a high printing speed, in which the induced prestrain is generated along with the sample and under the thermal stimulus changes its dynamic behavior and shows self-folding features. Therefore, by employing active and passive elements and arranging them into a periodic structure, two architected

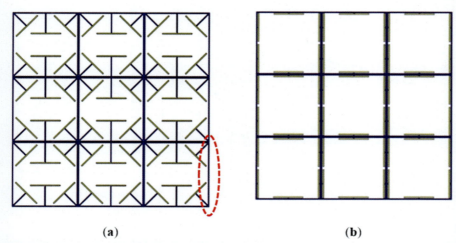

Fig. 7.20 Periodic adaptive structures. Periodic adaptive structures with active and passive elements: (A) diagonal structure and (B) parallel structure (the *dashed line* shows the fixed-fixed beam which is used for the frequency normalization).
From Noroozi, R., Bodaghi, M., Jafari, H., Zolfagharian, A., & Fotouhi, M. (2020). Shape-adaptive metastructures with variable bandgap regions by 4D printing. *Polymers*, *12*(3), 519. https://doi.org/10.3390/polym12030519.

structures have been designed. Structures consist of a mainframe and beam-like elements, which are made of passive and active elements, respectively. The first arrangement is such that the active elements in the diagonal and parallel form are connected to the passive frame. The second arrangement is such that the active elements are connected in parallel to the passive frame. In order to simplify, the first and second structures are called diagonal and parallel, respectively. Fig. 7.20 represents the designed periodic adaptive structure, in which yellow and blue colors indicate active and passive elements, respectively.

Adaptive diagonal structure

The wave propagation properties in the terms of dispersion curve of adaptive diagonal structures are studied. In this section, COMSOL-based numerical have been used to show how adaptive periodic structures can be designed with variable dynamic performance without adding additional resonant components.

To ensure the accuracy of results, the bandgap of a triangular structure was calculated and compared with the earlier results (Srikantha Phani, Woodhouse, & Fleck, 2006). As shown in Fig. 7.21, the COMSOL-based results are in good agreement with the earlier results, which indicates the accuracy of the method used in the COMSOL Multiphysics software.

Eigenfrequencies of adaptive periodic structures are calculated by applying periodic boundary conditions to different elastic wave vectors estimated based on the IBZ. The first natural frequency of the fixed-fixed beam, shown in Fig. 7.20 with a

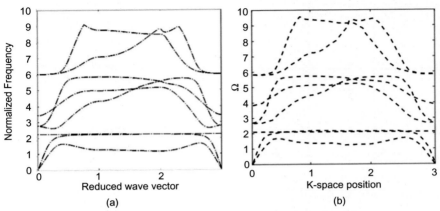

Fig. 7.21 Results validations. Comparison of COMSOL-based results with. From Noroozi, R., Bodaghi, M., Jafari, H., Zolfagharian, A., & Fotouhi, M. (2020). Shape-adaptive metastructures with variable bandgap regions by 4D printing. *Polymers, 12*(3), 519. https://doi.org/10.3390/polym12030519.

dashed line, is used to normalize eigenfrequencies in the form of $\Omega = \omega/\omega_0$, in which $\omega_0 = 22.4 \sqrt{EI/mL_0^4}$ expresses the first natural frequency of the fixed-fixed beam.

Fig. 7.22 represents dispersion curves and some of the mode shapes for diagonal structure before any external stimulus. It can be seen from the dispersion curves that there is a bandgap in the range of 1.902 to 2.043, which is shown in gray color. This bandgap region covers 4.68% of the total frequency range. By considering the mode shapes, it is determined that the cause of the bandgap is the resonation of active elements.

After stimulating the diagonal structure under the heating–cooling process, the active element shows their self-bending features and changes mass distribution and stiffness in the structures. Fig. 7.15 shows the changed configuration for three different printing speeds. As shown in Fig. 7.23, the structure with a higher 4D printing speed undergoes more change, and its 4D-printed elements have more curvature and deformation. Fig. 7.24 shows the dispersion curves for the diagonal structure at different printing speeds. The bandgap area for the print speed of 20 m/s is in the range of 1.751 to 2.043, which covers 4.68% of the total frequency range. For the print speed of 40 m/s (Fig. 7.25), the bandgap area covers $\Omega = 1.751$ to 1.812 of the frequency range, which covers a smaller range of the total frequency than the print speed of 20 m/s. Unlike the previous two print speeds, at a print speed of 70 m/s (Fig. 7.26), there is no bandgap area in the frequency range. The obtained dispersion curves for different printing speeds show that using 4D-printed elements can achieve adaptive structures which keep their performance in the various functional ranges.

The dispersion curves obtained from two printing speeds of 20 and 40 m/s indicate increasing printing speed has a little effect on bandgap width and only moves the

4D-printed shape memory polymer: Modeling and fabrication 219

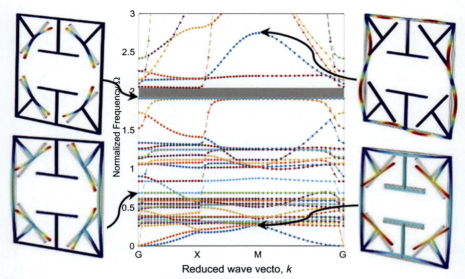

Fig. 7.22 Dispersion of the diagonal metastructure. Dispersion curves and mode shapes of the diagonal metastructure.
From Noroozi, R., Bodaghi, M., Jafari, H., Zolfagharian, A., & Fotouhi, M. (2020). Shape-adaptive metastructures with variable bandgap regions by 4D printing. *Polymers*, *12*(3), 519. https://doi.org/10.3390/polym12030519.

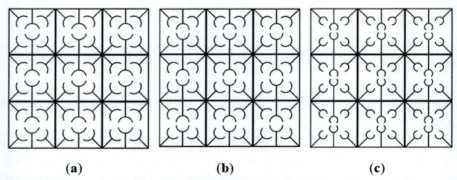

Fig. 7.23 The deformed configuration of the adaptive periodic diagonal structure. The deformed configuration of the adaptive periodic diagonal structure after the heating–cooling process for three different 4D printing speeds: (A) 20, (B) 40, and (C) 70 mm/s.
From Noroozi, R., Bodaghi, M., Jafari, H., Zolfagharian, A., & Fotouhi, M. (2020). Shape-adaptive metastructures with variable bandgap regions by 4D printing. *Polymers*, *12*(3), 519. https://doi.org/10.3390/polym12030519.

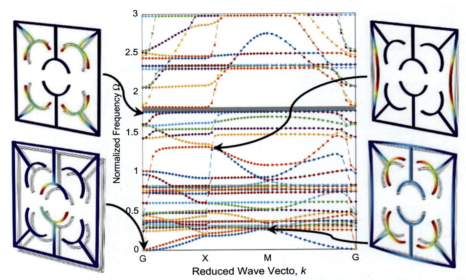

Fig. 7.24 Dispersion curves and mode shapes of the diagonal structure for printing speed 20 mm/s. Dispersion curves and mode shapes of the diagonal structure with the 4D printing speed at 20 mm/s after the heating–cooling process.
From Noroozi, R., Bodaghi, M., Jafari, H., Zolfagharian, A., & Fotouhi, M. (2020). Shape-adaptive metastructures with variable bandgap regions by 4D printing. *Polymers*, *12*(3), 519. https://doi.org/10.3390/polym12030519.

stopband region. However, increasing the print speed does not always change the stopband range. For example, when the printing speed increases to 70 mm/s, the bandgap region vanishes, and the structure transmits the whole wave in all frequency ranges. In general, the results show that by changing 4D printing speeds, different dispersion curves can be obtained, which by manipulating the appropriate and functional bandgap region can be found. Hence, the 4D printing elements have a high potential to fabricate adaptive structures with variable bandgap regions.

Adaptive parallel structure

As mentioned in the previous section, in the parallel structure, the active elements are connected to the passive frame in parallel. Fig. 7.27 shows the dispersion curves for parallel structures before any heating–cooling process. The dispersion behavior for parallel structures is different from the diagonal structure. As shown in dispersion curves, unlike the diagonal structure in which the bandgap region appears in the frequency range, no bandgap region has not appeared, and the whole elastic wave in the range of 1–3 passes through the structure. Fig. 7.28 indicates the parallel structure configuration after the heating–cooling process for three 4D printing speeds. As expected, the 4D-printed active elements under thermal stimulus change their configuration and activate their self-bending features. Hence, by changing the configuration

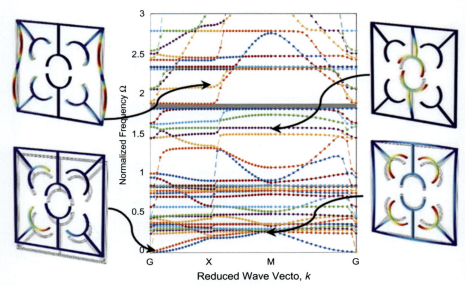

Fig. 7.25 Dispersion curves and mode shapes of the diagonal structure for printing speed 40 mm/s. The counterpart of Fig. 7.12 for 40 mm/s.
From Noroozi, R., Bodaghi, M., Jafari, H., Zolfagharian, A., & Fotouhi, M. (2020). Shape-adaptive metastructures with variable bandgap regions by 4D printing. *Polymers*, *12*(3), 519. https://doi.org/10.3390/polym12030519.

of structures, the mass distribution and stiffness of the structure change, and as a result, dynamic behaviors change. Figs. 7.29–7.31 show the dispersion curves and mode shapes of the parallel structure after the heating–cooling process for different printing speeds of 20, 40, and 70 m/s, respectively.

After the heating–cooling process, the 4D-printed beam-like elements change their configuration to a bent state. As shown in the dispersion curves, after the heating–cooling process, the parallel structure printed at a speed of 20 m/s, a narrow region of bandgap in the range of 1.945 to 1.989 appears in dispersion curves. The ratio of the bandgap range to the total frequency range is 1.48%. By changing the printing speed to 40 mm/s, the beam-like elements bend, and as a result, the dispersion curves change. At this printing speed, multiple bandgap regions appear in the dispersion curve. The multiple ranges of bandgap regions are in the range of 2.172 to 2.231, 2.371 to 2.505, 2.421 to 2.430, 2.441 to 2.569, and 2.594 to 2.765, which in the practical view has more performance than other structures and covers 12.32% of the total frequency range. The printed structure, at a printing speed of 70 m/s, does not experience any bandgap region, and all of the elastic waves pass through the structure. By considering the parallel structure dispersion curves, it could be concluded that the parallel structure shows a better performance than the diagonal structure. The parallel structure covers more and multiple bandgap regions so that it locally resonates filters in the broader range.

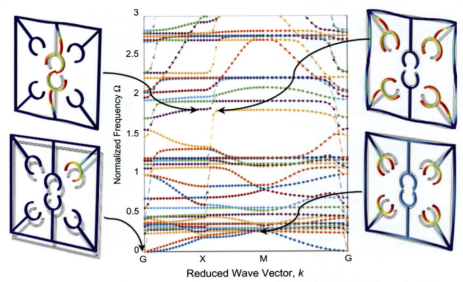

Fig. 7.26 Dispersion curves and mode shapes of the diagonal structure for printing speed 70 mm/s. The counterpart of Fig. 7.12 for 70 mm/s.
From Noroozi, R., Bodaghi, M., Jafari, H., Zolfagharian, A., & Fotouhi, M. (2020). Shape-adaptive metastructures with variable bandgap regions by 4D printing. *Polymers*, *12*(3), 519. https://doi.org/10.3390/polym12030519.

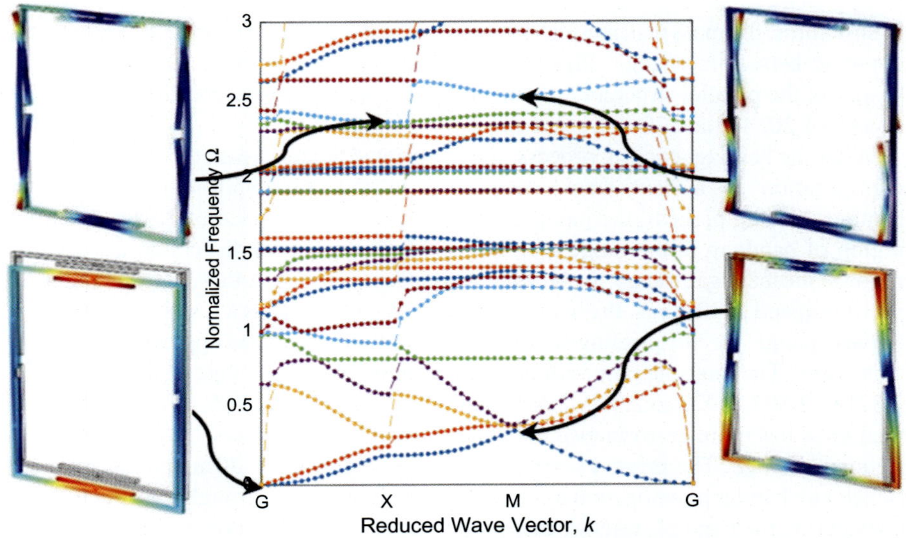

Fig. 7.27 Dispersion curves and mode shapes of the adaptive parallel structure. Dispersion curves and mode shapes of the adaptive parallel structure.
From Noroozi, R., Bodaghi, M., Jafari, H., Zolfagharian, A., & Fotouhi, M. (2020). Shape-adaptive metastructures with variable bandgap regions by 4D printing. *Polymers*, *12*(3), 519. https://doi.org/10.3390/polym12030519.

4D-printed shape memory polymer: Modeling and fabrication   223

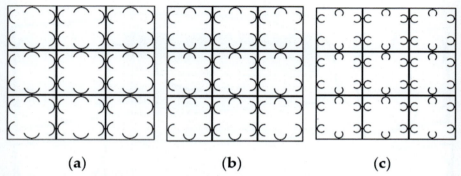

Fig. 7.28 The deformed configuration of the adaptive periodic parallel structure. The deformed configuration of the adaptive periodic parallel structure after the heating–cooling process for three different 4D-printing speeds: (A) 20, (B) 40, and (C) 70 mm/s.
From Noroozi, R., Bodaghi, M., Jafari, H., Zolfagharian, A., & Fotouhi, M. (2020). Shape-adaptive metastructures with variable bandgap regions by 4D printing. *Polymers, 12*(3), 519. https://doi.org/10.3390/polym12030519.

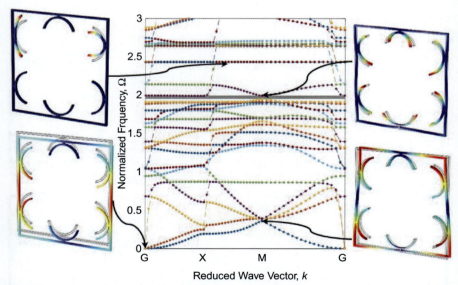

Fig. 7.29 Dispersion curves and mode shapes of the adaptive parallel structure for printing speed 20 mm/s. Dispersion curves and mode shapes of the adaptive parallel structure with 4D-printed active elements at 20 mm/s after the heating–cooling process.
From Noroozi, R., Bodaghi, M., Jafari, H., Zolfagharian, A., & Fotouhi, M. (2020). Shape-adaptive metastructures with variable bandgap regions by 4D printing. *Polymers, 12*(3), 519. https://doi.org/10.3390/polym12030519.

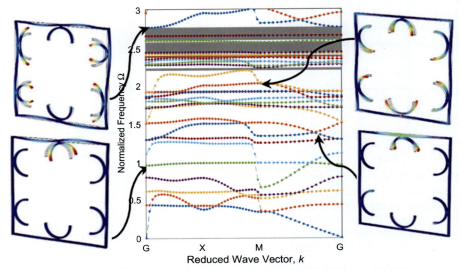

Fig. 7.30 Dispersion curves and mode shapes of the adaptive parallel structure for printing speed 40 mm/s. The counterpart of Fig. 7.17 for 40 mm/s.
From Noroozi, R., Bodaghi, M., Jafari, H., Zolfagharian, A., & Fotouhi, M. (2020). Shape-adaptive metastructures with variable bandgap regions by 4D Printing. *Polymers*, *12*(3), 519. https://doi.org/10.3390/polym12030519.

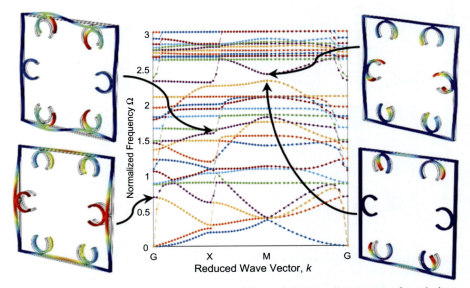

Fig. 7.31 Dispersion curves and mode shapes of the adaptive parallel structure for printing speed 70 mm/s. The counterpart of Fig. 7.17 for 70 mm/s.
From Noroozi, R., Bodaghi, M., Jafari, H., Zolfagharian, A., & Fotouhi, M. (2020). Shape-adaptive metastructures with variable bandgap regions by 4D printing. *Polymers*, *12*(3), 519. https://doi.org/10.3390/polym12030519.

Conclusion

In this chapter, first, the constitutive equations of SMPs based on the thermoviscoelastic and phase transformation approach were introduced. Then, by employing FDM technology, one of the most popular 3D printing methods, the concept of 4D printing technology was presented. It was shown, experimentally, that during the 3D printing technology, the sample is subjected to a thermomechanical cycle similar to shape memory programming. Moreover, it was shown that printing speed has many effects on the 3D-printed sample shape memory effect. Due to the importance of modeling the thermomechanical behavior of 4D-printed samples, based on the concept of FG 4D printing, simple to reproduce but the accurate method was presented to predict 4D-printed sample behavior. This method was validated and calibrated with experimental results and correctly predicted the self-folding feature of 4D-printed elements. As case studies, the thermomechanical behavior of the smart gripper and self-folding smart composite structures under external stimulus was investigated. Also, by using the 4D-printed elements, the adaptive dynamic structures were designed and modeled in finite element software and their bandgap region were analyzed for different printing speeds.

References

Ahmed, K., Shiblee, M. N. I., Khosla, A., Nagahara, L., Thundat, T., & Furukawa, H. (2020). Recent progresses in 4D printing of gel materials. *Journal of the Electrochemical Society*, *167*, 037563b.

Askari, M., Afzali Naniz, M., Kouhi, M., Saberi, A., Zolfagharian, A., & Bodaghi, M. (2021). Recent progress in extrusion 3D bioprinting of hydrogel biomaterials for tissue regeneration: A comprehensive review with focus on advanced fabrication techniques. *Biomaterials Science*, *9*(3), 535–573. https://doi.org/10.1039/d0bm00973c.

Attaran, M. (2017). The rise of 3-D printing: The advantages of additive manufacturing over traditional manufacturing. *Business Horizons*, *60*(5), 677–688. https://doi.org/10.1016/j.bushor.2017.05.011.

Bertoldi, K., & Boyce, M. C. (2008). Mechanically triggered transformations of phononic band gaps in periodic elastomeric structures. *Physical Review B*. https://doi.org/10.1103/physrevb.77.052105.

Bertoldi, K., Boyce, M. C., Deschanel, S., Prange, S. M., & Mullin, T. (2008). Mechanics of deformation-triggered pattern transformations and superelastic behavior in periodic elastomeric structures. *Journal of the Mechanics and Physics of Solids*, *56*(8), 2642–2668. https://doi.org/10.1016/j.jmps.2008.03.006.

Bodaghi, M., & Liao, W. H. (2019). 4D printed tunable mechanical metamaterials with shape memory operations. *Smart Materials and Structures*, *28*(4). https://doi.org/10.1088/1361-665x/ab0b6b, 045019.

Bodaghi, M., Serjouei, A., Zolfagharian, A., Fotouhi, M., Rahman, H., & Durand, D. (2020). Reversible energy absorbing meta-sandwiches by FDM 4D printing. *International Journal of Mechanical Sciences*, *173*. https://doi.org/10.1016/j.ijmecsci.2020.105451, 105451.

Bodaghi, M., Damanpack, A. R., & Liao, W. H. (2017). Adaptive metamaterials by functionally graded 4D printing. *Materials and Design*, *135*, 26–36. https://doi.org/10.1016/j.matdes.2017.08.069.

Bodaghi, M., Noroozi, R., Zolfagharian, A., Fotouhi, M., & Norouzi, S. (2019). 4D printing self-morphing structures. *Materials, 12*(8). https://doi.org/10.3390/ma12081353.

Damanpack, A. R., Bodaghi, M., & Liao, W. H. (2020). Contact/impact modeling and analysis of 4D printed shape memory polymer beams. *Smart Materials and Structures, 29*(8). https://doi.org/10.1088/1361-665x/ab883a, 085016.

D'Alessandro, L., Zega, V., Ardito, R., & Corigliano, A. (2018). 3D auxetic single material periodic structure with ultra-wide tunable bandgap. *Scientific Reports*. https://doi.org/10.1038/s41598-018-19963-1.

Dudek, P. (2013). FDM 3D printing technology in manufacturing composite elements. *Archives of Metallurgy and Materials, 58*(4), 1415–1418. https://doi.org/10.2478/amm-2013-0186.

El-Sayegh, S., Romdhane, L., & Manjikian, S. (2020). A critical review of 3D printing in construction: Benefits, challenges, and risks. *Archives of Civil and Mechanical Engineering, 20*(2). https://doi.org/10.1007/s43452-020-00038-w.

Felton, S. M., Tolley, M. T., Shin, B., Onal, C. D., Demaine, E. D., Rus, D., et al. (2013). Self-folding with shape memory composites. *Soft Matter, 9*(32), 7688–7694. https://doi.org/10.1039/c3sm51003d.

Honarpardaz, M., Tarkian, M., Ölvander, J., & Feng, X. (2017). Finger design automation for industrial robot grippers: A review. *Robotics and Autonomous Systems, 87*, 104–119. https://doi.org/10.1016/j.robot.2016.10.003.

Jebellat, E., Baniassadi, M., Moshki, A., Wang, K., & Baghani, M. (2020). Numerical investigation of smart auxetic three-dimensional meta-structures based on shape memory polymers via topology optimization. *Journal of Intelligent Material Systems and Structures, 31* (15), 1838–1852. https://doi.org/10.1177/1045389X20935569.

Khoo, Z. X., Teoh, J. E. M., Liu, Y., Chua, C. K., Yang, S., An, J., et al. (2015). 3D printing of smart materials: A review on recent progresses in 4D printing. *Virtual and Physical Prototyping, 10*(3), 103–122. https://doi.org/10.1080/17452759.2015.1097054.

Leng, J., Lu, H., Liu, Y., Huang, W. M., & Du, S. (2009). Shape-memory polymers—A class of novel smart materials. *MRS Bulletin, 34*(11), 848–855. https://doi.org/10.1557/mrs2009.235.

Levy, A., Miriyev, A., Sridharan, N., Han, T., Tuval, E., Babu, S. S., et al. (2018). Ultrasonic additive manufacturing of steel: Method, post-processing treatments and properties. *Journal of Materials Processing Technology, 256*, 183–189. https://doi.org/10.1016/j.jmatprotec.2018.02.001.

Li, Y., Wang, X., & Yan, G. (2021). Configuration effect and bandgap mechanism of quasi-one-dimensional periodic lattice structure. *International Journal of Mechanical Sciences, 190*. https://doi.org/10.1016/j.ijmecsci.2020.106017, 106017.

Li, J., Wang, Y., Chen, W., Wang, Y.-S., & Bao, R. (2019). Harnessing inclusions to tune post-buckling deformation and bandgaps of soft porous periodic structures. *Journal of Sound and Vibration, 459*. https://doi.org/10.1016/j.jsv.2019.114848, 114848.

Mao, Y., Yu, K., Isakov, M. S., Wu, J., Dunn, M. L., & Jerry Qi, H. (2015). Sequential self-folding structures by 3D printed digital shape memory polymers. *Scientific Reports, 5*. https://doi.org/10.1038/srep13616.

Meaud, J., & Che, K. (2017). Tuning elastic wave propagation in multistable architected materials. *International Journal of Solids and Structures, 122–123*, 69–80. https://doi.org/10.1016/j.ijsolstr.2017.05.042.

Ngo, T. D., Kashani, A., Imbalzano, G., Nguyen, K. T. Q., & Hui, D. (2018). Additive manufacturing (3D printing): A review of materials, methods, applications and challenges. *Composites Part B: Engineering, 143*, 172–196. https://doi.org/10.1016/j.compositesb.2018.02.012.

Olakanmi, E. O., Cochrane, R. F., & Dalgarno, K. W. (2015). A review on selective laser sintering/melting (SLS/SLM) of aluminium alloy powders: Processing, microstructure, and properties. *Progress in Materials Science*, *74*, 401–477. https://doi.org/10.1016/j.pmatsci.2015.03.002.

Penumakala, P. K., Santo, J., & Thomas, A. (2020). A critical review on the fused deposition modeling of thermoplastic polymer composites. *Composites Part B: Engineering*, *201*. https://doi.org/10.1016/j.compositesb.2020.108336.

Rayegani, F., & Onwubolu, G. C. (2014). Fused deposition modelling (fdm) process parameter prediction and optimization using group method for data handling (gmdh) and differential evolution (de). *International Journal of Advanced Manufacturing Technology*, *73*(1–4), 509–519. https://doi.org/10.1007/s00170-014-5835-2.

Schmied, J. U., Le Ferrand, H., Ermanni, P., Studart, A. R., & Arrieta, A. F. (2017). Programmable snapping composites with bio-inspired architecture. *Bioinspiration & Biomimetics*, *12*(2). https://doi.org/10.1088/1748-3190/aa5efd.

Shahrubudin, N., Lee, T. C., & Ramlan, R. (2019). An overview on 3D printing technology: Technological, materials, and applications. *Procedia Manufacturing*, *35*, 1286–1296. Elsevier B.V https://doi.org/10.1016/j.promfg.2019.06.089.

Shie, M. Y., Shen, Y. F., Astuti, S. D., Lee, A. K. X., Lin, S. H., Dwijaksara, N. L. B., et al. (2019). Review of polymeric materials in 4D printing biomedical applications. *Polymers*, *11*(11). https://doi.org/10.3390/polym11111864.

Shim, J., Wang, P., & Bertoldi, K. (2015). Harnessing instability-induced pattern transformation to design tunable phononic crystals. *International Journal of Solids and Structures*, *58*, 52–61. https://doi.org/10.1016/j.ijsolstr.2014.12.018.

Shirzad, M., Zolfagharian, A., Matbouei, A., & Bodaghi, M. (2021). Design, evaluation, and optimization of 3D printed truss scaffolds for bone tissue engineering. *Journal of the Mechanical Behavior of Biomedical Materials*, *120*. https://doi.org/10.1016/j.jmbbm.2021.104594, 104594.

Singh, B. K., & Pandey, P. C. (2018). Tunable temperature-dependent THz photonic bandgaps and localization mode engineering in 1D periodic and quasi-periodic structures with graded-index materials and InSb. *Applied Optics*, *57*(28), 8171–8181. https://doi.org/10.1364/AO.57.008171.

Soltani, A., Noroozi, R., Bodaghi, M., Zolfagharian, A., & Hedayati, R. (2020). 3D printing on-water sports boards with bio-inspired core designs. *Polymers*, *12*(1). https://doi.org/10.3390/polym12010249.

Srikantha Phani, A., Woodhouse, J., & Fleck, N. A. (2006). Wave propagation in two-dimensional periodic lattices. *Journal of the Acoustical Society of America*, *119*(4), 1995–2005. https://doi.org/10.1121/1.2179748.

Su, J. W., Tao, X., Deng, H., Zhang, C., Jiang, S., Lin, Y., et al. (2018). 4D printing of a self-morphing polymer driven by a swellable guest medium. *Soft Matter*, *14*(5), 765–772. https://doi.org/10.1039/c7sm01796k.

Sun, W., Ma, Y., Huang, W., Zhang, W., & Qian, X. (2020). Effects of build direction on tensile and fatigue performance of selective laser melting Ti6Al4V titanium alloy. *International Journal of Fatigue*, *130*. https://doi.org/10.1016/j.ijfatigue.2019.105260, 105260.

Sydney Gladman, A., Matsumoto, E. A., Nuzzo, R. G., Mahadevan, L., & Lewis, J. A. (2016). Biomimetic 4D printing. *Nature Materials*, *15*(4), 413–418. https://doi.org/10.1038/nmat4544.

Tibbits, S. (2014). 4D printing: Multi-material shape change. *Architectural Design*, *84*(1), 116–121. https://doi.org/10.1002/ad.1710.

Tobushi, H., Hara, H., Yamada, E., & Hayashi, S. (1996). Thermomechanical properties in a thin film of shape memory polymer of polyurethane series. *Smart Materials and Structures, 2716*, 483–491. https://doi.org/10.1088/0964-1726/5/4/012.

Wang, Q., Tian, X., Huang, L., Li, D., Malakhov, A. V., & Polilov, A. N. (2018). Programmable morphing composites with embedded continuous fibers by 4D printing. *Materials and Design, 155*, 404–413. https://doi.org/10.1016/j.matdes.2018.06.027.

Wong, K. V., & Hernandez, A. (2012). A review of additive manufacturing. *International Scholarly Research Notices, 2012*, 208760.

Yamamura, S., & Iwase, E. (2021). Hybrid hinge structure with elastic hinge on self-folding of 4D printing using a fused deposition modeling 3D printer. *Materials & Design, 203*. https://doi.org/10.1016/j.matdes.2021.109605, 109605.

Yan, Q., Dong, H., Su, J., Han, J., Song, B., Wei, Q., et al. (2018). A review of 3D printing technology for medical applications. *Engineering, 4*(5), 729–742. https://doi.org/10.1016/j.eng.2018.07.021.

Zadpoor, A. A., & Malda, J. (2017). Additive manufacturing of biomaterials, tissues, and organs. *Annals of Biomedical Engineering, 45*(1). https://doi.org/10.1007/s10439-016-1719-y.

Zhang, Z., Demir, K. G., & Gu, G. X. (2019). Developments in 4D-printing: A review on current smart materials, technologies, and applications. *International Journal of Smart and Nano Materials, 10*(3), 205–224. https://doi.org/10.1080/19475411.2019.1591541.

Zolfagharian, A., Denk, M., Kouzani, A. Z., Bodaghi, M., Nahavandi, S., & Kaynak, A. (2020). Effects of topology optimization in multimaterial 3D bioprinting of soft actuators. *International Journal of Bioprinting, 6*(2), 1–11. https://doi.org/10.18063/ijb.v6i2.260.

Zolfagharian, A., Kaynak, A., Bodaghi, M., Kouzani, A. Z., Gharaie, S., & Nahavandi, S. (2020). Control-based 4D printing: Adaptive 4D-printed systems. *Applied Sciences, 10*(9), 3020. https://doi.org/10.3390/app10093020.

Zolfagharian, A., Durran, L., Gharaie, S., Rolfe, B., Kaynak, A., & Bodaghi, M. (2021). 4D printing soft robots guided by machine learning and finite element models. *Sensors and Actuators A: Physical, 328*. https://doi.org/10.1016/j.sna.2021.112774, 112774.

4D textiles: Materials, processes, and future applications

8

David Schmelzeisen, Hannah Kelbel, and Thomas Gries
Institut für Textiltechnik of RWTH Aachen University, Aachen, Germany

Introduction

Known since about 1925, the first four-dimensional (4D) structures—bimetal stripes—are programmable only to a limited extent (Timoshenko, 1925). Multimaterial printing allows nowadays to print complex 4D structures. Materials with different coefficients of material expansion upon swelling or thermal change are combined to program macroscopic deformations due to residual stress. Common multimaterial printing processes are time-consuming and costly. Limited print bed dimensions limit the overall dimensions of the structures. This makes the scalability of 4D printed structures difficult in terms of geometry, weight, and cost-effectiveness. Table 8.1 summarizes the potentials and challenges of 4D printing technology (Narula et al., 2018).

In 2017, Skylar Tibbits, together with Christophe Guberan, presented textile-based 4D structures for the first time under the name "Programmable Materials" (Guberan, Clopath, & Tibbits, 2017). Prof. Thomas Gries extended this concept to a hybrid, time-varying 4D material consisting of generatively produced, heterogeneous microlayers and textiles: 4D Textiles (Simonis, Schmelzeisen, Gries, & Gesché, 2018). Tension energy can be introduced into textiles by stretching an elastic textile. The tension energy can be transferred to the 4D textile system via the form and material closure and can multiply the effect of previous 4D structures in the hybrid 4D textile system. Small changes in shape as a result of external stimuli can lead to large structural changes. In addition, the geometry, weight, and economy of the overall system can be scaled by the targeted use of 4D structures in textile carrier systems. For this purpose, the textile carrier is printed with 4D structures. The geometry of the printed 4D structures is composed of primitive surface geometries (tessellation). Fig. 8.1 shows the working principle of the hybrid-structured 4D textiles (Simonis et al., 2018).

The 4D textile consists of the elements' surfaces and reinforcements. The printed reinforcements maintain the prestress and provide the shape change. The system takes the most energetically efficient state upon relaxation, and paraboloidal shapes are created. The forces are tangential to the paraboloidal shape of the surface (Koch, Schmelzeisen, & Gries, 2021).

The interfaces between the surface and the reinforcement are crucial for the force transfer of prestressed surfaces to the reinforcements. In the manufacturing process, the fusible reinforcement material is introduced into the surface pores of the textile,

Table 8.1 Potentials and challenges of 4D printing.

Potentials	Challenges
Design freedom	Low productivity
Customizable	Limited area and volume
Reduction of complexity	Scalable work capacity
Simplified logistics	Reversibility
Material saving/recycling	Choice of production method

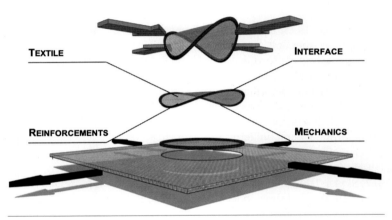

Fig. 8.1 Functional principle of the 4D textile hybrid structure.
From Schmelzeisen, D. (2020). *A design methodology for 4D textile.*

which must have a minimum size to be filled initially but not exceed a certain size since otherwise, adhesion will decrease. Once the reinforcing material hardens there, a positive bond is formed.

Depending on the mechanics of the composite, changes in shape occur. In 4D textile, the membrane is mechanically prestressed. The stiffness of the beams is activated by the prestressed, double-curved surface. The stiffness in 4D textile can be achieved by the prestressing, elastic stiffness, and construction of the reinforcements. Thus, the deformation of 4D textile is largely dependent on the prestress, its direction, and the design of the reinforcements. For the description of the mechanical behavior of 4D textile, continuum mechanics is suitable. For the continuum mechanics of 4D textile, the kinematic description is of great importance because large deformations occur. For the surfaces, hyperelasticity is important (Schmelzeisen, 2020).

Addressing existing issues of the production method in regard to prestressing of the textiles, a manufacturing support system using a magnetic tensioning frame is introduced. In addition to existing flat printed structures, this contribution also presents a novel circular printing method for 4D textiles. Furthermore, an adhesion model for the combination of textiles and polymer is presented. Building on material studies this contribution also presents developed applications to bring the technology closer to end-users and presents a method for other researchers and developers to create applications.

State of the art

Textile

Textiles are produced in processes that build on each other, from fiber to yarn and fabric production. In yarn formation, several short or continuous fibers are combined. Shape and mechanical properties can be adjusted via the choice of fibers and the yarn-forming process (Gries, Veit, & Wulfhorst, 2015). In surface formation, yarns are subsequently joined to form two or three dimensional objects. The type of joint can be adjusted locally in the process.

The technical properties of textiles can be adjusted on these successive scale levels from fiber to yarn, surface, semifinished product to finished product (Holme, 1999). Textiles are suitable for 4D hybrid materials because of their adjustable properties: elongation (structural and material elongation), porosity, and bending stiffness.

High direction-dependent strain values with almost complete recovery of the material make it possible to store and convert energy in the system in a targeted manner. The activation energy for the metastable states of the 4D textile structures can be set via prestress and elasticity. The porosity of the textile surface is crucial for the interface between textile and reinforcement. In textiles, porosity can be influenced by different parameters, such as fiber cross section, yarn structure, and type of textile surface.

High bending stiffness of textiles allows the reproduction of complex three-dimensional (3D) structures. The bending stiffness of textiles can be adjusted primarily via the textile structure. Elastane fibers have a high elongation of up to 800% on the material side. Elastane fibers are polyurethane elastomer fibers consisting of at least 85% by mass of segmented polyurethanes. Block copolymers with different properties alternate in the segmented polyurethanes. The crystalline, rigid portion of the polyurethane is joined by amorphous, elastic segments of a dihydroxyl compound (Gries, 2005). When the elastane is relieved after stretching, the high restoring forces in the material ensure an immediate and almost complete retraction to the initial length (Satlow, 2005). Therefore, this elasticity component is called material stretch (Gries, 2005; Satlow, 2005).

The porosity can be adjusted on fiber levels via the diameter, the cross sectional geometry, and the individual fiber length. Noncircular fibers can increase the surface area at the microlevel, while the fiber length has an influence on the porosity at the mesolevel.

Elasticity can be adjusted at the yarn level via the twist and structure of the yarns. For highly elastic products, elastane fibers are often combined with other fibers in the yarn formation process. Pure elastane fiber has poor clothing physiological properties due to its smooth surface and low water absorption.

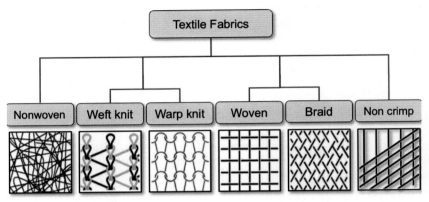

Fig. 8.2 Schematic classification of textile fabrics.
From Gries, T., Veit, D., & Wulfhorst, B. (2015). *Textile Technology–An Introduction.*

Fabric

Textile fabrics are divided into meshed, crossed, stretched thread systems or nonwovens (Gries, 2005; Grundbegriffe, 1969). Fig. 8.2 gives an overview of the textile fabrics.

In terms of process technology, textile surfaces can be produced very efficiently as continuous goods. In addition, textiles meet aesthetic design requirements. The combination of elastic continuous fibers with other fiber materials allows haptically appealing surfaces, such as underwear and swimwear (Gries, 2005; Schmelzeisen, 2020).

Meshed structures are considered for 4D textiles because they have structural elasticity and thus are suitable as substrate material for 4D textiles. In addition, the porosity of the surfaces is easily adjustable. Due to the production process, knitted fabrics have a structure with a high proportion of unbonded yarns. The nonbonded portions of the yarn can move well in relation to each other so that the structure has low bending stiffness. In addition, the knitted structure allows high structural elasticity. Knitted fabrics can be easily toughened and thus prepared for 3D printing (Schmelzeisen, 2020; Trümper, 2011).

Printing method

For 3D printing on textiles, material extrusion processes and specifically fused deposition modeling (FDM) are used (Sitotaw, Ahrendt, Kyosev, & Kabish, 2020). Grothe et al. used a Photon S 3D printer and thus depicted the first processes with the UV curing (digital light processing) method (Grothe, Brockhagen, & Storck, 2020). Due to its relevance, only FDM is further described.

FDM is the layer-by-layer application of plastics onto a building platform. In this process, one or more thermoplastics are melted in heated nozzles and then applied. The plastic cures at room temperature or at the adjustable print bed temperature. When

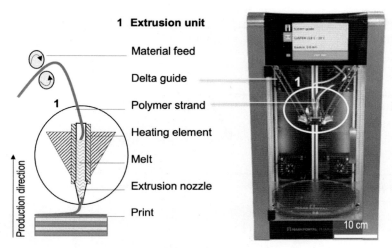

Fig. 8.3 Principle sketch (left) and photograph (right) of the Pharaoh XD20. No Permission Required.

a further layer is applied, the layer underneath is also melted so that a bond is formed. The extruder is moved in three-dimensional space with the help of a CNC unit. FDM printers differ in the number of extruders and the design of the CNC unit. Fig. 8.3 shows an FDM printer with a CNC unit in delta design (Gebhardt, Kessler, & Thurn, 2016).

Prestressing technologies

In addition to the already described plate with nails, a magnetic tensioning frame with an external pretensioning device is developed (Koch et al., 2021; Narula et al., 2018; Stapleton et al., 2019). A clamping device is specially designed and validated for uniform clamping. Fig. 8.4 gives an overview of the components of the newly designed clamping frame based on an extract of the morphological box (Schmelzeisen, 2020).

The tensioning device consists of a print bed that sits in an external tensioning unit. A magnetic ring is used to fix the prestressed textile. The clamping units are linked to the base plate by guide rods and threaded rods. The clamping units in x- and y-direction are perpendicular to each other. By means of independent threaded rods, it is possible to move the clamping units deviating from each other. Fig. 8.5 shows the overall structure of the clamping device (Schmelzeisen, 2020).

Clamping specimens using the external clamping device as shown in Fig. 8.6 are as follows:

1. Position pressure plate on base plate
2. Position textile on pressure plate and fix in clamping jaws
3. Stretch textile in x/y direction
4. Fix stretched textile with magnetic frame
5. Remove and print stretched textile with printing plate and clamping ring.

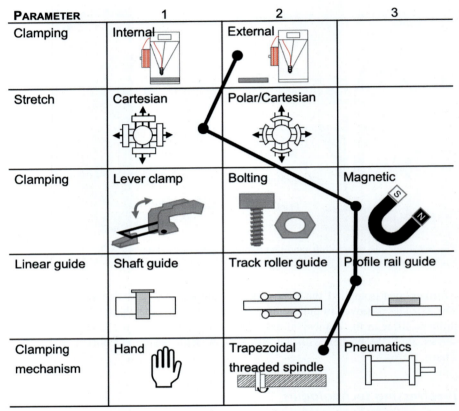

Fig. 8.4 Morphological box of the magnetic tensioning frame with external pretensioning device.
From Schmelzeisen, D. (2020). *A design methodology for 4D textile*.

With the external tensioning system, the regularity of the stress distribution in the edge areas of the clamping can be significantly improved. Fig. 8.7 shows the comparison between clamping the textile on the mandrel plate and using the magnetic clamping frame (Schmelzeisen, 2020).

Rotational symmetric substrate

For the production of rotational symmetric 4D textile applications, an FDM printing system is designed and validated. Fig. 8.8 gives an overview of the components of the newly designed printer for rotational symmetric parts based on an excerpt of the morphological box.

A Cartesian FDM printer concept is used for the printer setup. The Cartesian FDM printers of the Prusa product series (named after the inventor Josef Průša) are available at different suppliers as OpenHardware concepts starting at 160 €. The OpenHardware

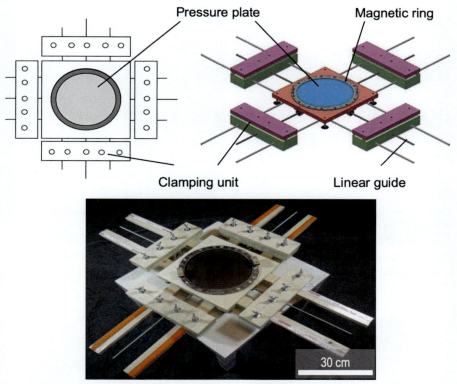

Fig. 8.5 The overall structure of the fixture: Principle sketch (up left), 3D representation (up right), and photograph (down middle).
From Schmelzeisen, D. (2020). *A design methodology for 4D textile*.

concept of the printers allows a comparatively simple change of the components. The motor necessary to move the printing table is connected via a toothed belt to a spring-mounted shaft made of aluminum. Shafts with different diameters can be used depending on the textile. Fig. 8.9 shows the overall design of the rotational symmetric printer.

The shaft is coated with an adhesion promoter so that the textile can be easily released after the printing process. The textile, a circular knit or knitted fabric, is drawn onto the shaft and thereby defined and uniformly pretensioned by stretching to the diameter of the shaft. For this purpose, the fastening of the motor must first be loosened and the drive belt removed. After the textile has been pulled onto the shaft, it is secured with cable ties against displacement by the extruder head. After that, the shaft must be fixed by means of the fastening screws. Printing can then be done in the FDM printing process according to the parameters listed in Table 8.2 (Schmelzeisen, 2020).

The general layer thickness of the print is set to 0.1 mm. The small layer thickness of 0.1 mm allows precise printing of reinforcing structures. The first layer when printing on textile determines the adhesion. The thickness of the first layer is

Fig. 8.6 The sequence of the clamping process using the external magnetic clamping frame. From Schmelzeisen, D. (2020). *A design methodology for 4D textile*.

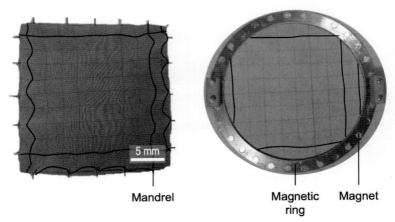

Fig. 8.7 Comparison of clamped knitwear on mandrel plate and magnetic tensioning frame. From Schmelzeisen, D. (2020). *A design methodology for 4D textile*.

doubled to 0.2 mm. Experience shows that this results in an improved form fit (Schmelzeisen, 2020).

Typically, fill structures are used in 3D printing for reasons of time and material savings. Fill structures often have similar strength values to a full fill. However, with the round printer, only a few layers are printed, so the savings can be neglected. Therefore, a density of 100% is used. The straight-line filling pattern is automatically set by the printer's software. An exemplary print with the dimensions 168 mm × 103.25 mm × 1 mm requires 1.19 m of filament and 80 min of printing time (Schmelzeisen, 2020).

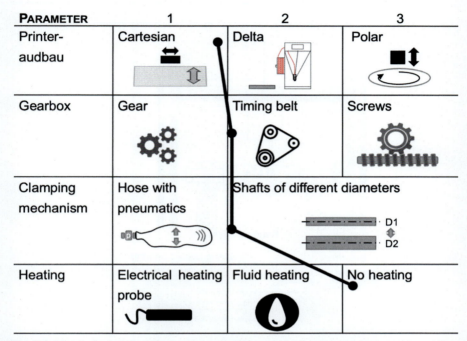

Fig. 8.8 Morphological box for the structure of the printer for rotational symmetric textiles. From Schmelzeisen, D. (2020). *A design methodology for 4D textile.*

Print parameters

The "Premium PLA" manufactured by the company FormFutura BV, Nijmegen, Netherlands, is used. By adjusting the viscosity via temperature changes, the penetration behavior of the polymer into the textile can be influenced. A processing temperature of 190 to 225°C is recommended by the manufacturer. Based on the results of Hashemi Sanatgar, Campagne, and Nierstrasz (2017) and the qualitative evaluation of our own tests, a nozzle temperature at the upper end of the manufacturer's specifications of 220°C is used. Table 8.3 summarizes the selected pressure settings. The temperature of the print bed is set to 40°C (Schmelzeisen, 2020).

The height of the base plane is called the offset. If the offset is set too high, the polymer melt will not bond with the textile. If the offset is too low, the print head is moved too close to the textile, resulting in textile damage. Up to now, suitable offset settings for material changes have been determined by time-consuming trial and error tests. In the future, laser triangulation methods will be used for the automated offset setting of the printer. The layer height influences both the print quality and the speed of the printing process. With low layer heights, higher quality parts can generally be printed. A layer thickness of 0.2 mm is used (Schmelzeisen, 2020).

The travel speed of the nozzle determines the application speed of the polymer. To achieve good adhesive interaction between the polymer and the textile, a high

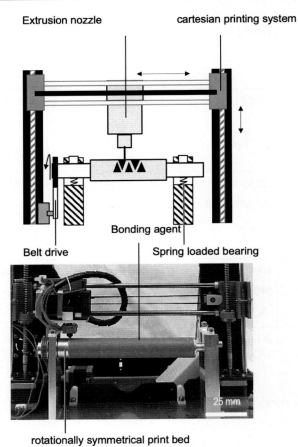

Fig. 8.9 The overall design of the rotational symmetrical printer. From Schmelzeisen, D. (2020). *A design methodology for 4D textile.*

Table 8.2 Pressure parameters for rotational symmetrical pressure (Schmelzeisen, 2020).

Parameter	Value	Unit
Offset	−0.15	[mm]
Nozzle diameter	0.40	[mm]
Minimum pressure head	0.07	[mm]
Layer thickness	0.10	[mm]
Layer thickness first layer	0.20	[mm]
Filling level	100.00	[%]
Print speed	30.00	[mm/s]

4D textiles: Materials, processes, and future applications

Table 8.3 Overview of the selected print settings (Schmelzeisen, 2020).

Parameter	Value
Nozzle temperature	220°C
Print bed temperature	40°C
Nozzle diameter	0.4 mm
Layer thickness	0.2 mm
Offset	Variable
Print speed	First part: 50 mm/s; second part: 65 mm/s
Infill	100%
Infill angle	In MSR

penetration depth of the polymer melt into the pores of the textile is required. To ensure that the highly viscous material is pressed deeper into the textile by the nozzle, the first layer is applied at a much slower traverse speed of 50 mm/s. The first layer is then applied at a speed of 0.5 mm/s. The second layer is applied at a speed of 0.5 mm/s. For the further printing process, a traverse speed of 65 mm/s is selected (Schmelzeisen, 2020).

The infill angle determines the direction of polymer application depending on the knitted fabric orientation. The application can take place in any direction. When selecting the infill angle, attention must be paid to the orientation of the fabric. The adhesion between reinforcement and surface is significantly dependent on the printing direction on knitted fabrics. Significantly higher forces can be absorbed in the mesh bar direction and then in the mesh row direction (Schmelzeisen, 2020).

The infill determines to what percentage the object is filled with polymer material. No infill settings are selected for 4D textile structures. The reinforcement structures are printed as solid material. For the experiments, the default value for nozzle diameter of 0.4 mm was used (Schmelzeisen, 2020).

Therefore, the textile is placed on the build plate and can be considered the new build surface (Koch et al., 2021).

Interfaces

The interface of surface and reinforcement is critical to the function of the 4D textile system. The interface between surface and reinforcement not only determines the static strength of the system but is also crucial for future studies on durability and reliability.

Different theories are known to explain adhesion. In general, a distinction is made between mechanical and specific adhesion. For the mechanical cohesion of the elements reinforcement and surface, both mechanical and specific adhesion processes play a role. Fig. 8.10 shows a schematic classification of adhesion theories (Brockmann et al., 2009).

In the specific adhesion subfield, adhesion is attributed to thermodynamic, physical, and chemical mechanisms (Brockmann et al., 2009). Mechanical adhesion is a

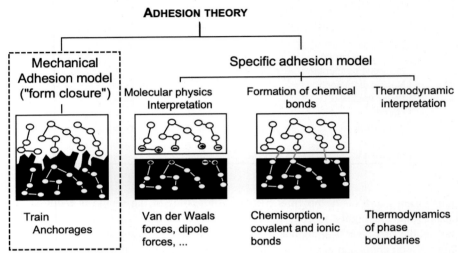

Fig. 8.10 Schematic classification of classical adhesion theories according to Brockmann, Geiß, Klingen, and Schröder (2009).
No Permission Required.

process in which the liquid adhesive is introduced into the surface pores of the solid. As soon as the adhesive has hardened there, a form-fit connection similar to a connection with screws or buttons is created. This is referred to as the "push-button principle." Fig. 8.11 shows the push-button principle (Bischof & Possart, 1982).

The pores in the adhesive surface must have a certain minimum size so that they can be filled with the initially liquid adhesive and a positive bond can be achieved.

Model

A comprehensive overview of experimental research on adhesion properties between 3D printed structures and fabric is given by Kozior, Blachowicz, and Ehrmann (2020). They identified three key parameters for the adhesion design of 3D printing on textiles: wetting, diffusion (material properties), and pressure (production property). An overview of mechanical testing methods for adhesion is given by Koch et al. (2021) including 3D printing on woven structures (Čuk, Bizjak, Muck, & Nuša Kočevar, 2020) and 3D printing on knitted structures (Narula et al., 2018).

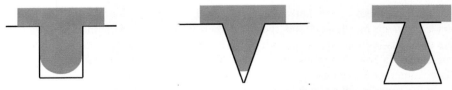

Fig. 8.11 Schematic representation of the "push-button principle" (Bischof & Possart, 1982).
No Permission Required.

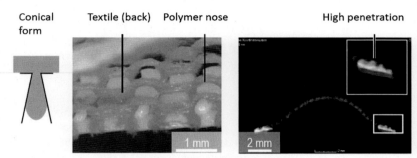

Fig. 8.12 Microscopy and CT image of the printed reinforcement.
From Schmelzeisen, D. (2020). *A design methodology for 4D textile.*

In the following, a model for FDM printing on textile fabrics as the impregnation of a porous surface is described. After printing, polymer anchors form on the back of the textile. Fig. 8.12 shows a microscopy and CT image of a printed reinforcement on a knitted fabric.

For the model, an ideal spherical anchorage is assumed. The peeling force Fs can be calculated as follows:

$$F_s = \left(\frac{E_A a h^3 b^4 F_Z^3}{4 d^6}\right)^{\frac{1}{4}} \quad (8.1)$$

where E_A: modulus of elasticity of the material A

a: distance of the center of the sphere from the interface
h: height of the material A
b: width of the material A
d: diameter of the balls
r: radius of the balls
σ_b: bonding strength

$$F_Z = 2\pi r \left(r - \sqrt{r^2 - a^2}\right) \sigma_b \quad (8.2)$$

Qualitatively, the peeling force F_s thus increases with increasing distance of the spherical center from the interface and decreasing diameter of the spherical anchorage.

The full factorial experimental design shows that knitting pattern and elastane addition have a direct influence on the peeling force. Knits with cross miss patterning and knits with elastane addition lead to significantly higher peel forces. Indirectly, the correlations are due to the pore geometry. Knitted fabrics with small, deep pores that allow melt penetration to improve adhesion. The observations correlate with the model of the "push-button principle" according to Bischof and Possart (1982) (Fig. 8.13) and with Formula (8.1) and (8.2).

The interface between textile and printed reinforcements can be considered with the help of the mechanical adhesion mechanism, form closure. The experimental results show that the adapted 180-degree peeling method is suitable for testing the

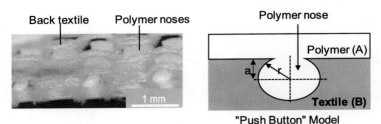

Fig. 8.13 Microscopy image and model description (Bischof & Possart, 1982). From Schmelzeisen, D. (2020). *A design methodology for 4D textile.*

adhesion properties of a 4D textile. When designing 4D structures, the orientation of the knitted fabrics must be taken into account. Significantly lower peel strengths are to be expected in the mesh row direction than in the mesh bar direction.

Form giving through surface tessellation

Repeating patterns are present in many man-made objects of daily life, such as textile production methods and designs on bolts of cloth, brickwork, or house tiles. They can also be seen in nature, in the patterns left behind by cracked mud.

The most basic tessellations are regular and semiregular tessellations. They both use regular polygons. There are exactly three regular tessellations, using the equilateral triangle, the square, and the regular hexagon only. There are exactly eight semiregular tessellations, using regular polygons ranging from the triangle to the dodecahedron.

For self assembly properties, the principle of body meshes needs to be considered. Body meshes are patterns that are created when a body is unfolded into a flat surface. A more modern extension of parqueting is spherical parqueting. Using a rotation pattern derived from the Platonic and Archimedean solids, curved surfaces can also be parquetized.

The research investigates tessellation in combination with membranes, where tessellation is used for repeatable shaping. Koch et al. show that different strategies are followed: Voronoi, auxetics, concentric rings or radial libs and geometric tessellations based on rectangles, triangles, or hexagons (Agkathidis, Berdos, & Brown, 2019; Grimmelsmann, Meissner, & Ehrmann, 2016; Koch et al., 2021; Kycia, 2019; Schmelzeisen, 2020).

Compliant mechanisms are flexible mechanisms that transmit motion or force via elastic deformation and in the case of 4D textiles appear in combination with membranes. The behavior depends on the design of the joints, which are applied between two beams on textiles and can be either textile, textile, and polymer or two different polymers (Koch et al., 2021).

Based on the classification of the structure made for 4D printing, according to Choi, Kwon, Jo, Lee, and Moon (2015), the evaluation of the structure of 4D textiles can be made. In Fig. 8.14, possible arrangement is shown. A change, its direction, and shape can be predicted and controlled by the arrangement of beams on the pretensioned textile.

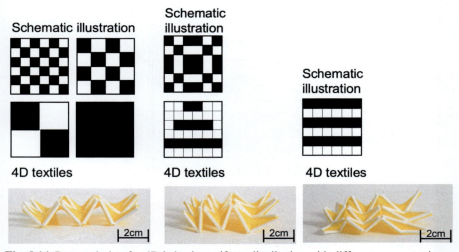

Fig. 8.14 Pattern design for 4D behavior uniform distribution with different concentrations (left), gradual distribution (middle), and special pattern (right).
From Koch, H. C., Schmelzeisen, D., & Gries, T. (2021). 4D textiles made by additive manufacturing on pre-stressed textiles—An overview. *Actuators*, *10*(2). https://doi.org/10.3390/act10020031.

Applications

Finger

The aim of the finger is to be able to automatically grasp objects with different geometries and dimensions. A soft robot gripper is modeled on the human finger. As a result, bionic motivated structures (honeycombs) are printed on an elastic knitted fabric on the printer for rotational symmetric surfaces. The actuator consists of two counterrotating shape memory metals. Fig. 8.15 summarizes the essential components based on the systematics developed. 1190 mm PLA-TPE composition, 200 mm × 110 mm polyamide-lycra composition, and 400-mm nitinol are used. For the production time, only the production steps of manufacturing the textile, parameter settings on the rotary printer, printing phase, manufacturing, and installation of the actuators are considered.

Today's gripping systems of industrial robots do not meet flexibility requirements. The derived goal is the development of an adaptive end effector based on the human index finger. The structure consists of three links and the middle and end joints. Fig. 8.16 sketches the bionic derivation of the structure on the human finger. This is suitable for gripping objects of different geometries. In addition, the gripper has a low weight and a high degree of customization, and it could be possible to manufacture it at a low cost (Schmelzeisen, 2020).

Strengths include the advantages of generative manufacturing processes in combination with the properties of textile fabrics. Conventional manufacturing processes limit the design of products, while 3D printing promotes design freedom to

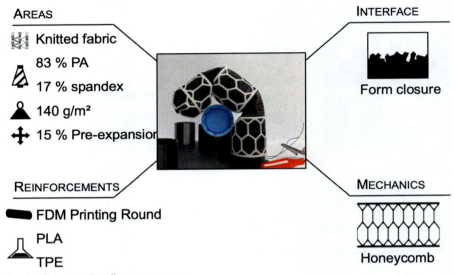

Fig. 8.15 4D textiles finger summary.
From Schmelzeisen, D. (2020). *A design methodology for 4D textile*.

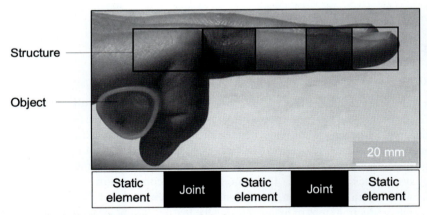

Fig. 8.16 Bionic derivation of the structural design.
From Schmelzeisen, D. (2020). *A design methodology for 4D textile*.

approximate humanoid actuators. Rapid prototyping enables timely response to specific user requirements and product demands (Schmelzeisen, 2020).

Weaknesses are the poor reproducibility and simulation of the structures. The round printer is currently unable to produce reproducible results. In addition, there is no validated test procedure for monitoring the serviceability of the actuator. Design decisions can only be roughly estimated and evaluated on the produced prototype (Schmelzeisen, 2020).

In the second step, two analogies are derived for the static elements: static elements are constructed as honeycombs with high stability. External forces are transferred to

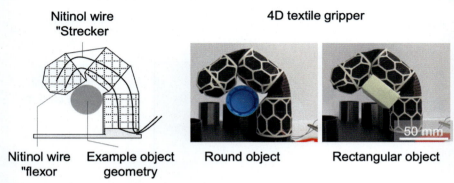

Fig. 8.17 Gripping behavior of soft robot gripper.
From Schmelzeisen, D. (2020). *A design methodology for 4D textile.*

the entire honeycomb structure via the honeycomb walls, and stress peaks are avoided (Bitzer, 1997).

The actuator converts the gripping movement of the soft robot gripper. The shape memory alloy nitinol with a diameter of 0.5 mm is selected as the actuator. Two actuators, flexor and extensor, are fabricated and controlled by a voltage source. Fig. 8.17 shows the installation of the actuators. According to empirical values, the best thermal shape memory effects are achieved at a heat treatment temperature of 410°C and a heat treatment duration of 9 min (Schmelzeisen, 2020).

The nitinol wires are placed at the position of the flexor and stretcher and connected to the voltage source. At 5.5 V and 1.5 A, the nitinol wires heat above the critical temperature, and the gripping of the prototype concept is initiated via the flexor. The gripper is opened by activating the extender when the flexor is deactivated (Schmelzeisen, 2020).

In preliminary tests, circular knitted fabrics were tested against sewn, bonded, and welded knitted fabrics for the production of the textile hose. The best results in printing are achieved with knitted fabrics that are ultrasonically welded using the cut and seal process. The seam does not wear and can be made repeatably with an ultrasonic welding machine. The textile tube with a diameter of 28 mm is pretensioned on the axle and secured with plastic rings to prevent lateral slippage. The exemplary gripping of a rotationally symmetric object which diameter is 25 mm and a rectangular object which edge lengths are 30 mm × 10 mm × 50 mm is shown in Fig. 8.17 (Schmelzeisen, 2020). The soft robot gripper has a total weight of 13 g and allows gripping objects of different geometries (Schmelzeisen, 2020).

Orthosis

The aim of the orthosis is to combine the compression of bandages with the targeted immobilization of orthoses. The approach is to realize the compression with the help of the circumferential elasticity of a polyester circular knitted fabric. Stabilization is achieved with the aid of printed structures in FDM printing. PLA is used as the printing material. The joint is modeled on a mechanical spring element. The geometry of

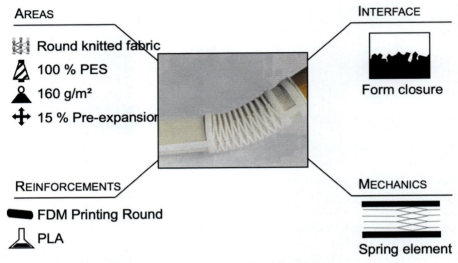

Fig. 8.18 Summary of the application orthosis in 4D systematics.
From Schmelzeisen, D. (2020). *A design methodology for 4D textile*.

the spring element can be designed individually for each patient according to the requirements. Fig. 8.18 summarizes the essential components on the basis of the systematics developed (Schmelzeisen, 2020). A piece of circular knitted polyester fabric with a diameter of 22 mm is used as the surface. The shaft diameter of 30 mm requires a preload of approx. 50%. PLA is used as a printing material.

Most of the benefits of custom orthoses are based on fit alone. Due to the high variance within the anthropometry, ready-made orthoses and supports are always only a "compromise solution" with the goal of being able to apply their effect to a maximum number of their users, and the assurance that this cannot be achieved for the entire population (Schmelzeisen, 2020).

Previous additively manufactured orthoses predominantly make use of the active principles: immobilization and stabilization. The advantages of the existing additively manufactured orthoses lie, among other things, in the variety of design options. In addition, a technical solution to combine the active principle of compression by textile support and the mentioned advantages of additive manufacturing of orthotic components is desirable for increasing the rehabilitation outcome. The structure of the orthosis is modeled on the structure of a ready-made orthosis. Two rigid elements stabilize the upper and lower arm, while a flexible element allows flexion (Fig. 8.19). Two prototypes with different tessellations are produced (Schmelzeisen, 2020).

In the first concept, scales are arranged to each other so that diffraction is allowed in the flexible zone, but rotation is prevented. The rigid elements are made from diamond-shaped geometries. In a second concept, a disk spring is simulated. The diamond-shaped reinforcement allows buckling of the structure but no rotation in the joint. The rigid elements are designed to be beam-shaped (Schmelzeisen, 2020).

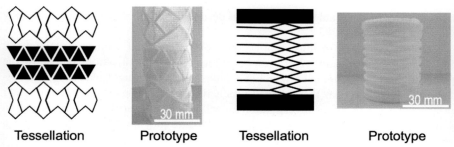

Fig. 8.19 Tessellation in CAD and on realized prototypes.
From Schmelzeisen, D. (2020). *A design methodology for 4D textile.*

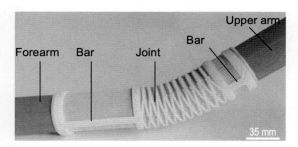

Fig. 8.20 Prototype of the overall concept of the orthosis made with circular 3D printing on textiles.
From Schmelzeisen, D. (2020). *A design methodology for 4D textile.*

The concept of the disk spring allows the angle of diffraction to be precisely adjusted via the geometry of the beams. Two bars hold the orthosis rigidly on the upper and forearm. The orientation of the disk spring can be adjusted so that only a defined movement of the joint is permitted. In this way, the desired freedom of movement can be adjusted by the doctor or orthopedist to suit the individual patient. The geometry can also be easily scaled to different overall dimensions. The reversibility of the system is achieved via the circumferential elasticity of the circular bend. Fig. 8.20 shows the overall concept of orthosis (Schmelzeisen, 2020).

The reproducibility of the prototypes produced is high. However, there is a risk of pinching the skin between the spring elements (Schmelzeisen, 2020).

Conclusion and outcomes

It could be shown that 4D textiles contribute to the development of 4D-printed structures by broadening the material usage and therefore especially the energy that can be stored in textiles. To significantly store, the energy in the textile production methods needs to be further developed of which two solutions have been described in this contribution, namely circular 3D printing on textiles and a magnetic prestressing unit. Since 4D textiles are multimaterial structures, the adhesion between the printed polymer and the textile is of major interest. A model based on the research on "push button" from Bischof and Possart (1982) is described and validated using microscopy (Bischof & Possart, 1982).

4D textiles have various application fields that need further development. Prototyping of applications has been performed for four applications that show the potential of the structure but production readiness is still a long way to go. Nevertheless, it could be shown that adapted production technologies such as the circular printer further open up the investigation as well as the application space. Tessellation strategies and modules that are adapted from nature play a major role in designing active structures with 4D textiles.

References

Agkathidis, A., Berdos, Y., & Brown, A. (2019). Active membranes: 3D printing of elastic fibre patterns on pre-stretched textiles. *International Journal of Architectural Computing, 17*(1), 74–87. https://doi.org/10.1177/1478077118800890.

Bischof, C., & Possart, W. (1982). *Adhäsion: Theoretische und experimentelle Grundlagen.* Akademie-Verlag.

Bitzer, T. (1997). *Honeycomb technology materials, design, manufacturing, applications and testing.* https://doi.org/10.1007/978-94-011-5856-5.

Brockmann, W., Geiß, P. L., Klingen, J., & Schröder, B. (2009). Adhesive bonding: Materials, applications and technology. In *Adhesive bonding: Materials, applications and technology* (pp. 1–414). John Wiley and Sons. https://doi.org/10.1002/9783527623921.

Choi, J., Kwon, O. C., Jo, W., Lee, H. J., & Moon, M. W. (2015). 4D printing technology: A review. *3D Printing and Additive Manufacturing, 2*(4), 159–167. https://doi.org/10.1089/3dp.2015.0039.

Čuk, M., Bizjak, M., Muck, D., & Nuša Kočevar, T. (2020). *3D printing and functionalization of textiles* (pp. 499–506). https://doi.org/10.24867/grid-2020-p56.

Gebhardt, A., Kessler, J., & Thurn, L. (2016). *Grundlagen und Anwendungen des Additive Manufacturing (AM).* 2. Aufl Münchner: Hanser.

Gries, T. (2005). *Elastische Textilien: Garne, Verarbeitung, Anwendung.* Frankfurt: Melliand Edition Textiltechnik.

Gries, T., Veit, D., & Wulfhorst, B. (2015). *Textile technology—An introduction.* Carl Hanser Verlag GmbH & Co. KG.

Grimmelsmann, N., Meissner, H., & Ehrmann, A. (2016). 3D printed auxetic forms on knitted fabrics for adjustable permeability and mechanical properties. In *Vol. 137, Issue 1. IOP Conference Series: Materials Science and Engineering* Institute of Physics Publishing. https://doi.org/10.1088/1757-899X/137/1/012011.

Grothe, T., Brockhagen, B., & Storck, J. L. (2020). Three-dimensional printing resin on different textile substrates using stereolithography: A proof of concept. *Journal of Engineered Fibers and Fabrics, 15.* https://doi.org/10.1177/1558925020933440.

Grundbegriffe. (1969). *DIN69/DIN6000 Textilien.*

Guberan, C., Clopath, C., & Tibbits, S. (2017). *Active shoes.* http://www.selfassemblylab.net/ActiveShoes.php.

Hashemi Sanatgar, R., Campagne, C., & Nierstrasz, V. (2017). Investigation of the adhesion properties of direct 3D printing of polymers and nanocomposites on textiles: Effect of FDM printing process parameters. *Applied Surface Science, 403*, 551–563. https://doi.org/10.1016/j.apsusc.2017.01.112.

Holme, I. (1999). Adhesion to textile fibres and fabrics. *International Journal of Adhesion and Adhesives*, *19*(6), 455–463. https://doi.org/10.1016/S0143-7496(99)00025-1.

Koch, H. C., Schmelzeisen, D., & Gries, T. (2021). 4D textiles made by additive manufacturing on pre-stressed textiles—An overview. *Actuators*, *10*(2). https://doi.org/10.3390/act10020031.

Kozior, T., Blachowicz, T., & Ehrmann, A. (2020). Adhesion of three-dimensional printing on textile fabrics: Inspiration from and for other research areas. *Journal of Engineered Fibers and Fabrics*, *15*. https://doi.org/10.1177/1558925020910875.

Kycia, A. (2019). Form finding of performative surfaces through 3D printing on prestressed textiles. In *CA2RE + trondtheim proceedings* (pp. 1–13).

Narula, A., Pastore, C. M., Schmelzeisen, D., El Basri, S., Schenk, J., & Shajoo, S. (2018). Effect of knit and print parameters on peel strength of hybrid 3-D printed textiles. *Journal of Textiles and Fibrous Materials*. https://doi.org/10.1177/2515221117749251, 251522111774925.

Satlow, G. (2005). Herstellung von Elastanfasern. In T. Gries (Ed.), *Elastische Textilien* Melliand Edition Textiltechnik.

Schmelzeisen, D. (2020). *A design methodology for 4D textile*. Shaker Verlag.

Simonis, K., Schmelzeisen, D., Gries, T., & Gesché, V. (2018). *Vierdimensionales Textilmaterial Deutsche Offenlegungsschrift DE 10*.

Sitotaw, D. B., Ahrendt, D., Kyosev, Y., & Kabish, A. K. (2020). Additive manufacturing and textiles-state-of-the-art. *Applied Sciences*, *10*(15). https://doi.org/10.3390/app10155033.

Stapleton, S. E., Kaufmann, D., Krieger, H., Schenk, J., Gries, T., & Schmelzeisen, D. (2019). Finite element modeling to predict the steady-state structural behavior of 4D textiles. *Textile Research Journal*, *89*(17), 3484–3498. https://doi.org/10.1177/0040517518811948.

Timoshenko, S. (1925). Analysis of bi-metal thermostats. *Journal of the Optical Society of America*, *233*. https://doi.org/10.1364/JOSA.11.000233.

Trümper, W. (2011). *Gestrickte Halbzeuge und Stricktechniken* (pp. 225–263). Springer Science and Business Media LLC. https://doi.org/10.1007/978-3-642-17992-1_6.

Closed-loop control of 4D-printed hydrogel soft robots

Ali Zolfagharian[a], Mahdi Bodaghi[b], Pejman Heidarian[a], Abbas Z Kouzani[a], and Akif Kaynak[a]
[a]School of Engineering, Deakin University, Geelong, VIC, Australia
[b]Department of Engineering, School of Science and Technology, Nottingham Trent University, Nottingham, United Kingdom

Introduction

Four-dimensional (4D)-printed soft robots are compliant polymeric structures that are assembled to function like living organisms and precisely controlled to perform specialized delicate tasks that cannot be performed by conventional robotics (Bodaghi, Damanpack, & Liao, 2016; Zolfagharian, Denk, et al., 2020; Zolfagharian, Kaynak, et al., 2020; Zolfagharian et al., 2016). Through access to 3D printing and smart materials, scientists can come up with creative designs by incorporating specific functions to develop biodegradable actuators and renewable biocompatible soft robots for accomplishing designated tasks. However, cost and fabrication flexibility are the primary factors to consider before pursuing the development of these designs. Recent developments in biodegradable soft robots driven by microbial fuel cells have led to further potential for developing autonomous biodegradable 4D-printed soft robots working with low electrical voltage (Rossiter, Winfield, & Ieropoulos, 2016; Winsberg, Hagemann, Janoschka, Hager, & Schubert, 2016). Also, edible actuators, made from biodegradable conductive hydrogels, have recently drawn attention due to potential applications in less invasive interactions with internal organs (Keller, Pham, Warren, & In het Panhuis, 2017). Furthermore, an application of such soft robotic systems has been demonstrated in controllable cancer drug release under a stimulus voltage of less than 20 V in a saline solution (Aybala & Ramazan, 2016). These studies suggest that electrically driven contactless soft actuators are evolving toward polyelectrolyte hydrogels with a low-voltage actuation capability.

Hydrogels being natural polymers have biocompatibility and biodegradability which are desirable properties for soft actuators (Rossiter et al., 2016). Biodegradable natural hydrogels are suitable materials for developing sustainable and renewable soft actuators. Conversely, the biocompatibility in compliance with some synthesized hydrogels limits their applications in the biomedical and pharmaceutical sectors (Zhao, Sun, Wang, Xu, & Muhammad, 2015). Chitosan hydrogel is a widely used biocompatible and biodegradable polyelectrolyte hydrogel with highly polar groups that can be cross-linked into a network of polymer chains to enhance the performance of actuators made from it (Altinkaya et al., 2016). The chitosan hydrogel exhibits controlled movement by an electrical stimulus that could be utilized in the production of soft actuators in a basic medium through monitoring the strength and duration of the input current (Zhao et al., 2015). These hydrogel actuators are often referred to as "throw-away" robots due to their biodegradable nature (Rossiter et al., 2016). Chitosan, a natural amphoteric hydrogel, has been used in drug delivery and tissue engineering (Altinkaya et al., 2016). A chitosan-based electroactive polymer actuator has exhibited increased ionic conductivity and mechanical performance compared to biopolymer-based actuators (Jeon, Cheedarala, Kee, & Oh, 2013).

The fabrication of soft actuators with intricate geometrical designs particularly in miniature scales faces traditional manufacturing constraints. However, this can be circumvented by 4D printing through the production of complex designs imported from a computer-aided design (CAD) model, which significantly reduced fabrication time, postprocessing effort, and material wastage. Also, customized geometry, specific functions, and control properties (Ali et al., 2017; Campbell, Tibbits, & Garrett, 2014) can be directly incorporated into the 4D printing of the morphing structure without needing to print each part individually and subsequently assemble them in a process involving multiple steps (Bodaghi et al., 2020; Cvetkovic et al., 2014; Mao et al., 2016; Ozbolat & Hospodiuk, 2016; Sydney Gladman, Matsumoto, Nuzzo, Mahadevan, & Lewis, 2016; Tibbits & Cheung, 2012). The 3D printing requires rapid solidification of the extrudate for reproducible prints with well-defined, intricate structures. Extensive intermolecular attractions in highly polar molecules of polyelectrolytes such as poly(acrylic acid) limit their processability since they are not ideally suited for extrusion-based fabrication of 4D printing. Chitosan, on the contrary, is more suitable for printing due to its favorable rheological properties and better mechanical properties in both dry and hydrated states.

In this study, 4D printing of polyelectrolyte soft actuators and their comparison with cast counterparts in terms of actuation performance are presented. The methodology of optimization of the 3D printing parameters is described. In addition, modeling, system identification, and closed-loop control of 4D-printed polyelectrolyte soft robot actuators are explained.

Motion mechanism of the soft actuator

In the presence of an electric field, polyelectrolytes swell and show an electro-osmotic response to electrolyte salt concentration due to the interaction of their polar groups with the electrolyte ions (Zolfagharian, Kaynak, Khoo, & Kouzani, 2018; Zolfagharian, Kouzani, Khoo, Noshadi, & Kaynak, 2018). The electrical potential applied across the electrolytic cell causes the asymmetric distribution of the mobile ions between the electrolyte and the hydrogel resulting in osmotic pressure differences that lead to the deformation of the gel (Zolfagharian, Kouzani, Khoo, Nasri-Nasrabadi, & Kaynak, 2017). Chitosan is a linear polysaccharide containing highly polar hydroxyl and amine groups along its chains. Alkaline solutions can induce negative polarity to the chains through deprotonation. An applied potential to the electrochemical cell causes movement of anions and cations (Na+ and OH-) toward the electrodes of opposite charges. The negatively charged impermeable anions in the chitosan film attract positive ions, Na⁺, while repelling negative ions, OH⁻, from the boundary, thus resulting in the increase of the hydroxide ion concentration on the side of the film facing the anode. This manifests as an ionic concentration gradient across the membrane and causes an osmotic pressure increase. The film then swells and bends toward the negative electrode while the number of counterions on either side of the gel-solution interface remains balanced, as per Gibbs-Donnan equilibrium, or the so-called Donnan equilibrium, to preserve electroneutrality within the system. Hence, the concentration gradients for each ion are balanced and there is no net flow across the membrane at equilibrium (Shang, Shao, & Chen, 2008).

The osmotic pressure (π) caused by ionic distribution in solutions is calculated by Flory's theory of polyelectrolytes as follows (Flory, 1953; Li et al., 2016; Shiga & Kurauchi, 1990):

$$\Pi_{ion} = RT \left(\sum_i C_i^{in} - \sum_i C_i^{out} \right) \quad (9.1)$$

where R is the gas constant, T is the absolute temperature, C_i^{in} is the concentration of mobile ion i in the gel, and C_i^{out} is the concentration of mobile ion i in the solution. The ionic concentration gradient across the thickness of the hydrogel causes an osmotic pressure difference $\Delta\pi$, generating the driving force for the bending deformation:

$$\Delta \Pi_{ion} = \pi_1 - \pi_2 \quad (9.2)$$

where π_1 and π_2 are the osmotic pressure differences between the hydrogel and the solution on the anode and cathode sides (Fig. 9.1). In the case of polyanionic hydrogel, $\Delta\pi > 0$, and the hydrogel swells on the anode side, in other words, bend toward the cathode electrode. This is different for the polycationic hydrogel where the osmotic pressure difference is negative, $\Delta\pi < 0$, the hydrogel bends toward the anode electrode.

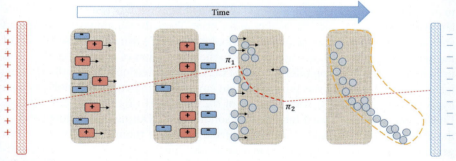

Fig. 9.1 Illustration of the bending motion in polyelectrolyte actuators; (A) counterion migration, (B) osmotic pressure on the anode side becomes greater than the cathode side, (C) more diffusion on the anode side, and (D) hydrogel bending toward the cathode. No permission required

In a system with i kinds of cations and j kinds of anions, $\Delta\pi$ is expressed as follows (Shiga & Kurauchi, 1990):

$$\Delta\pi = 2RT \sum_i \left[C_{i,c} + C_{i,B}\frac{V_B}{V_C}h_i t + C_{i,B}\frac{V_B}{V_C}h_i^2 t^2 - C_{i,A}(1 - h_i t) \right]$$
$$- 2RT \sum_j \left[C_{j,c} + C_{j,B}\frac{V_B}{V_C}h_j t + C_{j,B}\frac{V_B}{V_C}h_j^2 t^2 - C_{j,A}(1 - h_j t) \right] \qquad (9.3)$$

where R is the ion concentration of species i in the section x of the polyelectrolyte, $x = A, B, C$ defined as anode side electrolyte, gel, and cathode side electrolyte, respectively, and at time $t = 0$, h_i is the transport rate of species i between parts. The actuator deformation can be represented as a three-point flexural bending with the maximum bending stress, σ, related to the osmotic pressure differential, which can be determined as follows:

$$\Delta\pi = \frac{6DEY}{L^2} \qquad (9.4)$$

where L is the initial length, D is the thickness, E is the modulus of elasticity, and Y is the deflection of the actuator. From a chemical point of view, the bending could be determined by Eq. (9.4) knowing Eq. (9.3).

Materials and methods

Fabrication of the actuator

1.6 g of medium molecular weight chitosan with $75 - 85\%$deacetylation degree (Sigma-Aldrich, Australia) was added into 0.8 ml acetic acid (1 $v/v\%$) and stirred at $50\,°C$, for 2 h. A Petri dish was used to dry the solution in an oven at 50°C for 48 h to produce a gellified ink. The polyelectrolyte hydrogel ink was sonicated to remove air bubbles that can reduce the quality of the print. Then, the solution was centrifuged and poured into the low-temperature 3D-Bioplotter *30-CC* syringe. The CAD model of the actuator which had been drawn prior to the preparation of the hydrogel ink was imported into the 3D-Bioplotter EnvisionTEC GmbH (EnvisionTEC, Gladbeck, Germany). The CAD model was printed without any support materials. Ethanolic sodium hydroxide (EtOH-NaOH) with a recipe of $0.25\,M$ NaOH (Sigma-Aldrich), $70\,v/v\%$ EtOH ratio of $3\!:\!7$, was used for postprint solidification. During the printing process, the EtOH-NaOH is sprinkled on the printed chitosan at room temperature for fast gelation. A vernier caliper was used to measure the dimensions of the 3D-printed scaffolds. The average of three different readings of the length, width, and height of the dried samples was measured (K.D. et al., 1999). The chitosan samples were cut with precise dimensions of length, width, and thickness of $24\,mm \times 4\,mm \times 1\,m$. The procedure for the 3D printing of the chitosan is illustrated as a flowchart in Fig. 9.2.

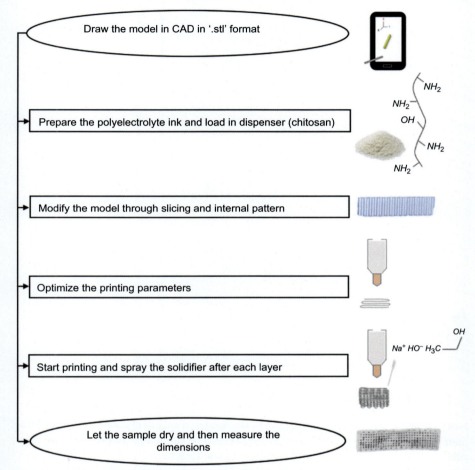

Fig. 9.2 The detailed procedure of 4D-printing the polyelectrolyte soft robot actuator. No permission required

Optimizing the printing parameters

The printing efficiency was optimized by an analysis of printing parameters with respect to the shape fidelity of the polyelectrolyte hydrogels. Print quality was achieved by carefully monitoring the accuracy of the width of the printed strands. A pneumatic system dispenses the ink in 3D-Bioplotter. The input pressure induces shear stress (τ) in the hydrogel in the nozzle, which could be expressed along the *y-axis* in terms of the force equilibrium on the fluid element as shown in Fig. 9.3 as follows:

$$\tau(\pi DL) = (P + \Delta P)\left(\pi D^2/4\right) - P\left(\pi D^2/4\right) \tag{9.5}$$

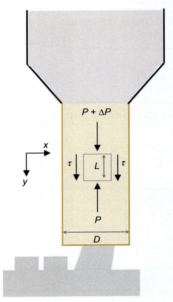

Fig. 9.3 Forces on an element in a nozzle.
No permission required

$$\tau = \frac{D \Delta P}{4L} \tag{9.6}$$

where D and L are diameter and length of the needle, respectively, and ΔP is the extrusion pressure at the nozzle. The shear stress could be calculated as a function of shear rate ($\dot{\gamma}$) knowing the viscosity of hydrogel (μ) as follows:

$$\tau = \mu \dot{\gamma} \tag{9.7}$$

Then, the volumetric flow rate ($\dot{q}$) in a shear-thinning fluid could be expressed as follows:

$$\dot{q} = K \frac{\dot{\gamma} \pi D^3}{32} \tag{9.8}$$

where K is the power law index coefficient that is assumed to be a constant value for a laminar flow. Using a cylindrical nozzle, the hydrogel strand diameter (d) could be assumed constant after extrusion to relate the shear rate to the velocity of nozzle (V) in the printing direction, x, as follows:

$$\dot{q} = V\left(\frac{\pi d^2}{4}\right) \tag{9.9}$$

Combining Eqs. (9.8) and (9.9), the shear rate could be obtained as follows:

$$\dot{\gamma} = K \frac{8V\,d^2}{D^3} \tag{9.10}$$

Using Eqs. (9.6), (9.7), and (9.10) and rewriting it in terms of the diameter of hydrogel strand, d, yield:

$$d = D^2 \sqrt{\frac{K\Delta P}{32V\mu L}} \propto \sqrt{\Delta P} \propto \frac{1}{\sqrt{V}} \tag{9.11}$$

The optimized design was developed for printing at a temperature of 20°C using a constant needle diameter. Only two key factors, the extrusion pressure (ΔP) and the nozzle speed (V), were investigated to control the quality of the extrudate according to Eq. (9.11). It is evident from Eq. (9.11) that the width of the printed strand is directly and inversely proportional to the square roots of the extrusion pressure and the speed of the nozzle, respectively. Experiments were conducted to study the relations in Eq. (9.11), in the range of 0.4 to 0.8 bar and 20 to 30 mm/s, in order to determine the desired hydrogel strand line size (closest to needle diameter) with the lowest extruder pressure and nozzle velocity.

Results and discussions

Optimization of the 3D printing parameters

Table 9.1 shows the width of the chitosan hydrogel strands, at different extrusion pressures and printing speeds, with the same needle size (0.250 mm). Based on the findings demonstrated in Fig. 9.4 and Table 9.1, extrusion pressure and nozzle velocity have an influence on the width of the printed strands; therefore, both variables must be adjusted to dispense the strand continuously. For instance, the diameter of the

Table 9.1 Strand width results from 3D-Bioplotter.

Sample #	Strand width (mm)	Extrusion pressure/ΔP (bar)	Nozzle velocity (mm/s)
1	0.420	0.8	25
2	0.400	0.6	30
3	0.450	0.8	20
4	Disjointed	0.4	20
5	Disjointed	0.4	25
6	Disjointed	0.4	30
7	0.340	0.6	20
8	0.320	0.8	30
9	0.290	0.6	25

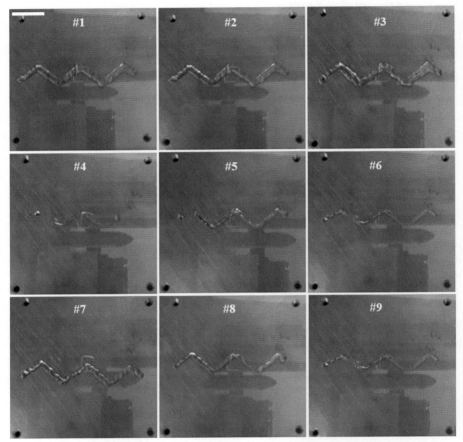

Fig. 9.4 Images of the printed chitosan strands during the processing parameters optimization. The length of scale bars is 2 mm.
No permission required

hydrogel strand becomes much greater than the needle size when there is excessive input pressure, and the hydrogels printed at higher nozzle speeds had more defects compared to those printed at lower speeds. Also, it was observed that at extrusion pressure and nozzle speed of *0.6 bar* and *25 mm/s*, respectively, the optimum width of the printed strand was *0.290 mm*.

Lattice-patterned chitosan beams of size $40 \times 10 \times 5\ mm^3$ were printed after determining the optimal 3D-Bioplotter parameters. Several printing experiments were conducted to determine the optimized values of 3D-Bioplotter settings (Table 9.2). The extrudates were consistent and predictable when the optimized settings were used. Using a set of optimized parameters from Table 9.2, parallel strands were printed on the first layer. The printing direction was then turned *90°* for the next layer, making a well-defined mesh with a mathematically defined porosity. The images of printed

Table 9.2 The optimized processing parameters in 3D-Bioplotter.

4D printing parameters	Value
Pressure (bar)	0.6
Speed (mm/s)	25
Transfer height (mm)	5
Postflow delay (s)	0.05
Preflow delay (s)	0.05
Layer thickness (mm)	0.2
Needle size (mm)	0.250
Temperature °C	20

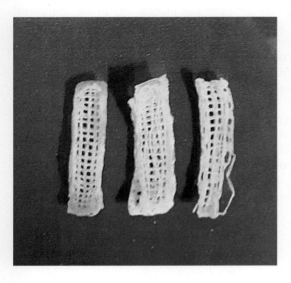

Fig. 9.5 Printed soft actuators with various processing parameters: (A) after extrusion and (B) after drying. The length of the scale bar is 10 mm.
No permission required

chitosan samples with various printer parameters are shown in Fig. 9.5. Clearly, printing with optimized parameters produced with better resolved samples and sharper grid pattern than those printed with higher pressure and nozzle velocity. Drying of the samples did not seem to have any significant effect on the overall print quality and the differences between the samples (Fig. 9.5).

Characterizations

3D printing parameters were optimized to print the chitosan hydrogels using a 3D-Bioplotter to design specifications. After producing both the 4D-printed and cast soft actuators, swelling ratio and mechanical property measurements were carried out on the samples.

Mechanical tests results

Polyelectrolyte hydrogels are ductile structures with large breaking strains and low elastic moduli (on the order of 100 kPa). However, at high strain rates, they exhibit brittle fracture (Czerner, Fasce, Martucci, Ruseckaite, & Frontini, 2016). Tensile testing was performed to investigate the fundamental mechanical properties of the dried hydrogel samples. Young's moduli and breaking strength of the samples of cast film chitosan and 3D-printed chitosan hydrogel were tested using a universal testing machine according to ASTM D638 (Type V) (Model: 5966, Instron, USA). The experiments were performed with a 10 kN load cell and a 2 mm/min crosshead speed at 20°C and 50% relative humidity. To achieve an average data value, three replicates of dried samples with measurements of *22 mm × 4 mm × 1 mm* were tensile tested. The effective area of the cross section of the tensile specimen was measured and used in the calculation of the tensile stress due to the wafer-like morphology of the 3D-printed chitosan.

Fig. 9.6 illustrates the stress–strain plots of the hydrogels tested under the same conditions. From these results, Young's moduli of the cast chitosan and 4D-printed chitosan were found 2.15 ± 0.32 MPa and 1.13 ± 0.20 MPa, respectively. The tensile strength of the cast chitosan and 4D-printed chitosan was also 1.62 ± 0.21 MPa and 0.37 ± 0.15 MPa, respectively.

The cast polyelectrolyte hydrogels displayed greater strength and ductility than the 4D-printed model. In the case of the printed chitosan sample, the percentage elongation was less than the cast one. The higher strength of the polyelectrolyte cast film samples is due to a more homogeneous film structure compared with the 4D-printed film. The grid pattern and printing defects may be caused by the irregular distributions of stress and local stress levels contributing to early failure.

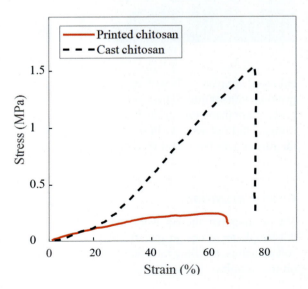

Fig. 9.6 Tensile test findings on the cast and printed chitosan films.
No permission required

Swelling measurements

Without dissolving into it, hydrogels can swell greatly in water. Measurement of fluid absorption is a significant consideration in estimating the swelling potential of the hydrogel, which, in turn, can be correlated with the bending performance of 4D-printed and cast film polyelectrolyte actuators in aqueous media. The swelling rate was measured by calculating the absorption of liquid as a function of time. The same dried samples for both cast and 4D-printed films were carefully weighted and submerged in 20 mL of 0.1 M NaOH at 20 ± 2°C before the swelling equilibrium was achieved (Czerner et al., 2016). Samples were taken out of the 0.1 M NaOH solution at certain intervals, measured, and reinfused with absorbent paper. Measurements of three replicates were carried out. The percentage of water intake was gravimetrically calculated as follows:

$$W_a\% = \frac{W_t - W_o}{W_o} \times 100 \tag{9.12}$$

where W_o and W_t are, respectively, the weights of the dry and wet samples.

Fig. 9.7 shows the quantity and pattern of the liquid uptake for the 3D-printed and cast chitosan actuators. After an immersion time of 120 min, the hydrogel samples reached an equilibrium state. Although the samples showed almost the same pattern of absorption prior to equilibrium, particularly after 60 min, the 4D-printed actuator showed a substantial increase in water absorption relative to the film cast actuators, suggesting a higher mass transfer rate in the 4D-printed hydrogels. This could be due to the larger mass transfer surface of the reticular 4D-printed structure with the same polymer and liquid environments.

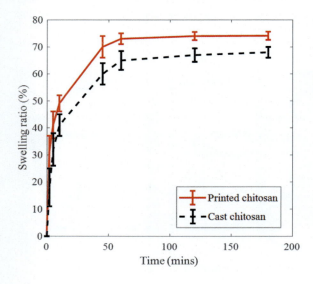

Fig. 9.7 Liquid uptake of the same dimension actuators for cast and 4D-printed chitosan. The standard deviation is shown by bars.
No permission required

Experimental setup and image processing

The bending magnitude was used to calculate the efficiency of the actuators and determine the effect of the 4D printing. Parametric modeling was used in response to electrical data to approximate and simulate the complex behavior of actuator bending. To control the applied voltage and orientation of the actuator according to the decisions taken by the key control algorithm, an Arduino-based motor driver is used. The Arduino communicates with the running MATLAB code via an RS232 serial adapter. The control algorithm outputs the appropriate voltage on a scale from 0 to 255, and the desired direction as 0 or 1. The serial port passes this information to the Arduino. The Arduino adapts its channel pulse width and output pins to the appropriate speed and direction for pulse width modulation (PWM). The driver produces voltages according to PWM and Arduino's direction pin values (Fig. 9.8).

MATLAB image processing is used to achieve the actuator's position in real time. The actuator was printed in natural chitosan color. The endpoint location of the actuator is red-marked to make image-processing effects. The endpoint coordinates represent the position of the red mark center on the actuator. This is used in the control algorithm to decide on the next command. HD PRO WEBCAM C920 was used to capture 30 fps images using an image resolution of 3280×1845 pixels/frame. The image processing steps for extracting the actuator's endpoint location can be described as following steps (Zolfagharian, Kouzani, Maheepala, Yang Khoo, & Kaynak, 2019; Zolfagharian, Kouzani, Moghadam, et al., 2018):

1. Camera pictures are imported in MATLAB in 0.1-s interval (Fig. 9.9(i)).
2. The imported RGB picture is transformed to gray scale (Fig. 9.9(ii)).
3. Red image portion is separated and subtracted (Fig. 9.9(iii)). To filter out noise, a median filter is used. The resulting image is translated into a binary image.
4. Unnecessary red components are removed from all pixels below 300 (Fig. 9.9(iv)). The label is then used for all connected components. Coordinates of the labeled area are obtained using "regionprops" command.

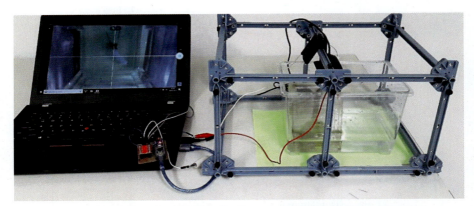

Fig. 9.8 Experimental setup for 4D-printed soft actuator data acquisition and control.
No permission required

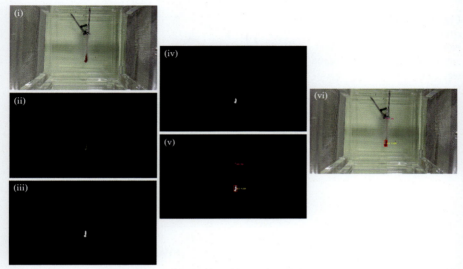

Fig. 9.9 Image processing steps for actuator endpoint position.
No permission required.

5. The reference point is transferred to the cantilever's origin (Fig. 9.9(v)) and the red element inside a rectangular box is bound.
6. For evaluation, the coordinates are superimposed on the original image (Fig. 9.9(vi)).

Ionic strength effect

The ionic concentration gradient driven by electric fields is the reason for the bending behavior of the hydrogel polyelectrolyte actuators. The Donnan effect manifests by creating an ionic concentration gradient along the direction of the electric field within the hydrogel that results in an osmotic pressure difference within the structure of the hydrogel, which therefore results in structural distortions. The maximum bending angle for the 4D-printed and cast actuators was measured with respect to the electrolyte concentration. The potential difference between the electrodes was set to 10 V while the NaOH solution concentrations varied as 0.1, 0.12, and 0.14 M. As Fig. 9.10 reveals, the optimum actuator bending angle was obtained at a concentration of 0.12 M. This can be due to the effective anionic gradient across the thickness of the actuator, which maximizes the difference in osmotic pressure.

Geometrical effects

In order to determine the effect of geometry, deflection tests were performed on different sample sizes at the same input voltage. Therefore, samples of the same weight but different dimensions were printed (Fig. 9.11). The input voltage of 10 V was applied, and their deflection was tracked when the equilibrium was achieved.

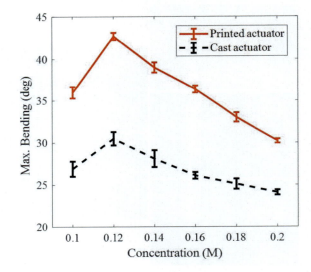

Fig. 9.10 The effect of ionic strength on the highest index bending in the 60s. Error bars are the standard error for three replicates.
No permission required.

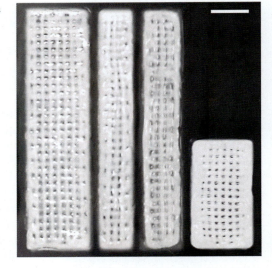

Fig. 9.11 Printed actuator samples with different dimensions before drying, from left to right, respectively,
S1: 64 mm × 16 mm × 4 mm,
S2: 64 mm × 16 mm × 2 mm,
S3: 64 mm × 8 mm × 4 mm, and
S4: 32 mm × 16 mm × 4 mm. The length of the scale bar is 10 mm.
No permission required.

Figs. 9.12 and 9.13 show the variation of the bending index with time for the printed actuators with different dimensions. The findings show a proportionate rise in the bending of the actuators with their length. As far as Fig. 9.12 is concerned, 3 of 4 samples are 64 mm long. We cannot then assume that bending is just proportional to the length. Because of its highest length-to-thickness ratio, S2 displays a considerably higher bending index profile than all other samples within the same time frame. With the smallest length-to-thickness ratio, S4 has the lowest bending index profile and S1

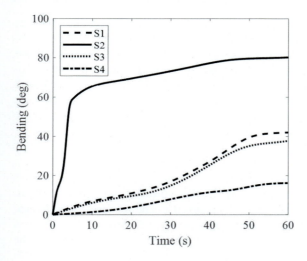

Fig. 9.12 Geometrical effects on bending behavior (S1: 64 mm × 16 mm × 4 mm, S2: 64 mm × 16 mm × 2 mm, S3: 64 mm × 8 mm × 4 mm, and S4: 32 mm × 16 mm × 4 mm). No permission required.

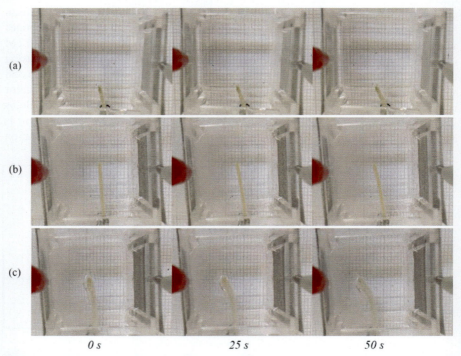

Fig. 9.13 Actuation sequences of the 3D-printed chitosan actuators: (A) S4, (B) S3, and (C) S2 samples.
No permission required.

and S3 have almost the same ratio, therefore displaying the same bending index variations. The bending index variations tend to be in line with the 4D-printed actuators' length-to-thickness ratio where the largest bending correlates to the highest length-to-thickness ratio, and vice versa. The results shown in Fig. 9.12 indicate a strong association of the displacement responses with the actuator's thickness, in accordance with the Shiga model (Shiga & Kurauchi, 1990). The photographic images of the sequence of bending motion of the 3D-printed samples S1, S2, and S3 at 0, 30, and 50s are shown in Fig. 9.13.

Actuation performance

The bending of the polyelectrolytes is, as stated earlier, a product of the Donnan effect. The rise in the counterion migration speeds increases the bending angle of the hydrogel and its actuation rate, which is why calculating the ideal electrolyte concentration for bending efficiency is a key factor for the actuator design as discussed earlier. The amplitude of the electrical voltage between the electrodes is another significant element in bending the actuator.

The increasing voltage accelerates the counterion migration rate, which consequently causes an increase in the bending angle and the actuation rate. The effect of input voltage magnitude on the actuation behavior of the 4D-printed chitosan hydrogel was tested in an attempt to validate this. The voltage between the electrodes was set at 5, 10, and 15 V while the NaOH concentration remained constant at 0.12 M (Fig. 9.14). It is noticeable in Fig. 9.15 that greater bending is achieved by increasing the electrical voltage applied to the actuator. Furthermore, in both 4D-printed and cast samples, the bending rate was initially high but declined with time as the actuator reached swelling equilibrium. This trend is further enhanced in the reverse cycle as it is apparent that with the increase in immersion time, the saturated condition is approached and the opposite voltage direction results in a decreased actuation

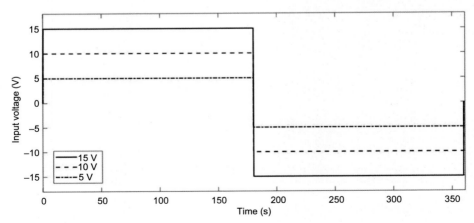

Fig. 9.14 Input voltage applied on electrodes.
No permission required.

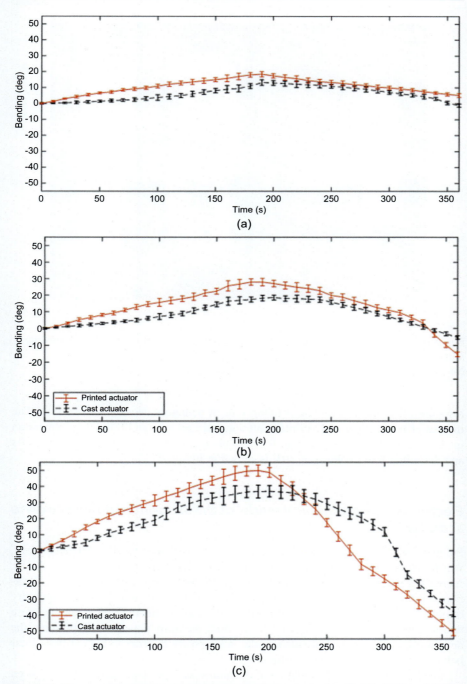

Fig. 9.15 The actuators bending in response to applied voltages (A–C) 5 V, 10 V, and 15 V, respectively.
No permission required.

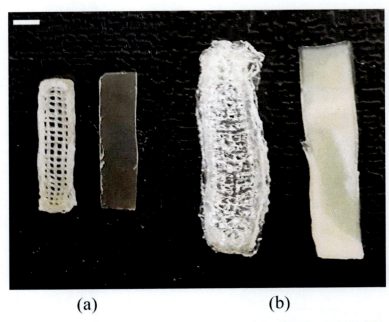

Fig. 9.16 Cast and 4D-printed and chitosan from right to left: (A) prior and (B) following immersion in solution. The bars are 5 mm in length.
No permission required.

response. This means that the bending rate of actuators decreases over time. The images of both cast and 4D-printed soft actuators prior to and following immersion in the solution are shown in Fig. 9.16. A photographic view of actuators at *120 s* and *320 s* in a fixed voltage of *10 V* is shown in Fig. 9.17.

Electro-chemo-mechanical model of the 3D-printed polyelectrolyte actuator

Polyelectrolyte soft actuators are known as an electro-chemo-mechanical mechanism, which makes their modeling somewhat complex (Zolfagharian, Kaynak, et al., 2019). Increasing uncertainties and time-dependent parameters resulting from back relaxation and mechanical property variations affect the bending and actuating capability of polyelectrolyte soft actuators. For further applications of such soft actuators, these undesirable aspects should be considered. With the introduction of 4D printing, the development of a control-oriented model is necessary in order to approximate their behavior in real-life implementations.

Several studies on the simulation of dynamics for typical soft actuators have been conducted. Black-box models were used to measure the actuator's curvature on the basis of input voltage in some works, but they were not scalable and unable to describe

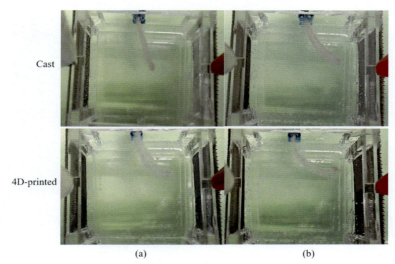

Fig. 9.17 Image sequences in order of maximal bending of chitosan actuators, cast (top) and 4D-printed (bottom): (A) after 120 s and (B) after 320 s.
No permission required.

the actuator function completely. Furthermore, advanced gray-box models, based on electric circuit models, such as RC (Attaran, Brummund, & Wallmersperger, 2015) and distributed transmission line models (Liu et al., 2012), have been built to relate the voltage with the actuator bending. More advanced white-box models have been developed in order to better understand more fundamental mechanics for precise dynamics simulation of polyelectrolyte actuation units (Liu et al., 2012), taking into account complicated electrochemical and mechanical concepts. However, these models were not ideal for the use of such actuators in real time. Hence, a mathematical gray-box relation of the 4D-printed polyelectrolyte soft actuator, combining the mechanical and electrical dynamics of the actuators, is developed in this study.

In control design implementations, Takagi-Sugeno (T-S) is a realistic approach to fuzzy modeling (Zolfagharian, Kaynak, et al., 2018). This work designs a reliable 4D-printed polyelectrolyte soft actuator model, based on the T-S modeling technique. The model proposed relates the input voltages of the actuator to an actuator bending through a universal T-S model that can be surfed between the submodels depending on the voltage range. This gives a scalable and functional model for further regulation of reversible 4D-printed mechanisms. Insert: ParagraphHeading 1Heading 2Heading 3BoxBulletFigureFormulaNumbered ListTable.

Lattice-patterned chitosan beams with different dimensions are 4D-printed using the bioprinter as shown in Fig. 9.18. This section introduces a scalable model of the polyelectrolyte actuator to address the complexity of multiphysics modeling. The actuator model involves electrochemical and electromechanical models, given the coupling of their dynamics (Bodaghi et al., 2016; Zolfagharian et al., 2016; Zolfagharian, Denk, et al., 2020). The electrochemical model relates the voltage

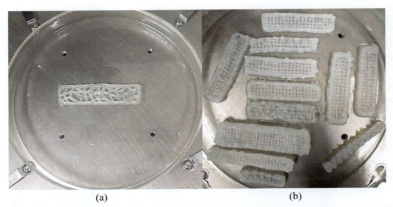

Fig. 9.18 (A) An arbitrary 4D printing pattern of polyelectrolyte actuator and (B) various printed actuator patterns and dimensions.
No permission required.

and the counterion gradient within the gel electrolyte. The electro-chemo-mechanical model is then linked by the counterion gradients in electrochemical phases, where the input voltage is proportional to the osmotic force. An approximation can be obtained for the bending curvature formed by the polymer actuator as follows:

$$\frac{Y(s)}{V(s)} = e^{\beta s} \frac{(s+z_1)\ldots(s+z_{n+1})}{(s+p_1)(s+p_2)\ldots(s+p_{n+1})(s+p_{n+2})}, \tag{9.13}$$

where β is a real number and z_i s and p_i s are, respectively, zeros and poles of the transfer function. The transfer function of input voltage to the actuator endpoint could be estimated via system identification approaches, such as ARMAX or weighted least squares. To do so, the optimal least squares and system identification toolbox in MATLAB could be used to construct the model of an arbitrary 4D-printed soft actuator by conducting experimental tests at different input voltages.

Controller design

Since its initial release (Young, Utkin, & Ozguner, 1999; Khawwaf et al., 2017; Khoo, Xie, & Man, 2009; Utkin, Guldner, & Shi, 2009), sliding mode control system (SMC) has been used in various robotic applications. An SMC is used here to regulate the sliding surface of the 4D-printed actuator for closed-loop control in the presence of external disturbances. SMC is employed as a global controller to deal with the actuator time-invariant characters representing the developed T-S model for endpoint estimation of the soft actuator and designed based on the fuzzy extreme subsystems concept for finding the uncertainties in lower and upper bounds of unknown parameters. Initially, the global fuzzy state space is broken down into m subspaces to form a fuzzy extreme subsystem. In every new subspace, the global fuzzy structure is governed by the dominant membership of its local fuzzy subsystem. In all the interactions among the local fuzzy subsystems, the worst stability case can be found in setting the subsystem's upper boundary. This information is then used in the SMC model to stabilize a dynamic nonlinear system (Ali et al., 2017).

T-S fuzzy system formulations

The T-S fuzzy method is a well-recognized model-based control that is described on the basis of IF-THEN fuzzy rules to describe local system dynamics with a combination of linear input/output relationships of a nonlinear system. This analysis uses the T-S fuzzy model. First, the entire state-space is split into subspaces. So, a linear model that is determined by the optimized least squares is used in each subspace to approximate the model. Finally, the fuzzy inference rule (here Pi membership) is used to fuzzify all subsystem parameter matrices to achieve a global dynamic fuzzy model.

A number of tests were performed on the actuator under real conditions to monitor the efficiency of the endpoint control of the 4D-printed actuator. First, multiple experiments have been performed to approximate the T-S actuator model via the least-square optimization technique. Different controllers were then implemented for robust position tracking of the actuator endpoint. Measuring all frequencies simultaneously within the bandwidth is impractical and ineffective. Therefore, the actuator models should be tested at frequencies of interest. Thus, a low-voltage spectrum of 2–8 V was applied throughout the thickness of actuators. The reference of the chirp signal as shown in Fig. 9.19 excited the actuator for 200 s at the sampling frequency of 25 Hz in the range of 0.01 to 11 Hz. Before starting the optimization method, it is necessary to remember that the desired order of the model must be determined. As polyelectrolyte actuators are usually restricted to a low frequency, the model order is tested from two to four. Model order four is eventually selected to ensure an accurate system modeling in the frequency spectrum of interest. The models obtained using the approaches suggested are similar to the test results obtained in Figs. 9.20 and

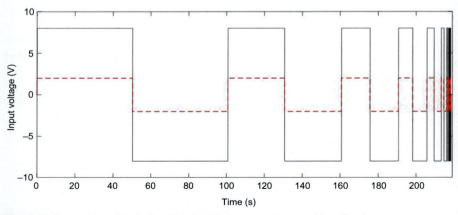

Fig. 9.19 Input chirp signals for 4D-printed actuator parameter identification. No permission required.

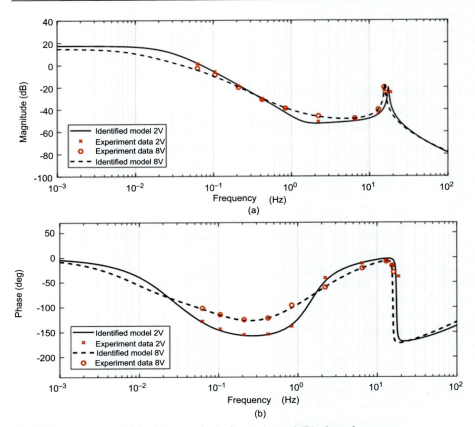

Fig. 9.20 Actuators models: (A) magnitude frequency and (B) phase frequency. No permission required.

9.21 for the frequency range of the system. The findings indicate that the magnitude shown in Fig. 9.20 decreases due to band-limited dynamics of the actuator in higher frequencies. At a low frequency, a low-pass response is observed when resonance behavior occurs near 11 Hz.

For the 4D-printed actuator, we identified two rule-based T-S models in this study as follows:

Rule 1: IF $|z(t)| \leq 5$, THEN $\begin{cases} \dot{x} = A_1 x(t) + B_1 u(t) \\ y = D_1 x(t) \end{cases}$.

Rule 2: IF $5 < |z(t)| \leq 8$, THEN $\begin{cases} \dot{x} = A_2 x(t) + B_2 u(t) \\ y = D_2 x(t) \end{cases}$.

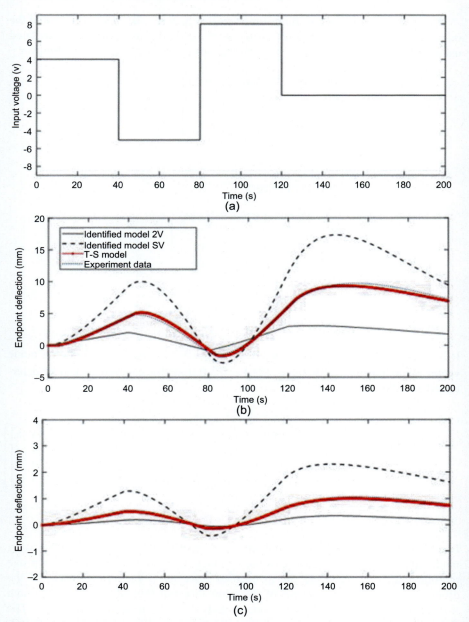

Fig. 9.21 (A) Signal for input voltage; (B) the random-patterned actuator's deflecting endpoint size S1, and (C) the deflection endpoint of the S4-style lattice pattern actuator.

The unknown transfer function of the actuator between the input voltage and the output bending angle corresponding to the two-zone linear systems is calculated using the least-square procedure for input signals between 2 V and 8 V as follows:

$$A_1 = \begin{bmatrix} -207 & -138.4000 & -7.5390 & -0.8623 \\ 256 & 0 & 0 & 0 \\ 0 & 8 & 0 & 0 \\ 0 & 0 & 1 & 0 \end{bmatrix}$$

$$B_1 = [1\,0\,0\,0]^T$$

$$C_1 = [2.9330\,3.9170\,0.4573\,1.0020]$$

$$A_2 = \begin{bmatrix} -0.4526 & -15.4300 & -1.9720 & -0.1145 \\ 16 & 0 & 0 & 0 \\ 0 & 1 & 0 & 0 \\ 0 & 0 & 0.1250 & 0 \end{bmatrix}$$

$$B_2 = [1\,0\,0\,0]^T$$

$$C_2 = [0.0094\,0.0488\,0.1442\,0.6250]$$

It is presumed the uncertain boundaries are $\varepsilon_m = 0.5$ and $\rho_m = 2$. The sliding mode vector is implemented as $S(t) = [52\,21\,1\,1]\ \tilde{x} = 0$. Extreme subsystems can be determined as follows:

$$E^1 = E^{11} = E^{21}$$
$$= \frac{1}{2}\left(\begin{bmatrix} -207 & -138.4000 & -7.5390 & -0.8623 \\ 256 & 0 & 0 & 0 \\ 0 & 8 & 0 & 0 \\ 0 & 0 & 1 & 0 \end{bmatrix} - \begin{bmatrix} -0.4526 & -15.4300 & -1.9720 & -0.1145 \\ 16 & 0 & 0 & 0 \\ 0 & 1 & 0 & 0 \\ 0 & 0 & 0.1250 & 0 \end{bmatrix}\right),$$

$$E^2 = E^{12} = E^{22} = [4\,0\,0\,0]^T$$

For 2 V and 8 V input signals, a comparison of experimental tests with the T-S fuzzy model and approximate specific voltage models is shown in Fig. 9.21. To prove reproducibility, the data are the average of three tests. The effectiveness of the established SMC based in terms of the scalability and flexibility for a 4D-printed soft actuator with an arbitrary pattern and size is shown in Fig. 9.22. The figure demonstrates the superiority of the T-S fuzzy prediction of the actuator end-point location relative to constant voltage versions even with the random-patterned 4D-printed actuator.

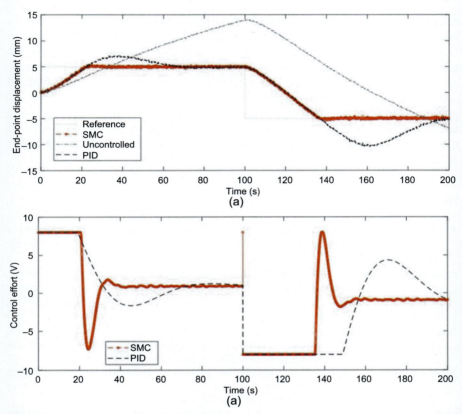

Fig. 9.22 Comparison of 4D-printed soft actuator control performance using closed-loop SMC and PID in terms of (A) endpoint displacement and (B) control effort.
No permission required.

Conclusion

This chapter presented a 4D printing study of a chitosan hydrogel soft actuator. To illustrate the principle of polyelectrolyte hydrogel actuator manufacturing using 4D printing, chitosan was chosen as a biocompatible and readily accessible hydrogel. To print the polyelectrolyte chitosan hydrogel, the printing parameters were optimized. 4D printing has been shown to increase the deflection frequency and the deflection magnitude relative to that of the cast film actuator. For a given sample size, the grid pattern formed by the 4D printing produced samples with a greater surface-to-volume ratio and was a significant factor in improving the efficiency of the actuation, so that a smaller electrical gradient could be applied through the sample thickness to provide the osmotic pressure gradient needed to initiate bending. The thickness and the length were found to be the most relevant geometrical considerations for the deflection of the printed actuator, while the width of the actuator had no noticeable impact. The driving force of bending originates from the gradient of ionic concentration through the actuator's thickness. Despite some loss of mechanical strength, the actuation output was enhanced through 4D printing. For the design of the actuator, all these factors should be considered. Due to its potential to impart customized and reproducible geometrical, mechanical, and control properties, the usage of 4D printing technology enables a new chapter in the additive manufacturing of soft robot actuators. These properties could be used to establish the time-related dynamics of these actuators, taking into consideration the parameters of geometry and materials in the printing process.

References

Ali, Z., Abbas, Z. K., Bijan, N.-N., Scott, A., Sui, Y. K., Michael, N., et al. (2017). 3D printing of a photo-thermal self-folding actuator. *KnE Engineering, 15*. https://doi.org/10.18502/keg.v2i2.590.

Altinkaya, E., Seki, Y., Yilmaz, Ö. C., Çetin, L., Özdemir, O., Şen, I., et al. (2016). Electromechanical performance of chitosan-based composite electroactive actuators. *Composites Science and Technology, 129*, 108–115. https://doi.org/10.1016/j.compscitech.2016.04.019.

Attaran, A., Brummund, J., & Wallmersperger, T. (2015). Modeling and simulation of the bending behavior of electrically-stimulated cantilevered hydrogels. *Smart Materials and Structures, 24*(3). https://doi.org/10.1088/0964-1726/24/3/035021.

Aybala, U., & Ramazan, A. (2016). Synthesis and analysis of electrically sensitive hydrogels incorporated with cancer drugs. *Journal of Pharmaceutics & Drug Delivery Research*. https://doi.org/10.4172/2325-9604.1000146.

Bodaghi, M., Damanpack, A. R., & Liao, W. H. (2016). Self-expanding/shrinking structures by 4D printing. *Smart Materials and Structures, 25*(10). https://doi.org/10.1088/0964-1726/25/10/105034.

Bodaghi, M., Serjouei, A., Zolfagharian, A., Fotouhi, M., Rahman, H., & Durand, D. (2020). Reversible energy absorbing meta-sandwiches by FDM 4D printing. *International Journal of Mechanical Sciences, 173*. https://doi.org/10.1016/j.ijmecsci.2020.105451, 105451.

Campbell, T. A., Tibbits, B., & Garrett, B. (2014). *The next wave: 4D printing programming the material world*. Atlantic.

Cvetkovic, C., Raman, R., Chan, V., Williams, B. J., Tolish, M., Bajaj, P., et al. (2014). Three-dimensionally printed biological machines powered by skeletal muscle. *Proceedings of the National Academy of Sciences of the United States of America, 111*(28), 10125–10130. https://doi.org/10.1073/pnas.1401577111.

Czerner, M., Fasce, L. A., Martucci, J. F., Ruseckaite, R., & Frontini, P. M. (2016). Deformation and fracture behavior of physical gelatin gel systems. *Food Hydrocolloids, 60*, 299–307. https://doi.org/10.1016/j.foodhyd.2016.04.007.

Flory, P. J. (1953). Molecular configuration of polyelectrolytes. *The Journal of Chemical Physics, 21*(1), 162–163. https://doi.org/10.1063/1.1698574.

Jeon, J. H., Cheedarala, R. K., Kee, C. D., & Oh, I. K. (2013). Dry-type artificial muscles based on pendent sulfonated chitosan and functionalized graphene oxide for greatly enhanced ionic interactions and mechanical stiffness. *Advanced Functional Materials, 23*(48), 6007–6018. https://doi.org/10.1002/adfm.201203550.

Keller, A., Pham, J., Warren, H., & In het Panhuis, M. (2017). Conducting hydrogels for edible electrodes. *Journal of Materials Chemistry B*, 5(27), 5318–5328. https://doi.org/10.1039/c7tb01247k.

Khawwaf, J., Zheng, J., Lu, R., Al-Ghanimi, A., Kazem, B. I., & Man, Z. (2017). Robust tracking control of an IPMC actuator using nonsingular terminal sliding mode. *Smart Materials and Structures*, 26(9). https://doi.org/10.1088/1361-665X/aa7d69.

Khoo, S., Xie, L., & Man, Z. (2009). Robust finite-time consensus tracking algorithm for multirobot systems. *IEEE/ASME Transactions on Mechatronics*, 14(2), 219–228. https://doi.org/10.1109/TMECH.2009.2014057.

Li, Y., Sun, Y., Xiao, Y., Gao, G., Liu, S., Zhang, J., et al. (2016). Electric field actuation of tough electroactive hydrogels cross-linked by functional triblock copolymer micelles. *ACS Applied Materials & Interfaces*, 8(39), 26326–26331. https://doi.org/10.1021/acsami.6b08841.

Liu, Y., Zhao, R., Ghaffari, M., Lin, J., Liu, S., Cebeci, H., et al. (2012). Equivalent circuit modeling of ionomer and ionic polymer conductive network composite actuators containing ionic liquids. *Sensors and Actuators, A: Physical*, 181, 70–76. https://doi.org/10.1016/j.sna.2012.05.002.

Mao, Y., Ding, Z., Yuan, C., Ai, S., Isakov, M., Wu, J., et al. (2016). 3D printed reversible shape changing components with stimuli responsive materials. *Scientific Reports*, 6. https://doi.org/10.1038/srep24761.

Ozbolat, I. T., & Hospodiuk, M. (2016). Current advances and future perspectives in extrusion-based bioprinting. *Biomaterials*, 76, 321–343. https://doi.org/10.1016/j.biomaterials.2015.10.076.

Rossiter, J., Winfield, J., & Ieropoulos, I. (2016, April). Here today, gone tomorrow: biodegradable soft robots. *In Electroactive polymer actuators and devices (EAPAD) 2016* (Vol. 9798, p. 97981S). International Society for Optics and Photonics.

Shang, J., Shao, Z., & Chen, X. (2008). Electrical behavior of a natural polyelectrolyte hydrogel: Chitosan/carboxymethylcellulose hydrogel. *Biomacromolecules*, 9(4), 1208–1213. https://doi.org/10.1021/bm701204j.

Shiga, T., & Kurauchi, T. (1990). Deformation of polyelectrolyte gels under the influence of electric field. *Journal of Applied Polymer Science*, 39(11 – 12), 2305–2320. https://doi.org/10.1002/app.1990.070391110.

Sydney Gladman, A., Matsumoto, E. A., Nuzzo, R. G., Mahadevan, L., & Lewis, J. A. (2016). Biomimetic 4D printing. *Nature Materials*, 15(4), 413–418.

Tibbits, S., & Cheung, K. (2012). Programmable materials for architectural assembly and automation. *Assembly Automation*, 32(3), 216–225. https://doi.org/10.1108/01445151211244348.

Utkin, V., Guldner, J., & Shi, J. (2009). *Sliding mode control in electro-mechanical systems.* CRC Press.

Winsberg, J., Hagemann, T., Janoschka, T., Hager, M. D., & Schubert, U. S. (2016). Redox-flow batteries: From metals to organic redox-active materials. *Angewandte Chemie, International Edition*, 56(3), 686–711.

Young, K. D., Utkin, V. I., & Ozguner, U. (1999). A control engineer's guide to sliding mode control. *IEEE Transactions on Control Systems Technology*, 328–342. https://doi.org/10.1109/87.761053.

Zhao, G., Sun, Z., Wang, J., Xu, Y., & Muhammad, F. (2015). Development of biocompatible polymer actuator consisting of biopolymer chitosan, carbon nanotubes, and an ionic liquid. *Polymer Composites*, 38(8), 1609–1615.

Zolfagharian, A., Denk, M., Kouzani, A. Z., Bodaghi, M., Nahavandi, S., & Kaynak, A. (2020). Effects of topology optimization in multimaterial 3D bioprinting of soft actuators. *International Journal of Bioprinting*, 6(2), 1–11. https://doi.org/10.18063/ijb.v6i2.260.

Zolfagharian, A., Kaynak, A., Bodaghi, M., Kouzani, A. Z., Gharaie, S., & Nahavandi, S. (2020). Control-based 4D printing: Adaptive 4D-printed systems. *Applied Sciences*, *10*(9). https://doi.org/10.3390/app10093020.

Zolfagharian, A., Kaynak, A., Khoo, S. Y., & Kouzani, A. Z. (2018). Polyelectrolyte soft actuators: 3D printed chitosan and cast gelatin. *3D Printing and Additive Manufacturing*, *5*(2), 138–150. https://doi.org/10.1089/3dp.2017.0054.

Zolfagharian, A., Kaynak, A., Yang Khoo, S., Zhang, J., Nahavandi, S., & Kouzani, A. (2019). Control-oriented modelling of a 3D-printed soft actuator. *Materials*, *12*(1), 71. https://doi.org/10.3390/ma12010071.

Zolfagharian, A., Kouzani, A. Z., Khoo, S. Y., Moghadam, A. A. A., Gibson, I., & Kaynak, A. (2016). Evolution of 3D printed soft actuators. *Sensors and Actuators, A: Physical*, *250*, 258–272. https://doi.org/10.1016/j.sna.2016.09.028.

Zolfagharian, A., Kouzani, A. Z., Khoo, S. Y., Nasri-Nasrabadi, B., & Kaynak, A. (2017). Development and analysis of a 3D printed hydrogel soft actuator. *Sensors and Actuators, A: Physical*, *265*, 94–101. https://doi.org/10.1016/j.sna.2017.08.038.

Zolfagharian, A., Kouzani, A. Z., Khoo, S. Y., Noshadi, A., & Kaynak, A. (2018). 3D printed soft parallel actuator. *Smart Materials and Structures*, *27*(4). https://doi.org/10.1088/1361-665X/aaab29.

Zolfagharian, A., Kouzani, A. Z., Maheepala, M., Yang Khoo, S., & Kaynak, A. (2019). Bending control of a 3D printed polyelectrolyte soft actuator with uncertain model. *Sensors and Actuators, A: Physical*, *288*, 134–143. https://doi.org/10.1016/j.sna.2019.01.027.

Zolfagharian, A., Kouzani, A., Moghadam, A. A. A., Khoo, S. Y., Nahavandi, S., & Kaynak, A. (2018). Rigid elements dynamics modeling of a 3D printed soft actuator. *Smart Materials and Structures*, *28*(2), 025003.

Hierarchical motion of 4D-printed structures using the temperature memory effect

10

Giulia Scalet[a], Stefano Pandini[b], Nicoletta Inverardi[b], and Ferdinando Auricchio[a]
[a]Department of Civil Engineering and Architecture, University of Pavia, Pavia, Italy
[b]Department of Mechanical and Industrial Engineering, University of Brescia, Brescia, Italy

Introduction

The name shape memory polymers (SMPs) refer to a class of stimuli-responsive materials, able to display significant dimensional changes upon the application of specific external stimuli (Huang et al., 2010; Lendlein & Langer, 2002; Otsuka & Wayman, 1999; Scalet, 2020). More precisely, their smart response consists in the possibility to deform them from an initial "permanent" shape to a "temporary" configuration, which can be maintained until the intervention of a triggering stimulus (typically a temperature rising above a characteristic value, called transformation temperature or T_{trans}). This stimulus activates the process in which the polymer reassumes the permanent shape. This specific behavior is typically classified as a one-way shape memory response (Leng, Lan, Liu, & Du, 2011).

Although this peculiar response is often associated with some intrinsic material aspects, early works on this subject strongly underlined that the shape memory response of thermally activated polymers derives from a combination of the material structure and of their thermomechanical history (Lendlein & Kelch, 2002; Yakacki et al., 2008). In fact, the shape memory effect (SME) may be seen as the response of the polymer to a certain cyclic history in which the material is first deformed in a temporary shape (this operation is typically called "programming"), and later brought to the original permanent shape upon the application of a temperature change.

On the material side, an adequate shape memory response finds its basis on a proper macromolecular architecture, typically involving the presence of a "hard" phase, consisting of either chemical or physical cross-links and responsible for the full recovery of the permanent shape, and of a "soft" phase, where softness is actually used to refer to the possibility of the polymer chains to highly gain mobility above certain temperatures, such as the glass transition or melting temperature, thus acting as a thermal switch between the two shapes. By modifying the architecture through chemical approaches, such as changing the cross-linking density (Song et al., 2010; Xie & Rousseau, 2009), monomer type, and/or comonomer ratio (Alteheld, Feng, Kelch, & Lendlein, 2005; Behl & Lendlein, 2007; Liu & Mather, 2002; Yakacki et al., 2008),

---

Smart Materials in Additive Manufacturing, Volume 2: 4D Printing Mechanics, Modeling, and Advanced Engineering Applications
https://doi.org/10.1016/B978-0-323-95430-3.00010-5
Copyright © 2022 Elsevier Inc. All rights reserved.

it is possible to tune the material transition temperature (Alteheld et al., 2005; Lendlein & Kelch, 2002; Liu & Mather, 2002; Song et al., 2010; Xie & Rousseau, 2009; Yakacki et al., 2008), to expand the broadness of the transition region (Kratz, Madbouly, Wagermaier, & Lendlein, 2011; Miaudet et al., 2007) or to provide more than one transition temperature (Behl, Bellin, Kelch, Wagermaier, & Lendlein, 2009; Bellin, Kelch, Langer, & Lendlein, 2006; Chen, Hu, Yuen, Chan, & Zhuo, 2010; Luo & Mather, 2010; Zotzmann, Behl, Hofmann, & Lendlein, 2010). Some approaches also use SMP-based composites or fillers to tune both the behavior and transition temperature of the polymer (Miaudet et al., 2007; Qi, Guo, Wei, Dong, & Fu, 2016).

Chemical modification represents the more investigated tailoring strategy, but it may be not so convenient from a practical point of view, due to difficulties related to polymer synthesis. Alternatively, it is possible to act with a thermomechanically based approach. In fact, it was also demonstrated that the application of specific thermomechanical histories, and in particular a proper adoption of parameters such as the deformation temperature (Gall et al., 2005; Khonakdar, Jafari, Rasouli, Morshedian and Abedini, 2007; Liu, Gall, Dunn and McCluskey, 2003; Miaudet et al., 2007; Xie, Page and Eastman, 2011; Yakacki et al., 2011), the strain rate (Hu, Ji, & Wong, 2005), and the time under load (Wong, Xiong, Venkatraman, & Boey, 2008), could be also a useful strategy to control the triggering temperature and the recovery rate.

Indeed, it is in the framework of thermomechanical tailoring that a particular effect, termed as "temperature memory effect" (TME), becomes relevant. In the next paragraph, an overview of this peculiar effect is presented, and the rest of the chapter will be focused on leveraging this effect for sequential and/or hierarchically structured shape changes.

Temperature memory effect: Basics and literature review

Description

The TME is displayed by certain SMP systems possessing a broad transition region. Interestingly, these materials are not just capable of actively undergoing a temporary-to-permanent shape transformation, but they also memorize the temperature at which they have been deformed in the temporary configuration (Kratz et al., 2011; Miaudet et al., 2007; Sun & Huang, 2010; Xie, 2010; Xie, Page, & Eastman, 2011).

In fact, various works confirmed that the temperature interval on which recovery occurs is roughly the same temperature at which programming was carried out (Fritzsche & Pretsch, 2014; Heuchel, Sauter, Kratz, & Lendlein, 2013; Kratz, Voigt, & Lendlein, 2012; Mirtschin & Pretsch, 2015; Nöchel et al., 2015; Wang et al., 2014; Xie, 2010). Therefore, for these systems, the TME may be defined as the possibility to activate the recovery process on the desired temperature region, imposed by carrying out the programming procedure at that specific temperature.

Such a response is again a combination of a peculiar polymer structure and proper thermomechanical history. In fact, while most of the standard programming strategies are carried out above T_{trans}, in the case of TME, programming takes place across the transition region, exploiting the fact that glass transition and melting are processes distributed across a thermal region. Indeed, the broader the transition region is, the wider would be the set of triggering temperatures. This particular thermomechanical tailoring also demands a proper structural, chemically tailored, condition to be effective, i.e., the polymer has to present a broad glass transition (Miaudet et al., 2007; Xie, 2010; Xie et al., 2011; Yu, Ritchie, Mao, Dunn, & Qi, 2015) or melting (Behl, Kratz, Noechel, Sauter, & Lendlein, 2013; Bodaghi, Damanpack, & Liao, 2018; Fritzsche & Pretsch, 2014; Mirtschin & Pretsch, 2015; Wang et al., 2014) region and may be obtained by using a single polymer (Xie, 2010; Xie et al., 2011), a polymer network (Bai, Zhang, Wang, & Wang, 2014; Li et al., 2012), or a supramolecular system (Ware et al., 2012).

The reason behind this peculiar behavior may be understood by considering the broad glass/melting transition as a consecutive distribution of a certain number of glass/melting transitions (Lendlein & Gould, 2019; Meng et al., 2013). Therefore, during programming within the transition region, a portion of the material is stiff while the rest is soft, and during the recovery process of the permanent shape from the different temporary shapes, only certain portions are subsequently activated corresponding to macromolecular segments which gain mobility in a sequential fashion, depending on the temperatures at which the material was deformed.

The TME may be also employed to obtain a programmable recovery from a temporary to a permanent shape not just in a single process, but also as a sequence of few intermediate steps, each one controlled in terms of strain change and triggering temperature through the programming procedure (Bellin, Kelch, & Lendlein, 2007; Heuchel et al., 2013; Samuel, Barrau, Lefebvre, Raquez, & Dubois, 2014; Sun & Huang, 2010). Such an approach has been largely employed to achieve multiple SMPs with a broad glass transition (Li, Pan, Zheng and Ding, 2018; Li et al., 2012; Miaudet et al., 2007; Samuel, Barrau, Lefebvre, Raquez and Dubois, 2014; Shao, Lavigueur and Zhu, 2012; Wang, Jia and Zhu, 2015; Ware et al., 2012; Xie, 2010; Zhang, Wei, Liu, Leng and Du, 2016; Zheng et al., 2017) and with a broad melting transition (Dolog & Weiss, 2013; Kratz et al., 2011; Yang et al., 2014). In fact, by deforming the SMP in a sequence of different steps carried out at various temperatures within the transition region, the subsequent recovery will occur in an inverse fashion with respect to the previous multistep programming. This ability has two main advantages with respect to other SMPs: first, it allows to design of a complicated sequence of motions in a continuous recovery process; secondly, it allows to program, for the same polymer, more than one switching temperature, with no need to chemically modify the material structure to introduce other transition temperatures. However, following this way, it is difficult to obtain a SMP with the capability of more than four or five shape transitions. Therefore, in order to achieve quintuple SME, an additional chemical tailoring is needed, and by way of example, Li et al. (2011) incorporated another melting transition into a SMP already possessing a broad glass transition.

Experimental testing

The most employed testing methodologies aimed at quantifying the material shape memory ability rely on the one-way SME, and therefore, they replicate its peculiar dual-step history. First, the change from the permanent to the temporary shape is applied during the programming step; afterward, the return to the permanent shape is carried out by subjecting the material to specific thermal conditions and measuring the shape evolution along time, in the so-called "recovery" or "reheating" step.

Programming is the most articulated step and consists of heating the specimen at a specific deformation temperature, T_{def}, which, in most of the cases, is well above T_{trans} (i.e., either T_g or T_m of the soft domains); deforming the material at given strain, or in a specific shape; fixing the shape by cooling the specimen, under fixed strain conditions, well below T_{trans}, and finally releasing the specimen, to remove any external load. This allows to easily deform and fix the temporary shape configuration, and, on the quantitative side, to evaluate the ability to fix the desired shape (also known as "strain fixity").

After programming, once the temporary shape is fixed, the "recovery" step takes place: this is typically carried out by reheating the specimen above T_{trans} along heating ramps at a constant heating rate (also known as "thermally stimulated recovery," or TSR, tests) while monitoring the recovery as a function of temperature; otherwise, recovery is promoted by exposing the polymer at a certain temperature and monitoring its isothermal recovery as a function of time. This step is usually carried out under stress-free conditions, allowing to evaluate how much the permanent shape is recovered (also known as "strain ratio"), and the temperature (timescale) required for the recovery for TSR (isothermal) tests.

A wider panorama of the testing geometries, procedure, and effect of parameters (holding time, strain rate, heating rate, stress-free and fixed-strain conditions during programming and recovery step) may be found in articles specifically focusing on a quantitative analysis of the SME of polymers in thermomechanical cycles (Liu, Gall, Dunn, & McCluskey, 2003; Sauter, Heuchel, Kratz, & Lendlein, 2013).

The testing approach developed in the case of the material with broad transition is generally more oriented in evaluating the TME abilities, and thus the possibility to simultaneously control both the deformation and the corresponding switching temperature, which roughly corresponds to the deformation temperature.

This effect may be used to obtain a dual-shape response, according to the aforementioned approach for SMPs: however, in this case, programming, instead of being carried out well above T_{trans}, may be performed on a larger set of deformation temperatures, spanning across the transition region, and it leads to direct control of the switching temperature in the recovery process. This experimental approach typically involves, as a programming step, a single deformation at a chosen T_{def}, followed by cooling well below the onset of the transition region; the subsequent recovery step typically shows a recovery process achieving its highest rate close to the deformation temperature, and it is typically employed in order to prove the effectiveness of the TME for a given polymer (Fritzsche & Pretsch, 2014; Kratz et al., 2011, 2012; Miaudet et al., 2007; Mirtschin & Pretsch, 2015; Nöchel et al., 2015; Wang et al., 2014; Xie, 2010; Xie et al., 2011; Yu & Qi, 2014).

In the case of a multiple shape memory response based on the TME, the material is programmed following a multistep deformation procedure, similar to a subsequent application of dual-shape programming cycles: the first one is carried out at $T_{\mathrm{def},1}$, corresponding to the highest desired switching temperature, and followed by cooling well below the onset of the transition region, T_{onset}; the second process will be carried out at $T_{\mathrm{def},2}$, freely chosen in the range $T_{\mathrm{onset}} < T_{\mathrm{def},2} < T_{\mathrm{def},1}$; and so on, the limits being represented by the extent of the transition region and by ensuring a certain range of separation between consecutive temperatures, so to avoid overlapping in the recovery processes. The recovery stage will activate the return to the original shape according to the previous cycle, following the deformation step but in an inverse manner. This type of experiments is also largely studied (Heuchel et al., 2013; Samuel et al., 2014; Sun & Huang, 2010), in particular when the TME is used for controlled sequential motions.

Modeling and simulation

The constitutive modeling and numerical simulations are fundamental steps to support the design of components or structures based on SMPs. On the practical side, they allow to reduce the high costs and times often associated with experimental testing, whereas on the theoretical side, they provide an understanding of complex phenomena that cannot be quantified experimentally and the mechanisms underlying this behavior.

Despite the importance, few contributions are currently available to describe the behavior of polymers featuring the TME and aim at modeling the behavior of polymers featuring a broad glass transition by adopting a thermoviscoelastic approach.

This choice is motivated by the fact that the physical mechanism behind the TME is associated with high chain mobility and multiple relaxation processes in the glass transition region. More specifically, from a micromechanical point of view, a polymer with such a transition region may be schematically imagined as composed of entangled domains with different thermoresponsive behavior; on a micromechanical scale, the broad glass transition corresponds to a continuous distribution of sub-elements, each one with specific relaxation times. For a deformation across the transition region, a portion of these elements, i.e., those having a long relaxation time at the deformation temperature, act as stiff elements, while the others have high mobility and may be easily deformed. After cooling the material under fixed strain conditions, a complex repartition of load occurs between these two sets of elements (Sun & Huang, 2010; Yu & Qi, 2014), which, interestingly, depend primarily on the deformation temperature, and this strongly influences the subsequent recovery process.

Consequently, the thermoviscoelastic approach is well suited to model such a mechanism, since it introduces rheological elements, consisting of springs and dashpots, that are able to describe this dependence of the molecular mobility and of the relaxation time on temperature.

In this regard, viscoelastic models with multiple nonequilibrium branches are generally applied (Sun & Huang, 2010; Xiao, Guo, & Nguyen, 2015; Yu & Qi, 2014; Yu, Xie, Leng, Ding, & Qi, 2012). Particularly, in the widely used generalized Maxwell model, the mechanical elements consist of an equilibrium branch and several non-equilibrium branches placed in parallel. The equilibrium branch is a linear spring with Young's modulus to represent the equilibrium behavior, while the nonequilibrium branches are Maxwell elements where an elastic spring and a dashpot are positioned in series to represent the viscoelastic response and thus the different relaxation modes of the material system. The spring and the dashpot of the ith Maxwell element are thus mechanically defined by Young's modulus and relaxation time, respectively. As the temperature is increased, different numbers of branches become shape memory active or inactive, leading to the observed TME, significantly altering the amount and nature of the internally stored energy. As anticipated, for a deformation across the transition region, the portion of branches with short relaxation time (at that deformation temperature) is easily deformed and minimally contribute to the polymer stiffness, while the remaining part of branches, which are not thermally activated due to their long relaxation time, requires a certain external load to be deformed. When cooling, all the domains are frozen in the deformed configuration, and the more rigid domains are also fixed due to the frozen soft portion they are tangled with. This leads to a repartition of the stored internal stress among the various branches, but the elastic stress accumulated in the more rigid set of branches cannot be totally released and will be stabilized as internal energy when the temperature is low. Only the heating of the polymer above temperatures able to reactivate these elements will promote the release of the stored energy, leading to that specific temperature dependence of the strain recovery process proper of the TME.

Starting from the qualitative discussion by Sun and Huang (2010), the paper by Yu et al. (2012) proposes a quantitative analysis for the TME using this modeling approach. Particularly, the authors employed the one-dimensional standard linear solid model under small strains, proposed by Qi and Dunn (2010), to illustrate the multiple relaxation processes of the polymer chains. Above the transition temperature, the Williams-Landel-Ferry (WLF) equation was used, while, below such temperature, the Arrhenius equation was adopted. Such a model was also used in Yu, Ge, and Qi (2014) and Yu and Qi (2014), while extensions to finite strain can be found in Diani, Gilormini, Frédy, and Rousseau (2012) and Westbrook, Kao, Castro, Ding, and Qi (2011). Xiao et al. (2015) applied a three-dimensional finite strain, nonlinear viscoelastic model to describe the behavior of Nafion, which has a broad glass transition region. Compared to Yu et al. (2012), who assumed an even distribution of relaxation times, the authors developed a method to obtain the parameters of the relaxation spectrum from the master curve of the relaxation modulus. The model was used to program various switchable pattern transformations in Nafion-based membranes using finite element simulations.

Exploitation of the temperature memory effect toward applicative examples

From the applicative point of view, the use of polymers capable of TME attracts particular interest by providing more flexibility in designing the shape memory response.

Until now, the applications of the TME have been limited to a few numbers of examples, mainly consistent in proofs of concept with multiple and/or sequential simple shape changes.

This is the case of the first works on the TME, as reported by Xie (2010) and Kratz et al. (2011), respectively.

Xie (2010) carried out an in-depth study of the thermomechanical conditions enabling the TME, and the successful result was proved with a simple test involving a bar shape, first stretched to the double of its initial length and then deformed into a helix-like shape. In this visual experiment, shown in Fig. 10.1A, the sequential recovery was triggered and led to the achievement of the permanent initial shape by passing through the two different programmed temporary shapes.

Similarly, Kratz et al. (2011) performed a careful chemical tailoring of the macromolecular structure of poly[ethylene-*ran*-(vinyl acetate)] (PEVA)-based networks with crystallizable controlling units in order to achieve TME properties also in semicrystalline networks. They proposed the newly synthesized SMP for use in prototypal fixtures, as shown in Fig. 10.1B. Specifically, the actuation of holders/releasers based on the same material was achieved at different temperatures, by simply modifying the temperature at which the deformation of the material occurs. With the advent of 4D printing, the consequent design flexibility hugely increased and the possibility of achieving more complex sequential shape transformations seemed more feasible and started to be explored with a wide body of scientific work based on the employ of multiple materials with different shape memory profiles (Ge, Dunn, Qi, & Dunn, 2014; Ge, Qi, & Dunn, 2013; Mao et al., 2015; Wu et al., 2016).

However, more recently, some works started to consider the opportunity to leverage the TME for designing complex actively moving structures on the basis of a single material. By way of example, Yu, Dunn, and Qi (2015) exploited inkjet printing to create a truss structure. Thanks to the TME, they were able to design a precise sequence of movements involving first a longitudinal contraction of the structure, followed afterward by its extension up to about 100% with respect to the first contracted recovered shape (Fig. 10.2).

Inverardi et al. (2020) performed an experimental investigation on a commercially available resin for 3D printing, identifying thermal and thermomechanical properties promising for the TME and proposing various thermomechanical procedures to evaluate, leverage, and model the TME. Further, by taking inspiration from the design challenge of building self-locking clamps and structures, they proposed the self-locking clamp which will be the subject of the case study later described in "Case study" section. The same group designed an auxetic-based metamaterial capable to sequentially change its shape between an out-of-plane folded tubular shape and a planar axially stretched shape (Pandini et al., 2020). More in detail, as shown in Fig. 10.3, they performed TME testing on the single unit cell, with the aim to decouple, on both time and temperature, the axial recovery from the out-of-plane folding. They showed, as an example, that by a specific two-step programming, it is possible to separate on the well-distinguished thermal region the unfolding motion of the cell (taking place at high temperature, or at longer times at a given temperature) from the axial contraction motion (occurring at low temperature, or for a shorter time at a given temperature).

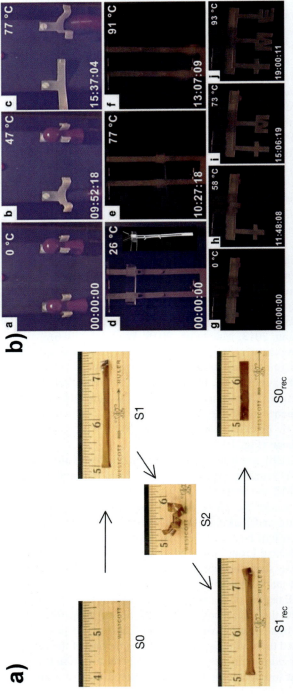

Fig. 10.1 (A) Visual demonstration of the triple SME. (B) Photograph series of three different TME demonstration devices. (A) Reprinted by permission from Springer Nature Customer Service Centre GmbH: Xie, T. (2010). Tunable polymer multi-shape memory effect. *Nature, 464*, 267–270. (B) Reprinted by permission from John Wiley and Sons: Kratz, K., Madbouly, S. A., Wagermaier, W., & Lendlein, A. (2011). Temperature-memory polymer networks with crystallizable controlling units. *Advanced Materials, 23*(35), 4058–4062.

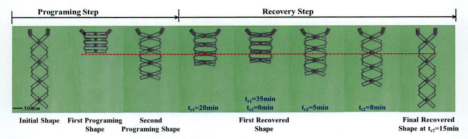

Fig. 10.2 Triple-SME in the printed SMP truss structure.
Reprinted from Yu, K., Dunn, M. L., & Qi, H. J. (2015). Digital manufacture of shape changing components. *Extreme Mechanics Letters*, *4*, 9–17, Copyright (2015), with permission from Elsevier.

Thanks to their potential coupling/decoupling of in-plane and out-of-plane deformations, other shape-morphing metamaterials targeting and approximating various shapes were investigated in Boley et al. (2019), Ding et al. (2017), Lei et al. (2019), Tibbits, Papadopoulou, and Laucks (2018), and Wagner, Chen, and Shea (2017).

Different approaches toward sequential motion are based also on polymers with more than one transition, having various phases acting as thermal switches (blends or copolymers (Lai, You, Chiu, & Kuo, 2017; Zhao et al., 2013)) or functionally graded structure of polymers differing for their T_g value (DiOrio, Luo, Lee and Mather, 2011; Ge et al., 2016; Hassan, Jo and Seok, 2018; Pei et al., 2017; Yu, Dunn, et al., 2015) or composition (Wu et al., 2016). For example, Peng, Yang, Gu, Amis, and Cavicchi (2019) interestingly proposed the achievement of a triple SME in resins for digital light processing 3D printing by obtaining two distinctly different glass transition temperatures at which activating different motions. They proposed the 4D-printed microfluidic device shown in Fig. 10.4 to prove different motions of the device when subjected to different temperatures for controlling and selecting the channel to deliver the liquid.

DiOrio, Luo, Lee, and Mather (2011) created a linear glass transition temperature gradient within one SMP leading to a glass transition distribution throughout the polymer and a recovery in the linear gradient direction. Such an approach (also known as "macroscale spatio-design"), consisting of endowing a material with spatially distributed transition temperatures, has attracted wide attention (Huang, Yang, An, Li and Chan, 2005; Li, Gao and Luo, 2016; Luo, Guo, Gao, Li and Xie, 2013; Zare, Prabhakaran, Parvin and Ramakrishna, 2019; Zheng et al., 2017).

However, polymers capable of the TME may reveal advantages compared to the aforementioned approaches. In fact, they may rely on common 3D printing technologies and avoid issues regarding poor compatibility and surface adhesion when compared to approaches based on multimaterial printing. Furthermore, they do not require posttreatments to generate property gradient distribution or structured layered depositions with respect to strategies based on the achievement of functionally graded structures. Finally, thanks to the TME, it may be possible to select a larger set of

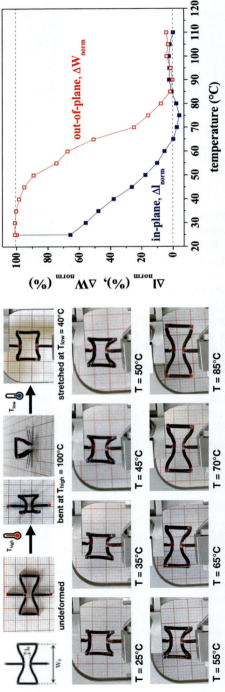

Fig. 10.3 Shape memory experiments decoupling out-of-plane flexural motions and in-plane contraction of an auxetic cell. The recovery process is visually described on the left and in the recovery curves representing how the in-plane elongation (Δl_{norm}) and the out-of-plane flexure (ΔW_{norm}) recover as a function of the increasing temperature.

Reprinted from Pandini, S., Inverardi, N., Scalet, G., Battini, D., Bignotti, F., Marconi, S., et al. (2020). Shape memory response and hierarchical motion capabilities of 4D printed auxetic structures. *Mechanics Research Communications*, *103*, 103463, Copyright (2020), with permission from Elsevier.

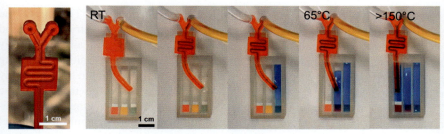

Fig. 10.4 Photographs of the 4D-printed microfluidic device and of the mixing and selective feeding process. At room temperature, the solutions were mixed and charged to the right groove; then, the channel pointed to the middle groove when heated to 65°C. At last, the channel became straight above 150°C and the mixture was charged to the left groove.
Reprinted (adapted) with permission from Peng, B., Yang, Y., Gu, K., Amis, E. J., & Cavicchi, K. A. (2019). Digital light processing 3D printing of triple shape memory polymer for sequential shape shifting. *ACS Materials Letters*, *14*, 410–417. Copyright (2019) American Chemical Society.

temperatures belonging to the broad transition to obtain the multiple shape memory response, instead of choosing the fixed transition temperatures of multiple SMPs, which may be changed only a priori, by playing on the chemistry of the system.

The field is still new for the employment of the TME to obtain sequential motions, so further works are expected to appear in the near future.

Experimental testing

Testing protocol

When aiming at designing an artifact displaying complex shape memory response through TME, experimental approaches are fundamental in order to identify whether material properties could fit a specific stimuli-responsive application and to evaluate the actual shape memory performance of an object based on that material. To fulfill these two objectives, two specific testing protocols may be conceived: first screening protocol, aimed at verifying the possibility to exploit the TME toward sequential motions, and second testing protocol to quantitatively describe the performances of objects, both for their evaluation and to generate input data for simulation.

Preliminary experimental activity to assess the possibility to exploit the temperature memory effect

This preliminary testing will focus on the properties of the materials. For this reason, tests have to be carried out on specimens with a simple shape, which is more representative of the material properties and easy to be characterized. A straightforward approach will be adopted, consisting of the following steps, whose results will eventually confirm a promising shape memory and temperature memory response.

Evaluation of thermal and thermomechanical properties: Testing of the material properties by means of differential scanning calorimetry (DSC) and dynamic-mechanical thermal analysis (DMTA) with the specific aim to identify the transformation temperature and the breadth of either the glass transition or melting region, depending on the type of polymer; the transition region should be sufficiently broad for optimal TME (i.e., at least 60–80°C from the onset to the end of the T_g/T_m region); this testing protocol may also provide information on the specific temperature (melting, crystallization, and glass transition temperature, employed to direct shape memory tests), and structural parameters (presence and amount of crystalline structure, effectiveness of the cross-linking). **Typical output:** *glass transition temperature, melting temperature, and breadth of the transition region.*

Evaluation of the mechanical behavior: Mechanical testing at the deformation temperature, via tensile/flexural/compression tests, depending on the type of deformation required, in order to identify the maximum deformation allowed under the various testing configurations. **Typical output:** *maximum strain at the break at a specific temperature.*

Evaluation of the shape memory and temperature memory behavior: Shape memory tests carried out via typical thermomechanical histories, consisting of a programming sequence (I. heating the specimen, II. deformation, III. cooling the specimen under fixed strain, and IV. unloading) and subsequent TSR (heating the specimen along a heating ramp while measuring the strain evolution due to the recovery process). This simple test allows quantifying the possibility to temporarily fix a given shape and to recover the original configuration. By carrying out programming at various temperatures, the TME is verified, and it is possible to identify, for each T_{def} value, the interval of temperature on which recovery is promoted as well as to evaluate strain fixity (i.e., the amount of applied strain effectively fixed by programming) and strain recovery (i.e., the amount of applied strain that may be recovered by heating). **Typical output:** *switching temperature, T_{trans}, at various deformation temperatures; strain recovery at various deformation temperatures; and strain fixity at various deformation temperatures.*

Experimental activity to evaluate and model shape memory response for sequential SMEs

The experimental response of objects capable of sequential motion and its modeling require advanced testing aimed at describing the peculiarity of the sequential response as a temperature and time-dependent process and at its simulation by means of finite element analysis. Toward such objectives, two sets of experimental tests are required.

Evaluation of the sequential shape memory response: An object is subjected to multiple-step programming and its recovery ability is evaluated. Programming a sequential motion clearly requires a multistep deformation history, applied as a sequence of deformations at (at least) two temperatures: the specimen is first deformed at a high temperature and fixed by cooling and then deformed at a lower temperature and again fixed by cooling. The higher and lower temperature may be optimally selected on the basis of TSR tests. The approach may also incorporate more than two deformations, provided the application of more than two deformation steps at various temperatures, in a sequence that starts from the highest temperature and then progressively moves to lower ones.

Evaluation of material parameters useful for implementing constitutive modeling: An experimental analysis is required in order to provide the necessary data to calibrate and validate the numerical predictions, based on a thermoviscoelastic constitutive model, implemented within a finite element framework, as detailed in "Constitutive modeling" section. DMTA is performed according to the multiple-frequency approach, scanning a region from well below the onset of the transition region (at least 40°C below the onset of the transition) to well above the end of the transition region (at least 40°C above the end of the transition), with a sweep at a discretized set of frequencies. In addition, quasistatic mechanical tests are carried out both under tensile and compressive conditions above the glass transition temperature (T_g), i.e., in the rubber-like region.

In the following, the aforementioned methodological workflow is applied to a 3D-printed SMP featuring the TME in order to approach the design of the target component described in the case study ("Case study" section). The workflow will cover the experimental characterization of the material, its description through a constitutive model, the identification of model parameters on experimental data, as well as the design of the target component.

Results of the screening protocol to assess the possibility to exploit the TME

The activity is aimed at determining the possibility to successfully leverage the TME to obtain sequential motion capability. The tests were carried out on rectangular plate-shaped specimens ($30 \times 80 \times 1 \, mm^3$) and on cubic specimens ($4 \times 4 \times 4 \, mm^3$), printed by means of a Formlabs Form 2, based on stereolithography (SLA) technology, and by employing the software PreForm (Formlabs). The resin employed is the Clear FLGPCL02, a commercial name of the Clear Photoreactive Resin for Formlabs 3D printers, formulated and provided by the company itself. From a chemical point of view, the material is made out of a mixture of methacrylic acid esters and a suitable photoinitiator.

Thermomechanical testing

3D-printed bar specimens were characterized by means of DSC tests and dynamic mechanical analysis (DMA) to determine the location of the glass transition region, to identify the best conditions for shape memory characterization, and the achievement of a cross-linked structure.

Typical testing conditions were adopted for these tests. DSC tests were run on a DSC Q100 (TA Instruments) on slices (10 mg) cut from the plates, along a heating–cooling–heating sequence between −80°C and 200°C at 10°C/min. DMA tests were carried out by means of a DMA Q800 analyzer (TA Instruments) on bar-shaped specimens (gauge length: 15–18 mm) cut from the plates. The tests were performed under tensile configuration, with an applied displacement amplitude equal to 10 μm and at a frequency equal to 1 Hz, scanning a region between −50°C and 200°C at 1°C/min.

The results of this characterization are reported in Fig. 10.5. The DSC trace in Fig. 10.5A reveals a widely distributed glass transition region, spreading over a temperature interval between about 10°C and 80°C. From the traces of storage and loss modulus, reported in Fig. 10.5B and C, it was possible to confirm the breadth of the glass transition region, approximate spanning on a similar interval (about 40–90°C according to the storage modulus trace; −10°C to 80°C according to loss modulus trace), with some reasonable differences ascribed to experimental peculiarities of the technique employed, testing condition, and parameter employed to estimate the

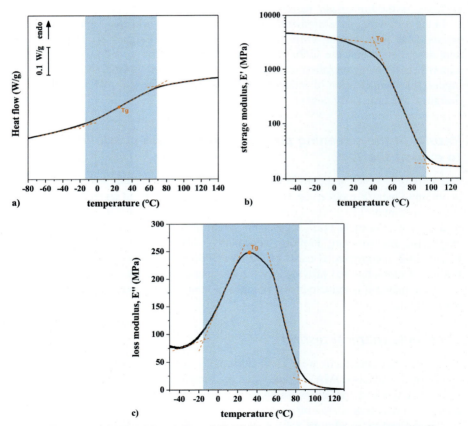

Fig. 10.5 Results of (A) DSC and (B and C) DMA characterization and construction lines to evaluate the extent of the transition region.
No Permission Required.

transition region breadth. It was possible to evaluate a nominal value of T_g equal to 32°C, by considering its most typical determination, i.e., as the peak of the loss modulus, E'', whereas the presence of a well-developed rubbery plateau in the storage modulus trace testifies the achievement of a proper cross-linking.

Mechanical testing

Preliminary mechanical tests were performed to identify the optimal deformation conditions to be used in the shape memory characterization.

Mechanical tests were carried out both in tensile and in compressive conditions between two extreme deformation temperatures: at room temperature, T_{room}, and at 100°C (about T_g+70°C, with T_g evaluated by DMA), two conditions that are representative of the minimum and maximum deformation temperature to be investigated in the programming stage. In particular, the tests allowed to identify the maximum safe strain to apply without the occurrence of specimen failure.

The tests were performed by means of an electromechanical dynamometer (Instron, Mod. 3366) equipped with a thermal chamber; for the tests at temperatures above T_{room}, the samples were equilibrated at T_{def} for at least 10 min before the testing. The tensile tests were carried out with a crosshead speed of 1 mm/min on 50 mm long bar-shaped samples (gauge length: 35 mm, cross section: 5×1 mm^2) cut from the plates, whereas for the compression tests, cubic specimens ($1.5 \times 1.5 \times 1.5$ mm^3), machined from the 3D-printed cubes, were used and the crosshead speed was set equal to 0.5 mm/min. Tensile tests revealed a brittle behavior at T_{room} (strain at break lower than 10%), not particularly mitigated by testing the materials above the glass transition with a strain at break never higher than 12%, as shown in Fig. 10.6A. Such behavior may be expected for this type of material and could be harshened by the probable

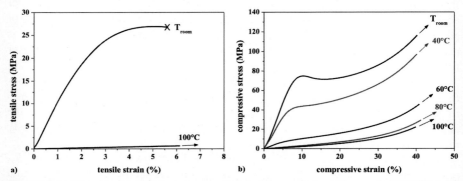

Fig. 10.6 (A) Results of tensile tests represented as stress vs strain at T_{room} and at 100°C (about T_g+70°C); (B) results of compression tests represented as stress vs strain curves for various deformation temperatures.

Reprinted by permission from John Wiley and Sons: Inverardi, N., Pandini, S., Bignotti, F., Scalet, G., Marconi, S., & Auricchio, F. (2020). Sequential motion of 4D printed photopolymers with broad glass transition. *Macromolecular Materials and Engineering*, *305*(1), 1900370.

presence of micrometrical voids and the inherent brittleness promoted by the highly cross-linked structure.

By contrast, the compressive stress vs strain curves shown in Fig. 10.6B clearly highlight two deformation regimes for the material. In fact, for deformation temperatures up to T_g (T_{def} equal to less than 40°C), an evident elastoplastic behavior is suggested, with a well-evidenced yield point and a softening region drop. As T_{def} increases, the occurrence of yielding is only slightly suggested. For T_{def} above 60°C, the mechanical response becomes similar to the typical response of rubber elasticity. At all temperatures, a compressive deformation up to 40% may be applied without failure.

Therefore, the shape memory behavior was studied under the more damage tolerant compressive and bending conditions, this latter allowing to apply a 180-degree bending of straight bars without any damage.

Shape memory tests were carried out according to the so-called TSR tests to measure the shape memory response as a function of temperature along a heating ramp. The tests were carried out on cubic specimens and consisted of early thermomechanical programming of a temporary shape, later followed by TSR tests.

The specimens were compressed at various deformation temperatures between T_{room} and 100°C up to a nominal strain value equal to 40%. Afterward, specimens were quickly cooled to T_{room} under fixed strain conditions. The compression was performed by means of an electromechanical dynamometer (Instron, Mod. 3366) equipped with a thermal chamber, and the cooling was carried out by maintaining fixed the crosshead at the applied displacement for at least 30 min, waiting for the specimen temperature to lower and for the applied stress to be fully relieved. Programmed specimens were then removed from the dynamometer and kept in a refrigerator at −20°C before measuring their shape memory behavior in TSR tests. It is noteworthy to remark that for this experimental setup, cooling under constant strain condition allowed to fix a large part of the applied deformation (for all the conditions tested higher than 65%), although a part of the applied strain is not effectively set but recovered due to the thermal contraction of the dynamometer fixture.

TSR tests were performed by means of the DMA machine, under compression configuration. The deformed specimens were subjected to a heating ramp at 0.5°C/min from T_{room} to well above the glass transition temperature (up to 120°C), under the application of a constant load of moderate entity (0.02 N), in order to continuously monitor the specimen height during the test.

Strain recovery during the TSR test is described in terms of "recovery ratio," R_r, representing the percentage of recovery taking place as temperature increases and defined as follows:

$$R_r(\%) = \frac{h - h_{prog}}{h_0 - h_{prog}} \times 100 \qquad (10.1)$$

where h is the actual height of the specimen, h_{prog} is the height after the programming step, and h_0 is the original height of the specimen.

Hierarchical motion of 4D-printed structures

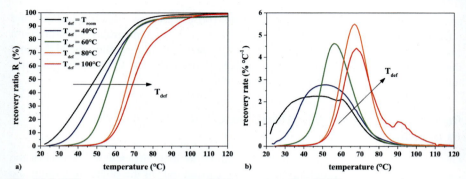

Fig. 10.7 TSR curves' results in terms of (A) recovery ratio and (B) recovery rate. Reprinted by permission from John Wiley and Sons: Inverardi, N., Pandini, S., Bignotti, F., Scalet, G., Marconi, S., & Auricchio, F. (2020). Sequential motion of 4D printed photopolymers with broad glass transition. *Macromolecular Materials and Engineering*, *305*(1), 1900370.

"Recovery rate" is the temperature derivative of the recovery ratio and allowed to effectively represent the distribution of the recovery process along the temperature scale, according to the following:

$$recovery\ rate\ (\%\,°C^{-1}) = \frac{dR_r}{dT} \qquad (10.2)$$

The results of the TSR tests are reported in Fig. 10.7. The results clearly show that all the samples, regardless of the deformation temperature, achieve high degrees of strain recovery, with an almost full recovery of the original shape (corresponding to $R_r = 98\%$–99%) at the end of heating. However, T_{def} strongly influences the distribution of the recovery process along the heating ramp as shown in Fig. 10.7A. The onset and the end of the recovery process are the parameters most influenced by T_{def}, moving to higher temperatures as T_{def} increases. For deformations in the sub-T_g region, a broad recovery process takes place, whereas at higher temperatures (T_{def} from 60°C), the recovery occurs with a similar narrow trend, shifted to higher temperatures as T_{def} increases. The broad recovery process found for sub-T_g deformations, onsetting close to T_{def} and finishing at about $T_g + 40$°C, may be justified by the fact that the recovery process also incorporates slower chain motions in the glassy region, further than those ascribed to the main relaxation process. At higher temperatures, recovery occurs on the basis of entropic forces acting on chains once the specific deformed domain gained enough mobility. Therefore, the dependence of the recovery process on T_{def} is related to the broad distribution of the segmental relaxation motions activating the recovery and, as a direct consequence, T_{def} may be used to control the switching temperature, accordingly to the TME. However, it is important to remark that for each T_{def} value, a specifically distributed recovery process takes place, whose curve may be distinguished but partially overlaps at different T_{def}. This is not unexpected, due to the slow recovery kinetics typical of SMPs whose recovery is triggered by T_g. However, the representation of the recovery rate in Fig. 10.7B offers the possibility to identify the T_{def} values leading to minimum overlapping of the recovery curves. In particular, specimens deformed at 40°C (blue curve) and 100°C (red curve) display the least, and only partial, overlapping of the recovery processes. These temperatures of minimum overlapping will be hereafter referred to as T_{high} and T_{low}, representing the higher and the lower temperature, respectively, employed in the multistep deformation programming for structures capable of sequential motion.

Results of the testing protocol based on the possibility to exploit the TME

Once identified the two temperatures providing the most efficient separation of the two peaks, a two-step programming can be carried out on the object to exploit the TME toward sequential motions. To provide a simple example, the two-step programming was carried out on a bar-shaped specimen (length: 80 mm, cross section: 5×1 mm^2), as outlined in Fig. 10.8A: the specimen was first heated at $T_{high} = 100$°C and deformed on one side by a 180-degree bending and cooled under this bent configuration to T_{room}; later, the same programming was carried out at $T_{low} = 40$°C on the other side, followed by cooling well below T_{room}, by placing the specimen in a refrigerator at about -20°C, where it was maintained until the shape memory test. Its self-deploying process was studied by placing the specimen in an oven with a clear glass window. The oven temperature was increased with a rate of about 1°C/min and the shape changes were monitored by means of a camera (Nikon D700) placed outside the oven (frame rate: 2 pic/min). On the collected pictures, the angle of the arm, α, defined as shown in Fig. 10.8B, was measured during unfolding by means of a digital image analysis software (Image J), allowing an accurate measurement of the angle (error lower than 1 degree on repeated measurements). Concurrent measurement of the temperature was carried out by means of a thermocouple lying close to the specimen.

For this specific geometry, a recovery ratio was evaluated on the angle α as depicted in Fig. 10.8B, according to the following:

$$R_{r,angle}(\%) = \frac{\alpha_{prog} - \alpha}{\alpha_{prog}} \times 100 \tag{10.3}$$

where α is the actual angle of the arm and α_{prog} is the angle after the programming step, all the angles being measured in rad. The results are reported as a function of temperature as shown in Fig. 10.8C.

The two processes are seen to the onset and occur, along this ramp, at different temperatures. The latter onsets with a lag of about 20°C with respect to the earlier one, so that for temperatures up to 45°C, the arm deformed at T_{low} recovers more than 70%, while that deformed at T_{high} displays negligible recovery. In addition, the recovery curves suggest that the lag of 20°C is approximately kept constant within the whole recovery process. This finding confirms the possibility to obtain well-distinguished opening effects along the temperature scale for sequential thermally triggered shape variations.

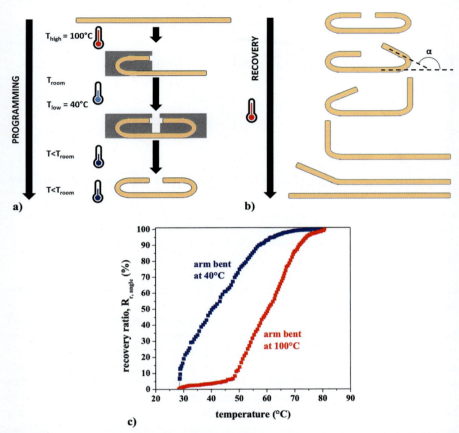

Fig. 10.8 (A) Sketch of the double-programming step on a bar-shaped geometry under flexure conditions; (B) sketch of the recovery and definition of the angle α; (C) TSR curves, depicted as recovery ratio vs temperature for the two programmed deformations at T_{low} (40°C) and T_{high} (100°C).
Reprinted by permission from John Wiley and Sons: Inverardi, N., Pandini, S., Bignotti, F., Scalet, G., Marconi, S., & Auricchio, F. (2020). Sequential motion of 4D printed photopolymers with broad glass transition. *Macromolecular Materials and Engineering*, *305*(1), 1900370.

Results of the experimental activity for the generation of input data for the numerical simulation

To provide useful data for the simulation on the basis of a thermoviscoelastic approach, DMA was performed by means of a DMA Q800 analyzer (TA Instruments) on a bar-shaped specimen (13–14 mm long). The specimen was tested under tensile configuration (displacement amplitude: 15 μm) and multifrequency mode (frequency sweeps from 0.4 Hz to 100 Hz), scanning a temperature range from −40°C to 150°C at

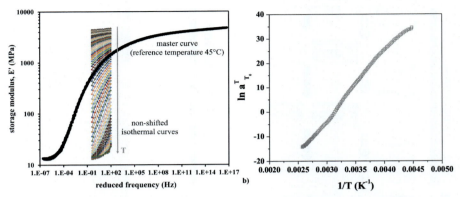

Fig. 10.9 (A) Non-shifted isothermal curves (*pale curves* between 0.4 Hz and 100 Hz) and master curve of the storage modulus (*black curve*) as a function of frequency, for a reference temperature T_0 equal to 45°C; (B) relationship between shift factors and temperature. No Permission Required.

0.5°C/min. The heating rate was kept low in order to recreate nearly isothermal conditions during each frequency sweep.

The results of the multifrequency DMA characterization are shown as a function of frequency in Fig. 10.9A in terms of isothermal frequency sweeps (colored curves). Adopting 45°C as reference temperature, T_0, and according to a time-temperature superposition scheme, the isothermal curves are rigidly shifted along the temperature axis until best superposition to finally obtain a storage modulus master curve (black curve). The shift factors used for the master curve construction are represented in Fig. 10.9B as a function of $1/T$ (being T the absolute temperature), showing a fairly linear dependence. Thus, the resulting master curve covers a wide range of frequencies and provides information and data to describe the viscoelastic behavior for the constitutive modeling of the material, while the shift factors vs T relationships allow for evaluation for the time–temperature dependence.

Constitutive modeling

Model formulation

The shape memory response of the commercial resin Clear FLGPCL02 studied in previous sections was modeled on the basis of the thermoviscoelastic approach discussed in "Modeling and simulation" section. Specifically, its viscoelastic behavior was captured by using the three-dimensional generalized Maxwell model. As discussed previously, this model is the most used model for describing the TME and it is also implemented within built-in libraries of commercial finite element software, such as Abaqus Standard finite element code (Simulia, Providence, RI). Accordingly, it was employed for our purposes in order to give a versatile approach that can be also reproduced for readers who approach this topic for the first time. Following the

Abaqus Standard nomenclature, the storage and loss moduli are defined as a function of frequency, respectively, as follows:

$$G'(\omega) = G_0 \left[1 - \sum_{i=1}^{N} \overline{g}_i^P \right] + G_0 \sum_{i=1}^{N} \frac{\overline{g}_i^P \tau_i^2 \omega^2}{1 + \tau_i^2 \omega^2} \tag{10.4}$$

$$G''(\omega) = G_0 \sum_{i=1}^{N} \frac{\overline{g}_i^P \tau_i \omega}{1 + \tau_i^2 \omega^2} \tag{10.5}$$

where $\overline{g}_i^P = G_i/G_0$ and τ_i are the Prony parameters of the ith Maxwell element, G_0 represents the unrelaxed shear modulus (i.e., the modulus at time $t=0$), G_i is the shear modulus, and N the number of Maxwell elements.

The functional time and temperature dependencies were expressed through the shift factors, $a_{T_0}{}^T$, obtained through the master curve construction and whose dependence on temperature was approximated by an Arrhenius-like law, as follows:

$$\ln a_{T_0}^T = \frac{E_0}{R} \left(\frac{1}{T} - \frac{1}{T_0} \right) \tag{10.6}$$

where E_0 typically represents the activation energy of the relaxation process, R is the universal gas constant, T is the actual temperature, and T_0 is the reference temperature, both expressed as absolute temperatures. The convenient choice of the Arrhenius expression was motivated on the basis of the overall good linear relation between $\ln a_{T_0}{}^T$ and $1/T$.

In addition, the mechanical behavior in terms of an elastic response needs to be described. In order to account for finite deformation and based on experimental evidence in Fig. 10.6A, the hyperelastic Neo-Hookean model was adopted under the hypothesis of incompressible behavior of the material. The corresponding strain energy potential function per unit of reference volume is defined as follows:

$$U = C_{10} (\overline{I}_1 - 3) \tag{10.7}$$

where $\overline{I}_1$ is the first deviatoric strain invariant and C_{10} is a material parameter.

Identification of model parameters

The parameters of the adopted model were all identified from the experimental curves presented in previous sections.

First, the Prony parameters, $\overline{g}_i^P$ and τ_i, in Eqs. (10.4) and (10.5) were determined from the storage and loss modulus master curves presented in "Results of the experimental activity for the generation of input data for the numerical simulation" section, as proposed in Diani et al. (2012). Accordingly, we implemented the optimization procedure proposed by Kraus, Schuster, Kuntsche, Siebert, and Schneider

Table 10.1 Prony parameters of the generalized Maxwell model: Normalized moduli and relaxation times, evaluated at $T_0 = 45°C$.

Prony parameters ($i = 1, ..., 11$)		Prony parameters ($i = 12, ..., 22$)		Prony parameters ($i = 23, ..., 33$)	
$\bar{g}_i^P$	τ_i	$\bar{g}_i^P$	τ_i	$\bar{g}_i^P$	τ_i
0.01289	1.00E-21	0.03769	1.00E-10	0.02195	10
0.01265	1.00E-20	0.04552	1.00E-09	0.00602	100
0.01093	1.00E-19	0.03764	1.00E-08	9.46730E-04	1000
0.04943	1.00E-18	0.05721	1.00E-07	1.45937E-04	10,000
0.04570	1.00E-17	0.06567	1.00E-06	8.51017E-05	100,000
0.02865	1.00E-16	0.07897	1.00E-05	3.75531E-05	1.00E+06
0.02596	1.00E-15	0.07933	0.0001	3.23778E-05	1.00E+07
0.03037	1.00E-14	0.08078	0.001	3.19557E-05	1.00E+08
0.02605	1.00E-13	0.07425	0.01	3.19143E-05	1.00E+09
0.03463	1.00E-12	0.06464	0.1	3.19102E-05	1.00E+10
0.02648	1.00E-11	0.04226	1	3.19098E-05	1.00E+11

(2017) in MATLAB to determine the optimal Prony parameters. Table 10.1 reports the parameters obtained using $N = 33$ Maxwell elements.

The obtained parameters ensured a good fit of the experimental curves as demonstrated in Fig. 10.10, where the tan δ curve is also reported (Fig. 10.10C) for model validation, in addition to the storage and loss moduli curves (Fig. 10.10A and B).

Then, the Arrhenius parameters in Eq. (10.6) were identified from the shift factors experimental shift factor versus $1/T$ correlation represented in Fig. 10.9. The parameter T_0 was assumed to be equal to the reference temperature, i.e., $T_0 = 45°C$, while the parameter E_0 was fitted on a region of temperature between $-20°C$ ($1/T = 0.00395 \,\mathrm{K}^{-1}$) and $80°C$ ($1/T = 0.00283 \,\mathrm{K}^{-1}$), i.e., the region where the curve in Fig. 10.9B exhibits a fairly regular linear correlation and because it offered a good coverage of the temperatures involved in the shape memory cycle. The parameter E_0 for the Arrhenius equation was taken equal to 247.6 kJ/mol. The comparison between the calibrated and experimental curves is reported in Fig. 10.11.

Finally, the parameter of the Neo-Hookean law (see Eq. 10.7) was identified on the tensile stress-strain curves of Fig. 10.6 at $100°C$, obtaining the results reported in Fig. 10.12.

Case study

The target component is the 4D-printed self-locking clamp studied in Inverardi et al. (2020). Self-locking mechanisms are widely employed in the 4D printing community in order to prove the sequential motion capability of a 3D-printed structure, which is usually made by multiple SMPs (Mao et al., 2015). In this case, the TME results effectively in providing multiple and sequential motions relying only on a single shape

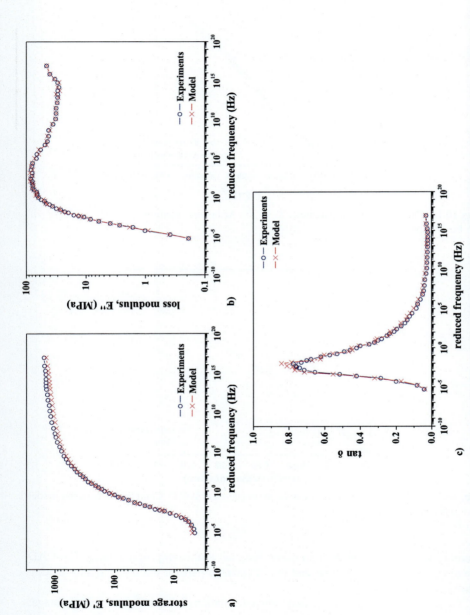

Fig. 10.10 Calibration of the Prony parameters of the generalized Maxwell model on (A) storage modulus, (B) loss modulus, and (C) tan δ curves. No Permission Required.

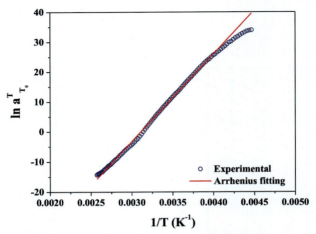

Fig. 10.11 Calibration of the parameter E_0 of the Arrhenius equation.
No Permission Required.

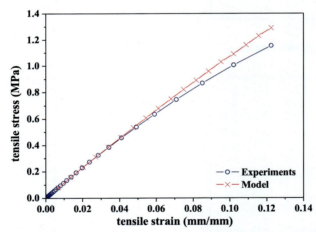

Fig. 10.12 Calibration on tensile stress vs strain curves at 100°C. Comparison between model and experimental data.
No Permission Required.

memory material and on the purposely designed multistep programming. The CAD drawing, the support structure, and the 3D-printed part for the selected self-locking clamp are shown in Fig. 10.13.

The clamp was subjected to a two-step deformation during the programming stage: first, after heating at T_{high}, a rotation was performed around hinge A (Fig. 10.14) to obtain a straight tab, and the structure was cooled at T_{room} under the imposed shape; then, after heating at T_{low}, a rotation was applied at the point called hinge B to obtain the second temporary shape (Fig. 10.14) and it was cooled at −20°C. The isothermal

Hierarchical motion of 4D-printed structures

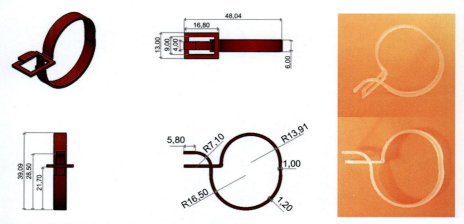

Fig. 10.13 Self-closing clamp under investigation.
No Permission Required.

recovery was studied by the immersion of the structure in a water bath at T_{rec}, equal to 60°C (Fig. 10.14). The choice of T_{rec} is important in order to promote the recovery of both the motions in a reasonable timescale, while still providing a good enough separation of the two motions to avoid any collision of the two sections of the structure undergoing recovery. This can be achieved by selecting a T_{rec} in between T_{low} and T_{high} and it can be optimized by either an experimental or a numerical campaign on simple shapes under isothermal recovery conditions at different temperatures.

Simultaneously, the programming and isothermal recovery were simulated within a finite element framework by means of the model described in "Constitutive modeling" section. Accordingly, the geometry was meshed by using 15,569 four-node linear isoparametric tetrahedron elements, hybrid with linear pressure. An analysis was performed by imposing appropriate boundary conditions and uniform temperature history to simulate the two-step programming as well as the isothermal recovery, as detailed in the following. First, the rotation around hinge A at T_{high} was ensured by applying proper pressure to the tab. Then, keeping the tab vertical, the sample was cooled up to T_{room} in 30min and heated up to T_{low} (0.5°C/min). Then, the rotation around hinge B at T_{low} was ensured by applying proper pressure. Keeping the applied deformations, the sample was cooled down to −20°C in 30min. Then, the applied loads were removed, and the sample heated up to T_{rec} in 1s. The clamp was finally kept at this temperature for 10min (isothermal recovery) and was left free to recover its initial closed shape.

As it can be seen in the sequence of images in Fig. 10.14, the recovery of the double-programmed clamp includes two separate motions along the timescale, allowing, through rotation around hinge B, the end-tab to enter in the hole before rotating to ensure the locking feature. Interestingly, the prediction of the simulation is in good agreement with the results of the experimental test. In particular, the first recovery, related to the deformation carried out at T_{low}, takes place in less than 10s of immersion in the hot water, both in the experiments and in the simulation. The recovery of the

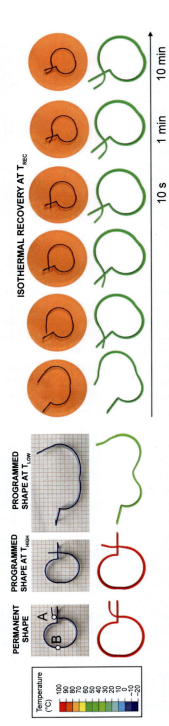

Fig. 10.14 Comparison between experimental and numerical results on the self-closing clamp. In the figure, $T_{high} = 100°C$, $T_{low} = 35°C$, and $T_{rec} = 60°C$.
No Permission Required.

second rotation (related to the programming at T_{high}) can be appreciated at least after 10s in the experimental case, while its starts in shorter times in the simulation. These differences may be ascribed to the difficult reproduction of the exact experimental steps, in particular those related to the actual change in temperature of the specimen during the immersion in the hot water bath, since a nonnegligible thermal lag may occur, but including also real cooling times and exact temperature distribution along with the sample (assumed uniform in the numerical model). However, the simulations still allow predicting the shape changes and their relevant timescale in a reliable fashion, resulting in a tool of utmost importance for the first design of sequentially moving devices to later be tested or employed in experimental conditions.

Conclusions and perspectives

This chapter has provided the reader with both analysis and discussion for a guided design of 4D-printed structures exhibiting the TME.

Specifically, the review of the current state of the art has revealed how this feature has been largely investigated from both the chemical and macroscopic material behavior point of view. Available results demonstrate the great potential of these polymers in promoting sequential shape changes that can be controlled by means of the deformation temperature. This potential can be exploited in numerous fields, from soft actuators and robotics to biomedical ones.

However, despite this potential, the application to real-world problems has been shown to be still limited, probably due to the complex behavior and difficult processing of these polymers. Only recently, the potentialities of such an effect have been coupled with 3D printing technologies to realize some applicative examples.

Obviously, this has to be accompanied by the formulation of appropriate material models to be used within numerical codes to support an efficient design of components featuring the TME, reducing costs and times of experimentation. In this regard, it has been discussed that few modeling contributions have been proposed, and they are mostly based on a thermoviscoelastic approach to catch the behavior of polymers with a broad glass transition. This approach, although simple and physical, has certain limitations in describing all the different types of polymers featuring the TME. Novel modeling works are thus required from the scientific community.

In order to promote the future use of this class of polymers in applications, a methodological workflow to approach the design of a target component based on a SMP featuring the TME, processed by means of 3D printing, has been presented. The workflow has covered the experimental characterization of the material, its description through a constitutive model, the identification of model parameters on experimental data as well as the design of a target component (i.e., a self-closing clamp). This component was realized by means of SLA 3D printing, based on a commercial resin, and it is thus easily reproducible.

The provided discussion as well as the obtained results clearly evidence how 4D printing of SMPs featuring the TME opens new possibilities in realizing complex structures capable of tunable shape morphing.

Acknowledgments

This work was partially supported by the Italian Minister of University and Research through the project "A BRIDGE TO THE FUTURE: Computational methods, innovative applications, experimental validations of new materials and technologies" (No. 2017L7X3CS) within the PRIN 2017 program. The authors kindly acknowledge Dr. Stefania Marconi for her contribution to the 3D printing of the self-locking clamp (details in Inverardi et al., 2020).

References

Alteheld, A., Feng, Y., Kelch, S., & Lendlein, A. (2005). Biodegradable, amorphous copolyester-urethane networks having shape-memory properties. *Angewandte Chemie International Edition, 44*(8), 1188–1192.

Bai, Y., Zhang, X., Wang, Q., & Wang, T. (2014). Shape memory property of microcrystalline cellulose–poly(ε-caprolactone) polymer network with broad transition temperature. *Journal of Materials Science, 49*(5), 2252–2262.

Behl, M., & Lendlein, A. (2007). Actively moving polymers. *Soft Matter, 3*, 58–67.

Behl, M., Bellin, I., Kelch, S., Wagermaier, W., & Lendlein, A. (2009). One-step process for creating triple-shape capability of AB polymer networks. *Advanced Functional Materials, 19*, 102–108.

Behl, M., Kratz, K., Noechel, U., Sauter, T., & Lendlein, A. (2013). Temperature-memory polymer actuators. *Proceedings of the National Academy of Sciences, 110*(31).

Bellin, I., Kelch, S., & Lendlein, A. (2007). Dual-shape properties of triple-shape polymer networks with crystallizable network segments and grafted side chains. *Journal of Materials Chemistry, 17*, 2885–2891.

Bellin, I., Kelch, S., Langer, R., & Lendlein, A. (2006). Polymeric triple-shape materials. *Proceedings of the National Academy of Sciences, 103*(48), 18043–18047.

Bodaghi, M., Damanpack, A. R., & Liao, W. H. (2018). Triple shape memory polymers by 4D printing. *Smart Materials and Structures, 27*, 065010.

Boley, J. W., van Rees, W. M., Lissandrello, C., Horenstein, M. N., Truby, R. L., Kotikian, A., et al. (2019). Shape-shifting structured lattices via multimaterial 4D printing. *Proceedings of the National Academy of Sciences, 116*(42), 20856–20862.

Chen, S., Hu, J., Yuen, C.-W., Chan, L., & Zhuo, H. (2010). Triple shape memory effect in multiple crystalline polyurethanes. *Polymers for Advanced Technologies, 21*, 377–380.

Diani, J., Gilormini, P., Frédy, C., & Rousseau, I. (2012). Predicting thermal shape memory of crosslinked polymer networks from linear viscoelasticity. *International Journal of Solids and Structures, 49*, 793–799.

Ding, Z., Yuan, C., Peng, X., Wang, T., Qi, H. J., & Dunn, M. L. (2017). Direct 4D printing via active composite materials. *Science Advances, 3*, e1602890.

DiOrio, A., Luo, X., Lee, K., & Mather, P. (2011). A functionally graded shape memory polymer. *Soft Matter, 7*, 68–74.

Dolog, R., & Weiss, R. (2013). Shape memory behavior of a polyethylene-based carboxylate ionomer. *Macromolecules, 46*, 7845–7852.

Fritzsche, N., & Pretsch, T. (2014). Programming of temperature-memory onsets in a semicrystalline polyurethane elastomer. *Macromolecules, 47*(17), 5952–5959.

Gall, K., Yakacki, C. M., Liu, Y., Shandas, R., Willett, N., & Anseth, K. S. (2005). Thermomechanics of the shape memory effect in polymers for biomedical applications. *Journal of Biomedical Materials Research Part A, 73*(3), 339–348.

Ge, Q., Sakhaei, A. H., Lee, H., Dunn, C. K., Fang, N. X., & Dunn, M. L. (2016). Multimaterial 4D printing with tailorable shape memory polymers. *Scientific Reports*, *6*, 31110.

Ge, Q., Dunn, C. K., Qi, H. J., & Dunn, M. L. (2014). Active origami by 4D printing. *Smart Materials and Structures*, *23*, 094007.

Ge, Q., Qi, H. J., & Dunn, M. L. (2013). Active materials by four-dimension printing. *Applied Physics Letters*, *103*, 131901.

Hassan, R. U., Jo, S., & Seok, J. (2018). Fabrication of a functionally graded and magnetically responsive shape memory polymer using a 3D printing technique and its characterization. *Journal of Applied Polymer Science*, *135*, 45997.

Heuchel, M., Sauter, T., Kratz, K., & Lendlein, A. (2013). Thermally induced shape-memory effects in polymers: Quantification and related modeling approaches. *Journal of Polymer Science Part B: Polymer Physics*, *51*, 621–637.

Hu, J. L., Ji, F. L., & Wong, Y. W. (2005). Dependency of the shape memory properties of a polyurethane upon thermomechanical cyclic conditions. *Polymer International*, *54*, 600–605.

Huang, W. M., Ding, Z., Wang, C. C., Wei, J., Zhao, Y., & Purnawali, H. (2010). Shape memory materials. *Materials Today*, *13*(7–8), 54–61.

Huang, W., Yang, B., An, L., Li, C., & Chan, Y. (2005). Water-driven programmable polyurethane shape memory polymer: Demonstration and mechanism. *Applied Physics Letters*, *86*, 114105.

Inverardi, N., Pandini, S., Bignotti, F., Scalet, G., Marconi, S., & Auricchio, F. (2020). Sequential motion of 4D printed photopolymers with broad glass transition. *Macromolecular Materials and Engineering*, *305*(1), 1900370.

Khonakdar, H. A., Jafari, S. H., Rasouli, S., Morshedian, J., & Abedini, H. (2007). Investigation and modeling of temperature dependence recovery behavior of shape-memory crosslinked polyethylene. *Macromolecular Theory and Simulations*, *16*, 43–52.

Kratz, K., Madbouly, S. A., Wagermaier, W., & Lendlein, A. (2011). Temperature-memory polymer networks with crystallizable controlling units. *Advanced Materials*, *23*(35), 4058–4062.

Kratz, K., Voigt, U., & Lendlein, A. (2012). Temperature-memory effect of copolyesterurethanes and their application potential in minimally invasive medical technologies. *Advanced Functional Materials*, *22*, 3057–3065.

Kraus, M. A., Schuster, M., Kuntsche, J., Siebert, G., & Schneider, J. (2017). Parameter identification methods for visco- and hyperelastic material models. *Glass Structures & Engineering*, *2*, 147–167.

Lai, S.-M., You, P.-Y., Chiu, Y. T., & Kuo, C. W. (2017). Triple-shape memory properties of thermoplastic polyurethane/olefin block copolymer/polycaprolactone blends. *Journal of Polymer Research*, *24*, 161.

Lei, M., Hong, W., Zhao, Z., Hamel, C., Chen, M., Lu, H., et al. (2019). 3D printing of Auxetic metamaterials with digitally reprogrammable shape. *ACS Applied Materials & Interfaces*, *11*, 22768–22776.

Lendlein, A., & Gould, O. (2019). Reprogrammable recovery and actuation behaviour of shape-memory polymers. *Nature Reviews Materials*, *4*, 116–133.

Lendlein, A., & Kelch, S. (2002). Shape-memory polymers. *Angewandte Chemie International Edition*, *41*, 2034–2057.

Lendlein, A., & Langer, R. (2002). Biodegradable, elastic shape-memory polymers for potential biomedical applications. *Science*, *296*(5573), 1673–1676.

Leng, J., Lan, X., Liu, Y., & Du, S. (2011). Shape-memory polymers and their composites: Stimulus methods and applications. *Progress in Materials Science*, *56*(7), 1077–1135.

Li, J., Liu, T., Xia, S., Pan, Y., Zheng, Z., Ding, X., et al. (2011). A versatile approach to achieve quintuple-shape memory effect by semi-interpenetrating polymer networks containing broadened glass transition and crystalline segments. *Journal of Materials Chemistry*, *21*, 12213–12217.

Li, J., Liu, T., Pan, Y., Xia, S., Zheng, Z., Ding, X., et al. (2012). A versatile polymer co-network with broadened glass transition showing adjustable multiple-shape memory effect. *Macromolecular Chemistry and Physics*, *213*, 2246–2252.

Li, H., Gao, X., & Luo, Y. (2016). Multi-shape memory polymers achieved by the spatio-assembly of 3D printable thermoplastic building blocks. *Soft Matter*, *12*, 3226–3233.

Li, X., Pan, Y., Zheng, Z., & Ding, X. (2018). A facile and general approach to recoverable high-strain multishape shape memory polymers. *Macromolecular Rapid Communications*, *39*, e1700613.

Liu, C., & Mather, P. T. (2002). Thermomechanical characterization of a tailored series of shape memory polymers. *Journal of Applied Medical Polymers*, *6*(2), 47–52.

Liu, Y., Gall, K., Dunn, M. L., & McCluskey, P. (2003). Thermomechanical recovery couplings of shape memory polymers in flexure. *Smart Materials and Structures*, *12*(6), 947–954.

Luo, X., & Mather, P. T. (2010). Triple-shape polymeric composites (TSPCs). *Advanced Functional Materials*, *20*, 2649–2656.

Luo, Y., Guo, Y., Gao, X., Li, B., & Xie, T. (2013). A general approach towards thermoplastic multishape-memory polymers via sequence structure design. *Advanced Materials*, *25*, 743–748.

Mao, Y., Yu, K., Isakov, M. S., Wu, J., Dunn, M. L., & Qi, H. J. (2015). Sequential self-folding structures by 3D printed digital shape memory polymers. *Scientific Reports*, *5*, 13616.

Meng, H., Mohamadian, H., Stubblefield, M., Jerro, D., Ibekwe, S., Pang, S. S., et al. (2013). Various shape memory effects of stimuli-responsive shape memory polymers. *Smart Materials and Structures*, *22*, 093001.

Miaudet, P., Derre, A., Maugey, M., Zakri, C., Piccione, P., Inoubli, R., et al. (2007). Shape and temperature memory of nanocomposites with broadened glass transition. *Science*, *318*, 1294–1296.

Mirtschin, N., & Pretsch, T. (2015). Designing temperature-memory effects in semicrystalline polyurethane. *RSC Advances*, *5*, 46307.

Nöchel, U., Reddy, C. S., Wang, K., Cui, J., Zizak, I., Behl, M., et al. (2015). Nanostructural changes in crystallizable controlling units determine the temperature-memory of polymers. *Journal of Materials Chemistry A*, *3*(16), 8284–8293.

Otsuka, K., & Wayman, C. M. (Eds.). (1999). *Shape memory materials* Cambridge University Press.

Pandini, S., Inverardi, N., Scalet, G., Battini, D., Bignotti, F., Marconi, S., et al. (2020). Shape memory response and hierarchical motion capabilities of 4D printed auxetic structures. *Mechanics Research Communications*, *103*, 103463.

Pei, E., Hsiang Loh, G., Harrison, D., de Amorim Almeida, H., Monzon Verona, M. D., & Paz, R. (2017). A study of 4D printing and functionally graded additive manufacturing. *Assembly Automation*, *37*(2), 147–153.

Peng, B., Yang, Y., Gu, K., Amis, E. J., & Cavicchi, K. A. (2019). Digital light processing 3D printing of triple shape memory polymer for sequential shape shifting. *ACS Materials Letters*, *14*, 410–417.

Qi, H., & Dunn, M. (2010). Thermomechanical behavior and Modeling approaches. In J. Leng, & S. Du (Eds.), *Shape-memory polymers and multifunctional composites* CRC Press.

Qi, X., Guo, Y., Wei, Y., Dong, P., & Fu, Q. (2016). Multishape and temperature memory effects by strong physical confinement in poly(propylene carbonate)/graphene oxide nanocomposites. *The Journal of Physical Chemistry B*, *120*, 11064–11073.

Samuel, C., Barrau, S., Lefebvre, J. M., Raquez, J. M., & Dubois, P. (2014). Designing multiple-shape memory polymers with miscible polymer blends: Evidence and origins of a triple-shape memory effect for miscible PLLA/PMMA blends. *Macromolecules, 47*, 6791–6803.

Sauter, T., Heuchel, M., Kratz, K., & Lendlein, A. (2013). Quantifying the shape-memory effect of polymers by cyclic thermomechanical tests. *Polymer Reviews, 53*(1), 6–40.

Scalet, G. (2020). Two-way and multiple-way shape memory polymers for soft robotics: An overview. *Actuators, 9*(1), 10.

Shao, Y., Lavigueur, C., & Zhu, X. (2012). Multishape memory effect of norbornene-based copolymers with cholic acid pendant groups. *Macromolecules, 45*, 1924–1930.

Song, L., Hu, W., Wang, G., Niu, G., Zhang, H., Cao, H., et al. (2010). Tailored (meth)acrylate shape-memory polymer networks for ophthalmic applications. *Macromolecular Bioscience, 10*, 1194–1202.

Sun, L., & Huang, W. M. (2010). Mechanisms of the multi-shape memory effect and temperature memory effect in shape memory polymers. *Soft Matter, 6*, 4403–4406.

Tibbits, S.J.E., Papadopoulou, A., & Laucks, J.S. (2018). US10549505B2.

Wagner, M., Chen, T., & Shea, K. (2017). Large shape transforming 4D auxetic structures. *3D Printing and Additive Manufacturing, 4*(3), 133–142.

Wang, Y., Li, J., Li, X., Pan, Y., Zheng, Z., Ding, X., et al. (2014). Relation between temperature memory effect and multiple-shape memory behaviors based on polymer networks. *RSC Advances, 4*, 20364.

Wang, K., Jia, Y. G., & Zhu, X. (2015). Biocompound-based multiple shape memory polymers reinforced by photo-cross-linking. *ACS Biomaterials Science & Engineering, 1*, 855–863.

Ware, T., Hearon, K., Lonnecker, A., Wooley, K. L., Maitland, D. J., & Voit, W. (2012). Triple-shape memory polymers based on self-complementary hydrogen bonding. *Macromolecules, 45*(2), 1062–1069.

Westbrook, K., Kao, P., Castro, F., Ding, Y., & Qi, H. (2011). A 3D finite deformation constitutive model for amorphous shape memory polymers: A multi-branch modeling approach for nonequilibrium relaxation processes. *Mechanics of Materials, 43*, 853–869.

Wong, Y. S., Xiong, Y., Venkatraman, S. S., & Boey, F. Y. C. (2008). Shape memory in un-cross-linked biodegradable polymers. *Journal of Biomaterials Science, Polymer Edition, 19*(2), 175–191.

Wu, J., Yuan, C., Ding, Z., Isakov, M., Mao, Y., Wang, T., et al. (2016). Multi-shape active composites by 3D printing of digital shape memory polymers. *Scientific Reports, 6*, 24224.

Xiao, R., Guo, J., & Nguyen, T. D. (2015). Modeling the multiple shape memory effect and temperature memory effect in amorphous polymers. *RSC Advances, 5*, 416.

Xie, T. (2010). Tunable polymer multi-shape memory effect. *Nature, 464*, 267–270.

Xie, T., & Rousseau, I. A. (2009). Facile tailoring of thermal transition temperatures of epoxy shape memory polymers. *Polymer, 50*(8), 1852–1856.

Xie, T., Page, K. A., & Eastman, S. A. (2011). Strain-based temperature memory effect for Nafion and its molecular origins. *Advanced Functional Materials, 21*, 2057–2066.

Yakacki, C. M., Shandas, R., Safranski, D., Ortega, A. M., Sassaman, K., & Gall, K. (2008). Strong, tailored, biocompatible shape-memory polymer networks. *Advanced Functional Materials, 18*(16), 2428–2435.

Yakacki, C. M., Nguyen, T. D., Likos, R., Lamell, R., Guigou, D., & Gall, K. (2011). Impact of shape-memory programming on mechanically-driven recovery in polymers. *Polymer, 52*(21), 4947–4954.

Yang, X., Wang, L., Wang, W., Chen, H., Yang, G., & Zhou, S. (2014). Triple shape memory effect of star-shaped polyurethane. *ACS Applied Materials & Interfaces, 6*, 6545–6554.

Yu, K., & Qi, H. J. (2014). Temperature memory effect in amorphous shape memory polymers. *Soft Matter, 10*, 9423.

Yu, K., Dunn, M. L., & Qi, H. J. (2015). Digital manufacture of shape changing components. *Extreme Mechanics Letters*, *4*, 9–17.

Yu, K., Ge, Q., & Qi, H. (2014). Reduced time as a unified parameter determining fixity and free recovery of shape memory polymers. *Nature Communications*, *5*(3066), 1–9.

Yu, K., Ritchie, A., Mao, Y., Dunn, M. L., & Qi, H. J. (2015). Controlled sequential shape changing components by 3D printing of shape memory polymer multimaterials. *Procedia IUTAM*, *12*, 193–203.

Yu, K., Xie, T., Leng, J., Ding, Y., & Qi, H. J. (2012). Mechanisms of multi-shape memory effects and associated energy release in shape memory polymers. *Soft Matter*, *8*, 5687–5695.

Zare, M., Prabhakaran, M. P., Parvin, N., & Ramakrishna, S. (2019). Thermally-induced two-way shape memory polymers: Mechanisms, structures, and applications. *Chemical Engineering Journal*, *374*, 706–720.

Zhang, Q., Wei, H., Liu, Y., Leng, J., & Du, S. (2016). Triple-shape memory effects of bismaleimide based thermosetting polymer networks prepared via heterogeneous crosslinking structures. *RSC Advances*, *6*, 10233–10241.

Zhao, J., Chen, M., Wang, X., Zhao, X., Wang, Z., Dang, Z.-M., et al. (2013). Triple shape memory effects of cross-linked polyethylene/polypropylene blends with cocontinuous architecture. *ACS Applied Materials & Interfaces*, *5*(12), 5550–5556.

Zheng, N., Hou, J., Xu, Y., Fang, Z., Zou, W., Zhao, Q., et al. (2017). Catalyst-free thermoset polyurethane with permanent shape reconfigurability and highly tunable triple shape memory performance. *ACS Macro Letters*, *6*, 326–330.

Zotzmann, J., Behl, M., Hofmann, D., & Lendlein, A. (2010). Reversible triple-shape effect of polymer networks containing polypentadecalactone- and poly(ε-caprolactone)-segments. *Advanced Materials*, *22*, 3424–3429.

Manufacturing highly elastic skin integrated with twisted and coiled polymer muscles: Toward 4D printing

Armita Hamidi[a] and Yonas Tadesse[b]
[a]David L. Hirschfeld Department of Engineering, Angelo State University, San Angelo, TX, United States
[b]Mechanical Engineering Department, The University of Texas at Dallas, Richardson, TX, United States

Introduction

Manufacturing of active materials or smart materials that are responsive to multiple stimuli is essential in many areas such as energy harvesting, energy storage, and soft robotics(Sydney Gladman, Matsumoto, Nuzzo, Mahadevan and Lewis, 2016). Active materials or smart materials are those that change their shapes or generate electrical charges upon the application of stimuli such as heat, vibration, electrical current, or diffusion of ions. Developing soft robots integrated with active materials has been gaining popularity due to the ability of these smart structures to interact safely with humans and the ability to function in an inaccessible environment (Case, White, & Kramer, 2015). Fabrication of these structures is extremely dependent on the overall design and the materials used during the fabrication process. Typically, soft materials in soft and biomimetic robots are commonly composed of hydrogels (Li, Go, Ko, Park, & Park, 2016; Morales, Palleau, Dickey, & Velev, 2014) and elastomers (Hines, Petersen, Lum, & Sitti, 2017; Lin, Leisk, & Trimmer, 2011; Onal, Chen, Whitesides, & Rus, 2017). Elastomers have lower elastic modulus (E < 1 MPa), similar to biological tissues and cells (Akhtar, Sherratt, Cruickshank, & Derby, 2011), and are suitable for stretchable electronics. They are mainly manufactured by conventional manufacturing technologies, namely casting or molding techniques. Lately, additive manufacturing techniques are partially or fully utilized through manufacturing soft robots (Zolfagharian et al., 2020; Zolfagharian, Kaynak, & Kouzani, 2020). The soft compartments of the designed structure can be directly 3D-printed or cast by 3D-printed molds. However, traditional casting these materials or conventional 3D printing is still time-consuming, as it requires multiple steps including preparing molds and coating with release agents as well as a long curing time (Calderon, Ugalde, Zagal, & Perez-Arancibia, 2016; Martinez, Fish, Chen, & Whitesides, 2012) or assembling the actuators into the soft structure. Moreover, none of these methods are suitable for producing parts having complex internal geometries.

New manufacturing methods are desired to replace traditional techniques and to realize complex and novel designs. The new manufacturing method should also accommodate the gradual transition from soft to hard components (Wallin, Pikul, & Shepherd, 2018).

Several additive manufacturing techniques and processing methods have been developed for printing silicone, active materials, and other soft materials. Stereolithography (SLA) (Odent et al., 2017; Plott, Tian, & Shih, 2018; Wallin et al., 2017), digital light processing (DLP) printing (Patel et al., 2017), selective laser sintering (SLS), photopolymer jetting, fused deposition modeling (FDM) (Christ, Aliheidari, Ameli, & Pötschke, 2017), and direct ink writing (DIW) (Plott et al., 2018) have been used for 3D printing soft materials. All these techniques have limitations either in printing resolution and mechanical properties as described in detail in our work (Hamidi & Tadesse, 2020). Among all these techniques, FDM and DIW are the most common methods due to their low material cost and ease of use. However, FDM has limitations in material compatibility. In the FDM process, the filament material is inserted into the machine in the form of a solid cylinder and only thermoplastic polymers can be used in this technology due to the heating and cooling process (Wallin et al., 2018). The most commonly used thermoplastics in FDM are acrylonitrile butadiene styrene (ABS), polylactic acid (PLA), polyethylene terephthalate glycol (PETG), and rubber-like thermoplastics such as thermoplastic elastomer (TPE). Liquid forms such as room-temperature-vulcanizing (RTV) silicones and gels cannot be used in FDM, which is one of the problems of this technology. DIW is another technique, which is widely used for printing hydrogel (Bakarich et al., 2017) and silicone (Morrow, Hemleben, & Menguc, 2017; Yirmibeşoğlu, Oshiro, Olson, Palmer, & Mengüç, 2019). This technique is also further developed for 3D printing other types of gels (Ren, Shao, Lin, & Zheng, 2016). Yet, preparing a customized ink (or solution) that has proper rheological properties for extrusion and that can provide expected mechanical properties after rapid solidification is challenging. Four-dimensional (4D) printing and multimaterial printing are some of the unique techniques that have been discussed to realize active materials or shape-morphing structures in recent years (Ge, Dunn, Qi, & Dunn, 2014; Ge et al., 2016; Momeni, Mehdi Hassani, Liu, & Ni, 2017; Sydney Gladman et al., 2016; Tibbits, 2014). However, these techniques did not utilize the benefits of twist insertion in the specialized technical fibers that can be manufactured with other structures to obtain unprecedented functional structures such as undulatory motion (two-wave or more) that resemble certain animals (Almubarak & Tadesse, 2017), crawling (Hamidi & Tadesse, 2020), and flapping-wing (Hamidi, Almubarak, Rupawat, Warren, & Tadesse, 2020; Hamidi, Almubarak, & Tadesse, 2019). The TCP muscles themselves discovered in 2014 (Haines et al., 2014) which have strain up to 50% and energy density 5.2 kW per fiber mass and presented as a new class of actuators. Such high-performance actuators or muscles are not utilized extensively in additive manufacturing systems as we presented in this chapter.

Continuous advancements in additive manufacturing (AM) enable the creation of autonomous robots through computer-aided design (CAD) models and digital fabrication. Thus, 3D printers are represented as smart factories, which shorten the product design and development time. AM is fundamentally different from conventional manufacturing such as subtractive and formative manufacturing techniques, as it employs digital design to build a structure into its designated shape using a layer-by-layer approach. This permits unprecedented freedom in fabricating complex, composite and hybrid structures with precision and control that cannot be achieved through traditional manufacturing routes.

The early use of AM is focused on preproduction visualization models. Rapid advances in technology result in the increasing demand for customized products with more complex designs and smaller features. Consequently, the AM application is broadening from building conceptual models and testing prototypes to the end-user products used in aerospace, automotive, healthcare, wearable devices, robots, and rehabilitation devices. This requires further development of AM machines and systems. Currently, 3D printing complex structure with versatile materials is one of the greatest challenges. So far, the AM has been used with a limited selection of compatible materials relative to the number of materials available in traditional manufacturing methods. Hence, the functionality of the parts produced by a 3D printer is also limited. 3D-printed parts generally suffer from anisotropic mechanical properties due to building orientation and interlayer bonding (Ahn, Montero, Odell, Roundy, & Wright, 2002). Moreover, the material and equipment cost make this technology expensive. The price of industrial-grade 3D printers is greater than $100k, which is high for the initial expenses of utilizing the technology.

Despite the benefits of 3D printing in realizing complex shapes, the inability of shape changes with time led to another technology known as 4D printing. Thus far, most of the works in 4D-printed systems are limited to direct 3D printing of stimuli-responsive materials with the emerged response(s) after the fabrication process, for example, hydrogel additive manufacturing (Shafranek et al., 2019). In another related work, computational models are proposed for 4D printing (Bodaghi, Noroozi, Zolfagharian, Fotouhi, & Norouzi, 2019; Zolfagharian, Denk, Bodaghi, Kouzani, & Kaynak, 2020). A review of 3D-printed sensors and 3D-printed actuators integrated into the various structures is discussed in Zolfagharian, Kaynak, and Kouzani (2020). We can say that soft polymeric materials are generally used in this method to keep the viscoelastic nature of the structure for deformation in applications related to soft robotics. In the context of this chapter, 4D printing refers to the ability of 3D printing some passive elastic material structural configuration and incorporating active materials during printing, so that the final structures can achieve morphed shapes when electrical current is provided. This is different than other approaches that directly print the active materials through additive manufacturing, because of the unique use of relatively new polymer muscles as active elements and the method of insertion in the 3D printing process.

Here, we describe an embedded 3D printing technique by means of flexible twisted and coiled polymer (TCP) fibers as embedded actuators. These actuators are well developed to integrate within the three-dimensional-printed flexible silicone matrix during the printing process, which is the important step. The printed structures acquired high flexibility and customization, can achieve morphed shapes by actuation, and then retained their original shape after actuation. We will describe the materials used, which are twisted and coiled polymer (TCP) muscles, custom-made 3D printer setup, the manufacturing process parameters, and the resulting structure/products and discussion.

Materials

TCP

New actuator technologies take advances from the properties of those materials which significantly changed in a controlled fashion by small changes in their environment. Materials that respond to multiple environmental stimuli are often called smart materials or active materials. These smart actuators include materials that demonstrate similar physical properties (stiffness, elasticity, strength, and density) as biological counterparts. Twisted and coiled polymer actuator (TCP) is one of the new types of smart actuators that is attracting for application as an artificial muscle in biomimetic robots due to its ability to reproduce the important features of natural muscle such as power, stress, strain, and speed. TCP muscle invented by Haines et al. (Haines et al., 2014; Saharan & Tadesse, 2019) is a polymer actuator that responds to the temperature gradient.

TCPs are made out of semicrystalline polymer fibers such as nylon 6,6; silver-coated nylon 6,6; and polyethylene. Typically, the polymer fibers are chosen from commercially inexpensive nylon fibers, namely fishing line or silver-coated sewing threads. Several companies provide the material, one example is manufactured by Statex Produktions & Vertriebs GmbH based in Germany. Silver-coated thread is a nylon thread that is coated with silver; therefore, it is conductive. The coating is around 100 nm (Saharan, De Andrade, Saleem, Baughman, & Tadesse, 2017), whereas the fiber diameter of the nylon is 0.85 mm for 272 fiber bundles in 2-ply TCP. TCP muscles made from silver-coated threads can carry current, heat up, and actuate by directly connecting to the electrical power source due to silver particles. As shown in Fig. 11.1, the TCP actuator is fabricated by suspending a deadweight (m_f) at the bottom end of a highly drawn polymer fiber such as silver-coated nylon and twisting it by a motor, which is connected to the top end of the fiber (Saharan & Tadesse, 2019). Using a similar setup as shown in Fig. 11.1, by folding or adding the fiber threads, different actuator structures such as 2-ply, 4-ply, and 6-ply TCP can be produced. The number of required threads can be determined based on the load and desired the deformation of the structure where TCP is used. Based on a silver-coated nylon precursor fiber of diameter 0.2 mm, the dead weight used for the fabrication process is also proportional to the number of fibers ($N*175/2$ g for N-ply TCP) (Saharan et al., 2017). After crimping twisted fibers on both ends, they are annealed and trained by suspending a deadweight (m_{tr}) and applying a sequence of input current ($N*0.75/2$ A for N-ply TCP). The annealing and training, in this case, are done for 18 and 15 heating/cooling cycles, respectively, until the actuation strain reaches a steady state.

A video of the fabrication process for 2-ply TCPs can be found from HBS Lab YouTube Channel https://youtu.be/S4-3_DnKE9E. The fabrication process consists of few steps, twisting, coiling, annealing, training, and finally actuation test. A similar procedure can be followed to make different structures such as 4-ply and 6-Ply TCP. The twisting process is typically set to occur until the entire length of the thread is coiled which results in N-ply coiled polymer. Various structures are used for practical applications. For example, 2-ply and 3-ply have been used for robotic hands

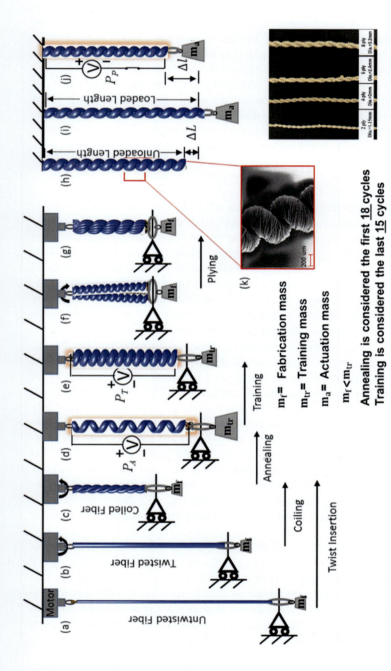

Fig. 11.1 TCP Fabrication setup and process (A) twisting under a load mf, (B) twisted fiber, (C) coiled fiber, (D) annealing under power (PA) and load m_{tr}, (E) training under power (PT) and m_{tr}, (F) plying 1-ply to form 2-ply, (G) 2-ply structure, (H) final structure before testing, (I) applying a test load, and (J) testing the actuator at a load m_a.

This figure is edited from our group's original publication, Wu, L., Jung De Andrade, M., Saharan, L. K., Rome, R. S., Baughman, R. H., & Tadesse, Y. (2017). Compact and low-cost humanoid hand powered by nylon artificial muscles. *Bioinspiration & Biomimetics*, *12*(2), 026004. The corresponding author for the aforementioned paper is myself, coauthor in this chapter, Yonas Tadesse. We edited and made several changes, so I think, we do not need permission from the publisher IOP science.

(Wu et al., 2017). Recently, a jellyfish-like robot is illustrated using 6-ply silver-coated TCP actuators made of 3 threads of equal length that are tied together at both ends and then coiled (Hamidi et al., 2020). The fabricated 6-ply actuator is then trained and annealed at an input current of 2.2A producing an average actuation strain of 19%. In general, TCPs are low-cost actuators and have several advantages. For instance, the cost of 6-ply TCP is \$4.5/m. Twisted and coiled polymer actuators/muscles made out of silver-coated nylon 6,6 (TCP_{Ag}) are relatively expensive than the TCP made out of monofilament nylon fishing line (TCP_{FL}) that are utilized for actuating a robotic finger (Wu et al., 2015). They cost \$5 per kg as the heating source is cold and hot water, they do not need heating elements such as nichrome.

Silicone

An elastomer is a polymer that can recover to its original shape after extensive deformation, i.e., elastic deformation. The elastomer primarily consists of a network of polymer chains, with the polymer being above its glass transition temperature and being amorphous, it has a high molecular weight (Mazurek, Vudayagiri, & Skov, 2019). Due to low Young's modulus and high strain, an elastomeric matrix is suggested for fabricating active composite materials by directly imprinting shape memory polymer fibers or other active materials into the elastomer template. In the context of the 4D printing method in this chapter, a commercial room-temperature-vulcanizing (RTV) silicone elastomer (Ecoflex Series, Smooth-On, Inc.) is used for embedding the TCP fibers as an active material. Based on experimental results, the prepared matrix has an adequate Young's modulus to overcome the problem of lateral deflections and twisting and can provide the printed structure with higher degrees of freedom where complex movement pattern is needed. The RTV silicone cures at room temperature upon exposure to air with negligible shrinkage. The material generally comes in two components. One part contains the base polymer with the catalyst (platinum), whereas the second part includes the cross-linker and together with the polymer to dilute the cross-linker. However, in the second part, the cross-linker and polymer are unreactive and stable since no catalyst is present. Mixing of the two parts causes the curing to take place at room temperature without any by-products; however, heating the mixture will increase the curing rate. Fig. 11.2 shows the chemical structures, the chemical reaction, and the resulting cured structure of the RTV silicone elastomer.

Due to the relatively low density ($1040.2\,kg/m^3$) and good flowability of the mixture (viscosity in the range of 140 Poise) (Gopinathan & Noh, 2018), such silicone is a good candidate for extrusion-based 3D printing. For obtaining lower viscosity and easier extrusion, oftentimes silicone thinner is added immediately after mixing two parts of RTV silicone before it is cured. Silicone thinner is a nonreactive silicone fluid that lowers the viscosity of both base polymer and cross-linker components during mixing (Tadesse, Moore, Thayer, & Priya, 2009) and allows the mixture to flow easily through the nozzle.

Fig. 11.3 presents the mechanical properties of RTV silicone. Maximum elongation versus strain is presented for different methods of curing elastomers along with

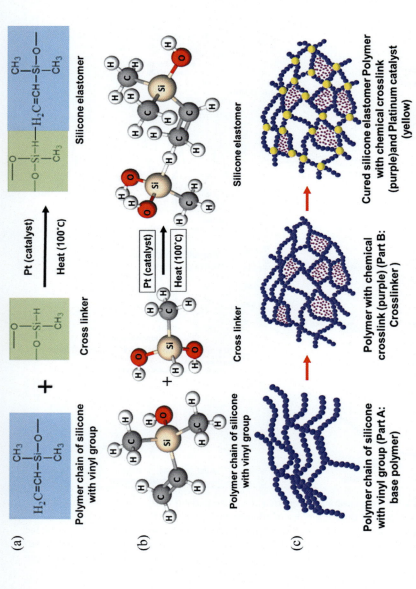

Fig. 11.2 Elastomer chemical reaction: (A) Chemical reaction between silicone polymer chain and cross-linker by adding heat and using a catalyst, (B) molecular structure of the silicone polymer chain, cross-linker and the product (after curing), and (C) chemical reaction between two premixes of RTV silicone, from left to right: base silicone polymer chain before mixing with cross-linker (part A), silicone polymer chain with a cross-linker (right after adding part B), cured silicone polymer chain (after the mixture of part A and B cured). No Permission Required.

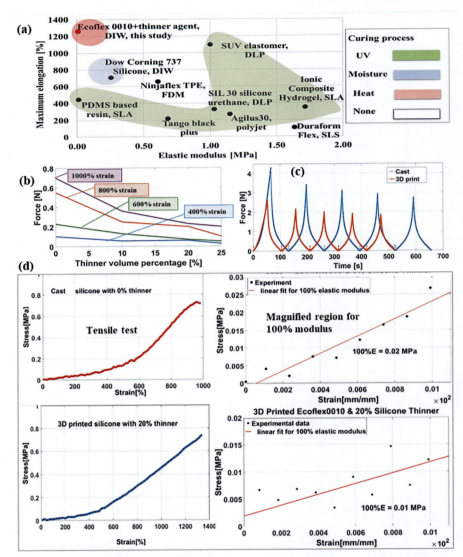

Fig. 11.3 RTV silicone elastomer properties: (A) Comparison of the elastic modulus and maximum elongation of different types of silicone that are utilized within additive manufacturing technology. *Color map* represents different curing processes, (B) RTV silicone elastic properties changed with adding different volume percentages of thinner for dog bone samples, (C) cyclic tensile test results of cast and 3D-printed RTV silicone, and (D) tensile test results and 100% modulus of cast and 3D-printed RTV silicone.
No Permission Required.

our sample. Cast samples of RTV silicone have a tensile strength of 0.62 ± 0.1 MPa and a strain of 700%. These are determined experimentally using a tensile testing machine. The tensile strength of the 3D-printed elastomer is lower than the cast ones, while the strain is increased to 800% (Hamidi & Tadesse, 2019). These results are comparable with other studies on cast Ecoflex 0030 and found 600% strain under 8 MPa stress (Case et al., 2015). Based on the tensile test results of 3D-printed samples, the elasticity of 800% at a load below 0.4 MPa is significant, which is due to the addition of silicone thinner. Cyclic testing of the silicone is also shown, where the strain magnitude for 3D-printed silicone is lower than the cast samples.

Manufacturing

3D printing is a technology that has been around for a while; therefore, it has been widely developed and had made its way from industry to household. Desktop 3D printers bring a new era of consumer 3D printing by offering an affordable tool for rapid prototyping, tooling, and end-part production. In this section, a novel approach for adopting an open-source desktop DIY kit 3D printer to fabricate 4D-printed structures is explained. As a case study, an existing open-source 3D printer (Prusa 8″ i3v kit, Maker Farm) is modified to print silicone elastomers with embedded TCP actuators. Later, it is shown that printed structure by actuating TCPs embedded within the elastomer matrix, the printed structure converts to a morphed structure. A paste extruder setup (DISCOV3RY, Structur3D Printing Inc. based in Canada) is modified to 3D print Ecoflex 0010 as shown in Fig. 11.4A. A motorized paste extruder shown in the schematic diagram in Fig. 11.4B uses a 60-cc syringe, plastic connectors, and a 305-mm-long plastic feeding pipe with an inner diameter of 3.2 mm. The motorized driver setup is preferred as it is easily adapted to the 3D printer control board without a need for any additional component such as a compressed air source (compressor) used in pneumatic drivers. A plastic nozzle is attached to the end of the tube for controlling the resolution. The material in the syringe is pushed through the tube due to the pressure of the plunger that is driven by a stepper motor at the top. To facilitate a better flow, the paste extruder is installed in a way that the syringe head faced the printer bed to utilize the gravity effect to achieve a more consistent flow. For 3D printing samples, 0.8, 0.6, and 0.4 mm (inner diameter) nozzles are used successfully. For this study, the smallest nozzle size (0.4 mm) is used to print the structures with high precision. A single layer thickness of a fabricated sample is 0.43 mm. For measuring the dimensional tolerances, a 100 mm × 100 mm × 15 mm structure is 3D-printed, and each side is measured and compared to the initial design. The measured error was 0.4 mm in the x- and y-directions and 0.1 mm in the z-direction, which is less than 1% error in all directions. The process parameters for silicone extrusion are shown in Table 11.1. The bed temperature is set at 100°C for printing with a 0.4-mm nozzle and the extruded layer height of the printed structure is measured as 0.43 mm. The curing rate of RTV silicone is heavily dependent on the temperature and amount of the material. For shortening the curing time and avoiding self-collapsing while maintaining enough bonding between the layers, the printer bed temperature was

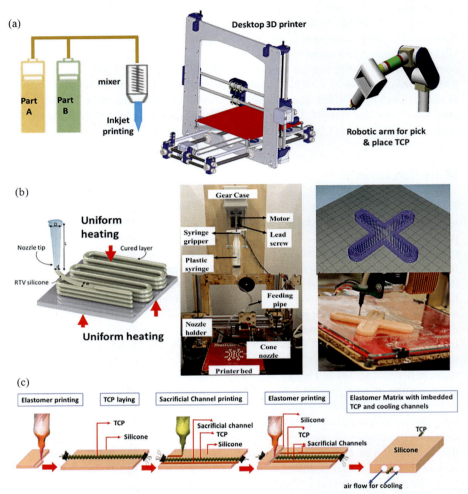

Fig. 11.4 4D additive manufacturing process: (A) Proposed concept: a desktop 3D-printed (middle) with RTV silicone elastomer setup (left) and automated pick and place robotic manipulator; (B) actual prototype as a preliminary result, RTV silicone printing process (left) and a simple set up with sliced sample before and after printing (right), and (C) proposed process steps for highly elastic skin integrated with TCP muscle.
No Permission Required.

set to maintain at 100°C with the other mentioned parameters in Table 11.1 to successfully print structures up to 25 mm height. This temperature is optimized based on the fastest curing time that is achieved for the silicone by experiment. After several trial-and-error experiments, it is obtained that a silicone layer extruded from a 0.4-mm-diameter nozzle will cure in 1–2 s. For printing structures with heights more than 25 mm, an external heating source should be installed for curing the top layers that are further distanced from the bed heating source.

Table 11.1 Process parameters.

Nozzle diameter	Layer thickness after curing	Nozzle temperature	Print bed temperature	Head translation speed	Layer height	Infill density
0.4 mm	0.5 mm	25°C (room temperature)	100°C	15 mm/s	0.43 mm	100%

Results and discussion

The development of high-performance artificial muscles with low-cost fabrication, complex actuation, easy operation, and scalable process is still a major scientific challenge. Advanced manufacturing methods are needed to utilize active fibers in the 3D printing process and obtain fully functional robotic systems. For example, the capabilities of conventional robots are often restricted as compared to bioinspired robots or in structures found in nature. The technical reason for this performance discrepancy is the limitations in the design, fabrication, sensing, control, and motion planning of the robot. Mismatch in material properties is one of the significant barriers inhibiting rigid robots from being incompatible with designs. Natural systems frequently benefit from a high degree of freedom owing to their body elasticity and their soft muscles. Taking inspiration from nature, a high degree of freedom in structural organization with embedded artificial muscles will result in a new class of smart devices. A functionalized network of flexible, responsive fibers embedded in a single conformable substrate allows sensing, actuation, and stiffness control of complex systems. Therefore, a well-established manufacturing technique, namely 4D printing, with embedded artificial muscles has a high potential to make breakthroughs into commercial products. Combination of active and responsive artificial muscles in any number of arrangements, yielding an immense design space for this concept.

To examine the functionality of embedded TCP in RTV silicone, a fish fin-like structure is fabricated using the traditional casting method (Fig. 11.5). The mold of the flapping structure is prepared by 3D printing ABS material. In the initial design, some slots are added all around the mold for placing and holding the TCPs. The depth of the slots designed halfway through the thickness of the flap so that the TCP would be covered completely with silicone. A spring steel with a thickness of 0.1 mm is mounted under the TCP where it is also embedded in the silicone matrix. Spring steel, the black color shown in Fig. 11.5, is also embedded to behave as passive membranes to help the soft matrix relax back to its original position after actuation, similar to the robotic jellyfish design presented in Hamidi et al. (2020). After the silicone is cured, the deformation of the soft structure is examined by applying electrical power (11 V) through the muscles, which is shown in Fig. 11.5 in the last row. However, this process of casting silicone is time-consuming and inconvenient.

As a preliminary result and demonstration, the potential of manufacturing embedded TCP in the soft matrix using an additive manufacturing technique, a starfish-like structure inspired from natural design (Fig. 11.6A–C) with a very soft body is printed

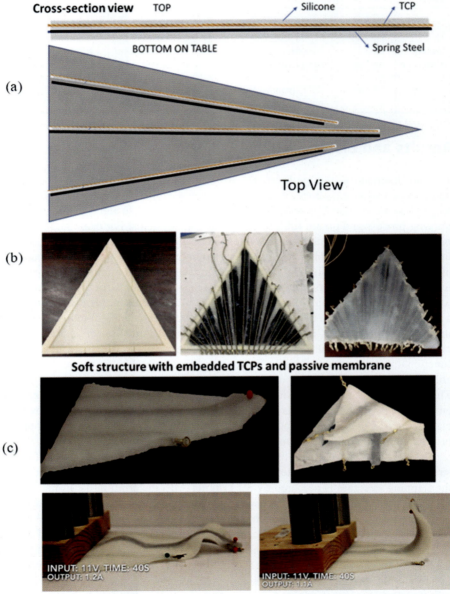

Fig. 11.5 3D-printed robotic biomimetic structure and preliminary initial results performed by casting: (A) An initial design of the soft structure with embedded TCPs and passive membrane (spring steel), (B) steps for fabricating a soft structure with embedded TCP muscles using a 3D-printed mold and cast RTV silicone, and (C) deformation of the soft structure during the actuation.
No Permission Required.

and actuated for evaluation. The final structure actuated successfully and obtained a morphed shape as shown in Fig. 11.6D when electrical power is provided to the active materials using DC power supply. The body structure is 3D-printed with RTV silicone to mimic the body of the starfish as shown in Fig. 11.6B. Multimaterial structures including a 3D-printed plastic (PLA) ring for fixing the TCPs, a fine nylon net fabric (tulle) for keeping the ring attached to the silicone, and the 2-ply TCPs are inserted in the structure during the fabrication process. One of the arms of the final structure with a thickness around 4.5 mm is actuated by providing a 3.4 W electrical DC power (4 V and 0.85 A) through the arm's TCP muscle for 28 s to obtain a morphed shape by respectively tilting the arm 2 mm and 5 mm in horizontal and vertical directions (8-degree bending angle), as shown in Fig. 11.6D.

Therefore, the method introduced in this chapter is versatile and can be employed to fabricate active materials and other printable passive materials for creating novel systems as shown in Fig. 11.6D (step by step). For instance, TCP muscle based on nylon 6,6 is chosen as one example of active materials, which has good compatibility in terms of design, size, and compliance to fit in a bioinspired design. Polylactic acid (PLA), silicone (highly elastic), and fabrics can be manufactured by using a custom-made printer setup that combines the approach from both DIW and FDM for printing a soft structure actuated with heat or resistive heating. The fabric was attached for a smooth transition from elastomer to thermoplastics by using a manual process (step 3). This will be replaced by a robotic manipulator and will be fully automated in the future. The process in this section is essential for the automated manufacturing system that is in line with the notion of 4D printing. The biomimetic system is created by embedding TCP actuators during the printing process of the elastomer as shown in Fig. 11.6. This approach presented in this chapter is unique and novel, and it will contribute to the method of manufacturing similar bioinspired structures. Pausing a printer and adding active materials such as SMAs has been described in Meisel et al. in 2015 (Meisel, Elliott, & Williams, 2015). However, the results and processes as shown in Fig. 11.6 are different due to the TCP materials used, the elastomer materials, and the fabric as well as processing parameters.

Conclusion

Soft robots are inherently deformable to pass through difficult environments, extensible to take on dynamic tasks, and resilient to outstand high loads. Inspired by nature, bioinspired robots have been presented recently which consist of multifunctional material systems. Typically, these systems can consist of both hard joints covered by soft silicone, to mimic tissues and smooth skin in biological counterparts. They may also require interfaces providing a smooth transition from hard to soft components. These kinds of structures will be able to replicate the dynamic motion of different creatures such as flapping, crawling, or swimming robotic systems. Thus, the process of building bioinspired robots is important for developing these robots. The manufacturing process should be able to combine the deposition of different materials, matching them and building a smooth transition at the interfaces. Moreover, the integration of actuators or artificial muscles and control methods are still challenging for

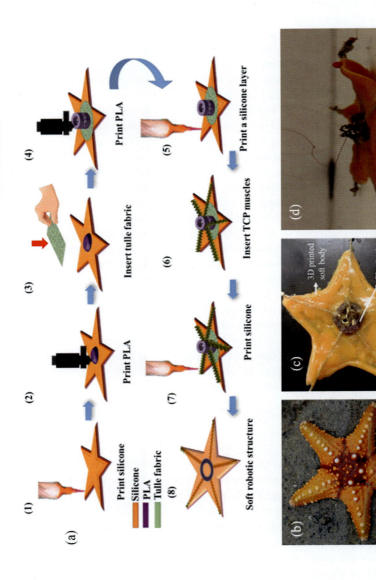

Fig. 11.6 Proposed 3D-printed robotic biomimetic structure and initial results: (A) Steps for fabricating a soft robot using TCP muscles embedded in a 3D-printed silicone structure, (B) natural starfish species, (C) 3D-printed starfish structure with inserted TCP muscles, and (D) deformation of the arm during the actuation.
No Permission Required.

making the robots, since bulky-powered driving mechanisms and electrical drives cannot easily fit in the biologically inspired structure. 4D printing technologies have the potential to fabricate these complex systems in a single process by embedding functional components and actuators into 3D structures. By introducing compliant active materials into the 3D printer, it will be able to build fully functional designs that can be activated by electricity, light, pressure, or shape memory effect. Despite several AM processes that are developed for 3D printing soft or hard materials, few of them have been applied for building robots. This study shows an inexpensive AM technique for integrating actuators in soft elastomers and identifies important material processing parameters to fabricate structures that mimic the flexible appendage of animals and develop a complete robot. The approach presented in this work is projected toward 4D printing of materials directly from CAD files. Here, we showed the integration of the active materials by embedding them in the middle of the printing process in a 3D printing elastomer. More work can be considered to fully automate the process which will be the future work.

References

Ahn, S. H., Montero, M., Odell, D., Roundy, S., & Wright, P. K. (2002). Anisotropic material properties of fused deposition modeling ABS. *Rapid Prototyping Journal, 8*(4), 248–257. https://doi.org/10.1108/13552540210441166.

Akhtar, R., Sherratt, M. J., Cruickshank, J. K., & Derby, B. (2011). Characterizing the elastic properties of tissues. *Materials Today, 14*(3), 96–105. https://doi.org/10.1016/S1369-7021 (11)70059-1.

Almubarak, Y., & Tadesse, Y. (2017). Twisted and coiled polymer (TCP) muscles embedded in silicone elastomer for use in soft robot. *International Journal of Intelligent Robotics and Applications, 1*. https://doi.org/10.1007/s41315-017-0022-x.

Bakarich, S. E., Gorkin, R., Gately, R., Naficy, S., in het Panhuis, M., & Spinks, G. M. (2017). 3D printing of tough hydrogel composites with spatially varying materials properties. *Additive Manufacturing, 14*, 24–30. https://doi.org/10.1016/j.addma.2016.12.003.

Bodaghi, M., Noroozi, R., Zolfagharian, A., Fotouhi, M., & Norouzi, S. (2019). 4D printing self-morphing structures. *Materials, 12*(8). https://doi.org/10.3390/ma12081353.

Calderon, A. A., Ugalde, J. C., Zagal, J. C., & Perez-Arancibia, N. O. (2016). Design, fabrication and control of a multi-material-multi-actuator soft robot inspired by burrowing worms. In *2016 IEEE international conference on robotics and biomimetics, ROBIO 2016* (pp. 31–38). Institute of Electrical and Electronics Engineers Inc. https://doi.org/10.1109/ROBIO.2016.7866293.

Case, J. C., White, E. L., & Kramer, R. K. (2015). Soft material characterization for robotic applications. *Soft Robotics, 2*(2), 80–87. https://doi.org/10.1089/soro.2015.0002.

Christ, J. F., Aliheidari, N., Ameli, A., & Pötschke, P. (2017). 3D printed highly elastic strain sensors of multiwalled carbon nanotube/thermoplastic polyurethane nanocomposites. *Materials and Design, 131*, 394–401. https://doi.org/10.1016/j.matdes.2017.06.011.

Ge, Q., Dunn, C. K., Qi, H. J., & Dunn, M. L. (2014). Active origami by 4D printing. *Smart Materials and Structures, 23*(9). https://doi.org/10.1088/0964-1726/23/9/094007.

Ge, Q., Sakhaei, A. H., Lee, H., Dunn, C. K., Fang, N. X., & Dunn, M. L. (2016). Multimaterial 4D printing with tailorable shape memory polymers. *Scientific Reports, 6*. https://doi.org/10.1038/srep31110.

Gopinathan, J., & Noh, I. (2018). Recent trends in bioinks for 3D printing. *Biomaterials Research*. https://doi.org/10.1186/s40824-018-0122-1.

Haines, C. S., Lima, M. D., Li, N., Spinks, G. M., Foroughi, J., Madden, J. D. W., et al. (2014). Artificial muscles from fishing line and sewing thread. *Science, 343*(6173), 868–872. https://doi.org/10.1126/science.1246906.

Hamidi, A., Almubarak, Y., Rupawat, Y. M., Warren, J., & Tadesse, Y. (2020). Poly-Saora robotic jellyfish: Swimming underwater by twisted and coiled polymer actuators. *Smart Materials and Structures, 29*(4). https://doi.org/10.1088/1361-665X/ab7738.

Hamidi, A., Almubarak, Y., & Tadesse, Y. (2019). Multidirectional 3D-printed functionally graded modular joint actuated by TCPFL muscles for soft robots. *Bio-Design and Manufacturing, 2*(4), 256–268. https://doi.org/10.1007/s42242-019-00055-6.

Hamidi, A., & Tadesse, Y. (2019). Single step 3D printing of bioinspired structures via metal reinforced thermoplastic and highly stretchable elastomer. *Composite Structures, 210*, 250–261. https://doi.org/10.1016/j.compstruct.2018.11.019.

Hamidi, A., & Tadesse, Y. (2020). 3D printing of very soft elastomer and sacrificial carbohydrate glass/elastomer structures for robotic applications. *Materials & Design, 187*, 108324. https://doi.org/10.1016/j.matdes.2019.108324.

Hines, L., Petersen, K., Lum, G. Z., & Sitti, M. (2017). Soft actuators for small-scale robotics. *Advanced Materials, 29*(13). https://doi.org/10.1002/adma.201603483.

Li, H., Go, G., Ko, S. Y., Park, J. O., & Park, S. (2016). Magnetic actuated pH-responsive hydrogel-based soft micro-robot for targeted drug delivery. *Smart Materials and Structures, 25*(2). https://doi.org/10.1088/0964-1726/25/2/027001.

Lin, H. T., Leisk, G. G., & Trimmer, B. (2011). GoQBot: A caterpillar-inspired soft-bodied rolling robot. *Bioinspiration & Biomimetics, 6*(2). https://doi.org/10.1088/1748-3182/6/2/026007.

Martinez, R. V., Fish, C. R., Chen, X., & Whitesides, G. M. (2012). Elastomeric origami: Programmable paper-elastomer composites as pneumatic actuators. *Advanced Functional Materials, 22*(7), 1376–1384. https://doi.org/10.1002/adfm.201102978.

Mazurek, P., Vudayagiri, S., & Skov, A. L. (2019). How to tailor flexible silicone elastomers with mechanical integrity: A tutorial review. *Chemical Society Reviews, 48*(6), 1448–1464. https://doi.org/10.1039/c8cs00963e.

Meisel, N. A., Elliott, A. M., & Williams, C. B. (2015). A procedure for creating actuated joints via embedding shape memory alloys in Poly Jet 3D printing. *Journal of Intelligent Material Systems and Structures, 26*(12), 1498–1512. https://doi.org/10.1177/1045389X14544144.

Momeni, F., Mehdi Hassani, M. N. S., Liu, X., & Ni, J. (2017). A review of 4D printing. *Materials and Design, 122*, 42–79. https://doi.org/10.1016/j.matdes.2017.02.068.

Morales, D., Palleau, E., Dickey, M. D., & Velev, O. D. (2014). Electro-actuated hydrogel walkers with dual responsive legs. *Soft Matter, 10*(9), 1337–1348. https://doi.org/10.1039/c3sm51921j.

Morrow, J., Hemleben, S., & Menguc, Y. (2017). Directly fabricating soft robotic actuators with an open-source 3-D printer. *IEEE Robotics and Automation Letters, 2*(1), 277–281. https://doi.org/10.1109/LRA.2016.2598601.

Odent, J., Wallin, T. J., Pan, W., Kruemplestaedter, K., Shepherd, R. F., & Giannelis, E. P. (2017). Highly elastic, transparent, and conductive 3D-printed ionic composite hydrogels. *Advanced Functional Materials, 27*(33). https://doi.org/10.1002/adfm.201701807.

Onal, C. D., Chen, X., Whitesides, G. M., & Rus, D. (2017). Soft mobile robots with on-board chemical pressure generation. In *Vol. 100. Springer tracts in advanced robotics* (pp. 525–540). Springer Verlag. https://doi.org/10.1007/978-3-319-29363-9_30.

Patel, D. K., Sakhaei, A. H., Layani, M., Zhang, B., Ge, Q., & Magdassi, S. (2017). Highly stretchable and UV curable elastomers for digital light processing based 3D printing. *Advanced Materials*, *29*(15). https://doi.org/10.1002/adma.201606000.

Plott, J., Tian, X., & Shih, A. J. (2018). Voids and tensile properties in extrusion-based additive manufacturing of moisture-cured silicone elastomer. *Additive Manufacturing*, *22*, 606–617. https://doi.org/10.1016/j.addma.2018.06.010.

Ren, X., Shao, H., Lin, T., & Zheng, H. (2016). 3D gel-printing—An additive manufacturing method for producing complex shape parts. *Materials and Design*, *101*, 80–87. https://doi.org/10.1016/j.matdes.2016.03.152.

Saharan, L., De Andrade, M. J., Saleem, W., Baughman, R. H., & Tadesse, Y. (2017). IGrab: Hand orthosis powered by twisted and coiled polymer muscles. *Smart Materials and Structures*, *26*(10). https://doi.org/10.1088/1361-665X/aa8929.

Saharan, L., & Tadesse, Y. (2019). Novel twisted and coiled polymer artificial muscles for biomedical and robotics applications. In *Materials for biomedical engineering: Nanobiomaterials in tissue engineering* (pp. 45–75). Elsevier. https://doi.org/10.1016/B978-0-12-816909-4.00003-8.

Shafranek, R. T., Millik, S. C., Smith, P. T., Lee, C. U., Boydston, A. J., & Nelson, A. (2019). Stimuli-responsive materials in additive manufacturing. *Progress in Polymer Science*, *93*, 36–67. https://doi.org/10.1016/j.progpolymsci.2019.03.002.

Sydney Gladman, A., Matsumoto, E. A., Nuzzo, R. G., Mahadevan, L., & Lewis, J. A. (2016). Biomimetic 4D printing. *Nature Materials*, *15*(4), 413–418. https://doi.org/10.1038/nmat4544.

Tadesse, Y., Moore, D., Thayer, N., & Priya, S. (2009). Silicone based artificial skin for humanoid facial expressions. In *Electroactive polymer actuators and devices (EAPAD) 2009* International Society for Optics and Photonics.

Tibbits, S. (2014). 4D printing: Multi-material shape change. *Architectural Design*, *84*(1), 116–121. https://doi.org/10.1002/ad.1710.

Wallin, T. J., Pikul, J. H., Bodkhe, S., Peele, B. N., Mac Murray, B. C., Therriault, D., et al. (2017). Click chemistry stereolithography for soft robots that self-heal. *Journal of Materials Chemistry B*, *5*(31), 6249–6255. https://doi.org/10.1039/c7tb01605k.

Wallin, T. J., Pikul, J., & Shepherd, R. F. (2018). 3D printing of soft robotic systems. *Nature Reviews Materials*, *3*(6), 84–100. https://doi.org/10.1038/s41578-018-0002-2.

Wu, L., Jung De Andrade, M., Saharan, L. K., Rome, R. S., Baughman, R. H., & Tadesse, Y. (2017). Compact and low-cost humanoid hand powered by nylon artificial muscles. *Bioinspiration & Biomimetics*, *12*(2). https://doi.org/10.1088/1748-3190/aa52f8.

Wu, A., Rome, R. S., Haines, C., Lima, M. D., Baughman, R. H., & Tadesse, Y. (2015). Nylon-muscle-actuated robotic finger. In *Vol. 9431. Active and passive smart structures and integrated systems* SPIE.

Yirmibeşoğlu, O. D., Oshiro, T., Olson, G., Palmer, C., & Mengüç, Y. (2019). Evaluation of 3D printed soft robots in radiation environments and comparison with molded counterparts. *Frontiers in Robotics and AI*, *6*(May). https://doi.org/10.3389/frobt.2019.00040.

Zolfagharian, A., Denk, M., Bodaghi, M., Kouzani, A. Z., & Kaynak, A. (2020). Topology-optimized 4D printing of a soft actuator. *Acta Mechanica Solida Sinica*, 418–430. https://doi.org/10.1007/s10338-019-00137-z.

Zolfagharian, A., Kaynak, A., Bodaghi, M., Kouzani, A. Z., Gharaie, S., & Nahavandi, S. (2020). Control-based 4D printing: Adaptive 4D-printed systems. *Applied Sciences*, *10*(9), 3020. https://doi.org/10.3390/app10093020.

Zolfagharian, A., Kaynak, A., & Kouzani, A. (2020). Closed-loop 4D-printed soft robots. *Materials & Design*, *188*, 108411. https://doi.org/10.1016/j.matdes.2019.108411.

Multimaterial 4D printing simulation using a grasshopper plugin

12

Germain Sossou, Hadrien Belkebir, and Frédéric Demoly
ICB UMR 6303 CNRS, University Bourgogne Franche-Comté, UTBM, Belfort, France

Introduction

The range of materials that are processable with additive manufacturing (AM) has been drastically expanded since the advent of AM in the 1980s (André, Méhauté, & Witte, 1984; Hull, 1984). Indeed, originally demonstrated with a (photocurable) polymeric material, various AM techniques can now process metals (as varied as aluminum alloys, copper, and gold), ceramic, concrete, food, biomaterials, etc. This expansion of additively manufacturable materials has even reached the stimulus-responsive materials realm. Despite evidence of AM with smart materials (SMs) dating as early as 2009 (Meier, Zarnetta, & Frenzel, 2009), it was in 2013 that this new material/process interaction has been popularized (both for the laypeople and the scientific community) with the wording "4D printing" after a TED talk by Tibbits. Since then, 4D printing has been a hot research topic as illustrated by the exponentially growing number of publications related to the topic (Fig. 12.1). There have even been reports such as (End User (Aerospace, Automotive, Clothing, Construction, Defense, Healthcare & Utility) & Geography—Global Trends & Forecasts to, 2015) forecasting a 4D printing market worth $537 million by 2025.

Worth noticing is that most of the research is dedicated to shape change-based 4D printing, with drug release, soft robotics, self-assembly, smart textiles, among others, as envisioned applications. There are three main ways by which shape change has been achieved through 4D printing:

- Shape memory effect (Yu, Dunn, & Qi, 2015; Zarek et al., 2016): a shape memory material (mostly a polymer) is printed in a permanent shape and thermomechanically trained to assume a temporary shape (or many). Upon exposure to heat, shape recovery occurs.
- Homogeneous and reversible shape change (Bakarich, Gorkin, Panhuis, & Spinks, 2015). In this case, a reversible shape-shifting material such as hydrogel or liquid crystal elastomer is used to print a part that deforms in unison with the stimulus.
- Material heterogeneity (Chen & Zheng, 2018; Ding, Weeger, Qi, & Dunn, 2018; Ge, Qi, & Dunn, 2013; Raviv et al., 2014; Sun, Wan, Nam, Chu, & Naguib, 2019; Sydney Gladman, Matsumoto, Nuzzo, Mahadevan, & Lewis, 2016; Zhang, Guo, Wu, Fang, & Zhang, 2018): two or more compliant materials, including at least one inert material and one material that shrinks, expands, or exhibits shape memory effect, are carefully combined in such a way that the overall shape change is more complex than simple shrinking, expansion, or shape recovery.

Fig. 12.1 Publications with the keyword "4D printing" on Elsevier database as of mid-June 2021.
No Permission Required.

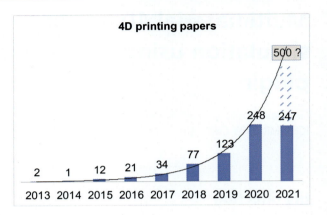

While among these works, a lot has been achieved in understanding the physics of 4D printing (namely regarding manufacturing, materials, and mechanisms yielding shape change) (Momeni & Ni, 2020), a relatively smaller research effort has been targeted at 4D printing from the computational perspective, that is, design, modeling, and simulation. The novelty of the proposed work consists of developing a voxel-based computational design approach to simulate multimaterial active structures and to identify material distributions by connectionist AI in an inverse problem logic. More specifically, the purpose of this chapter is to somewhat fill that gap in the 4D printing endeavor, especially regarding printed heterogeneous shape-changing objects. Given the capability of some AM machines to print objects with nearly continuously varying material properties, we posit that the potential of heterogeneity-based 4D printing is—in the current state of the art—largely untapped. A reason for this could be the difficulty in both designs (there is no provision in most of today's computer-aided design (CAD) software to model material variations in 3D objects) and simulation of such heterogeneous objects. Furthermore, predicting how a given heterogeneous object made among others of shape-changing SMs would deform is way more tedious and less intuitive than predicting how a homogeneous part could deform under load. This chapter is dedicated to alleviating this obstacle to tapping into the potential of heterogeneous 4D printing. It presents a computational tool for modeling, simulating, and generating heterogeneous objects.

The tool is described in "Computational design for 4D printing" section, and then, "Case studies" section is dedicated to showing some examples. Finally, conclusions are drawn, and work yet to be done is stressed out in "Conclusion and future work" section.

Computational design for 4D printing

Rationales and theoretical background

There have been many proofs-of-concept of 4D printing with various AM processes and techniques and various materials. Most of them consider shape-changing SMs.

Among these materials, there are those (a majority) which react to a stimulus by shrinking/expanding; this is the case with 3D-printed hydrogel (Han, Lu, Chester, & Lee, 2018), hydrophilic materials (Raviv et al., 2014), another subgroup is one of the shape memory materials (Zhao et al., 2018) whose shape-change path is between an as printed (memorized) shape and a deformed shape. The former type of SM does exhibit a rather simple shape change when taken alone; however, they can yield shape change more complex than their original shrinking/expanding when mixed with unresponsive (conventional) materials. An illustration of this is shown in Fig. 12.2 (according to Westbrook & Qi, 2008) in which hydrogel (which shrinks with heat) is combined with another inert hydrogel. In these examples, depending on how these materials are combined, the shape change triggered by heat is totally different. As such, the spatial arrangement of the various materials within an object becomes the key to designing such a shape-changing object.

Therefore, there is the need for a modeling technique that provides room for specifying such material heterogeneity in any object. To allow for easily exploring various materials arrangements of the same object, such representation should be intuitive (at least from a geometrical or a spatial reasoning standpoint) and easily editable. As shown in the introduction, knowing how a given spatial arrangement of SMs would behave upon exposure to the right stimulus is not as intuitive as knowing how a homogenous part would deform when submitted to a given load. For this reason and to ease the exploration of several materials' arrangements of the same object, an appropriate modeling technique should also allow for a rapid simulation of the object's mechanical behavior, namely how it would deform upon exposure to stimulus should be quickly found out.

To meet the aforementioned requirements, the authors elected to use a modeling scheme based on voxels. As essentially a volumetric pixel, a voxel can be used as a base element for representing three-dimensional objects in a discrete manner (Hiller & Lipson, 2010; Kaufman, Cohen, & Yagel, 1993). In the proposed modeling scheme, voxels are cubic and aligned with the Cartesian coordinate system. Therefore, any object can be represented as a 3D stack of equally sized voxels (Fig. 12.3).

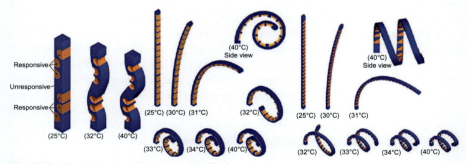

Fig. 12.2 The influence of material distribution.
From Westbrook, K. K., & Qi, H. J. (2008). Actuator designs using environmentally responsive hydrogels. *Journal of Intelligent Material Systems and Structures*, *19*(5), 597–607. https://doi.org/10.1177/1045389X07077856.

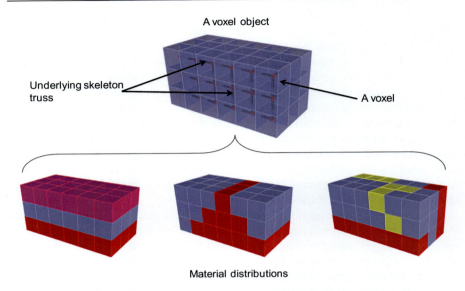

Fig. 12.3 A voxel object and various of its material distributions.
No Permission Required.

The collection of an object's voxels forms its *voxel object* (VO) model. The proposed tool (named VoxSmart) works on VOs. Any voxel has a material assigned to it. A VO includes also an underlying *skeleton truss* (as shown in Fig. 12.3), which is made of beams connecting the voxels' centers. The beams inherit their material properties from the two neighboring voxels they connect. The voxels' centers are the nodes of the truss. The main computation occurs on this truss. The direct stiffness method (Okereke & Keates, 2018) is used to compute the truss' degrees of freedom (DOFs) and subsequently its deformed shape. Each node has six DOFs. The deformation maps of the beams are used to drive the deformation of vertices making up the voxel shapes using a computational graphics method *skinning* (Jacobson, Deng, Kavan, & Lewis, 2014).

A VO can be made of voxels made of different materials. Any makeup of a VO's materials is called a material distribution (Fig. 12.3). VoxSmart is imbued with the vision that materials' heterogeneity is paramount to engineer functionality in an object. A vision similar to how the arrangement of 4 chemical bases defines an organism's deoxyribonucleic acid (DNA) and ultimately its traits (Fig. 12.4). As such, any material distribution is essentially represented as a collection of numbers where the ith number represents the material of the ith voxel in the VO as shown in Fig. 12.4.

It is one thing to know the behavior of given material distribution, but it is quite different to find an appropriate material distribution for the desired behavior. Given the desired shape change (from a source shape to a target one)—the behavior or "traits," a set of available materials—the "A, C, G, T bases"—and a stimulus state, finding a good material distribution—the "DNA"—achieving that behavior, can expand further the design space around material heterogeneity-based 4D printing.

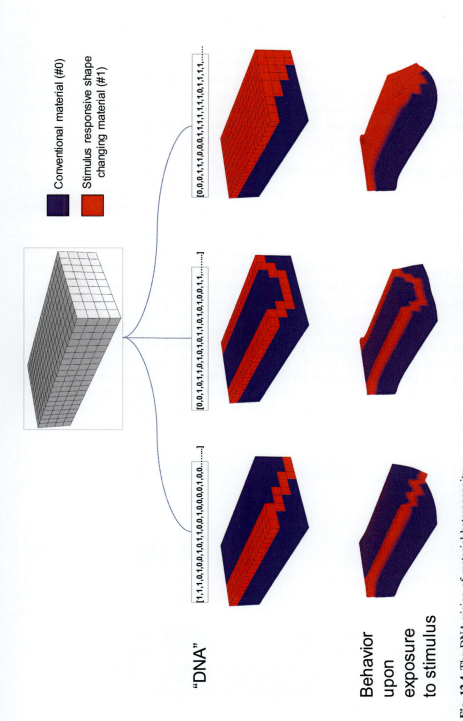

Fig. 12.4 The DNA vision of material heterogeneity. No Permission Required.

VoxSmart provides solutions to such inverse problems by harnessing the power of genetic algorithm to it, as successfully addressed in Hamel et al. (2019). Essentially, the desired shape change in the original object is first described as required DOFs, and then, the material distribution is optimized to minimize the difference between required DOFs and computed DOFs. More details regarding the theoretical background of the proposed tool can be seen in Sossou et al. (2019).

The proposed tool: VoxSmart

VoxSmart has been developed as a plugin of Grasshopper (Rutten, n.d.), a graphical algorithm editor of the CAD software Rhinoceros. In its current version, it is organized into five groups of components as shown in Fig. 12.5.

The *Voxel edition* group basically gathers all the components that create and visualize a VO. Materials (including various kinds of shape-changing SMs) can be defined and assigned to a VO with components from the *Material definition* group. Any boundary conditions (anchoring, prescribed degrees of freedom, and loads) can be applied to the modeled VO with the *Boundary conditions* group's components. Stimuli including heat, magnetic/electric field, and light can be set with the components from the *Stimulus definition* group. The *Simulation* group is made of a single component that takes as input the VO (with all the materials assigned and boundary conditions set), the stimulus, and the solver to be used for computing the DOFs and ultimately the deformed shape. Finally, the *Distributions* group includes components for retrieving/showing material distribution in an array of numbers form, preparing the material distribution computation problem, and solving it. Worth mentioning is the developed interoperability with MATLAB for this latter problem. The problem is

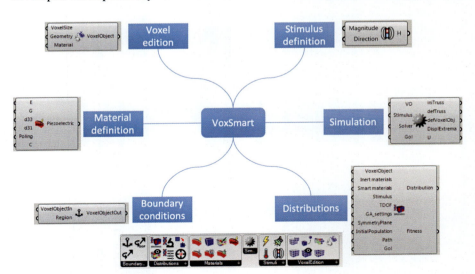

Fig. 12.5 VoxSmart components.
No Permission Required.

initiated in the plugin, and then, it is sent to MATLAB (in matrices form) where it is solved using the genetic algorithm toolbox. Results are sent back to the plugin upon completion of the genetic algorithm. All this is done without opening the MATLAB software. It does not even require MATLAB to be installed on the computer, but MATLAB runtime, a self-contained set of shared libraries for running compiled MATLAB applications or components, is required.

The typical workflow for a VoxSmart definition is as follows:

1. Defining the VO either by voxelizing an existing geometry or by crafting one from scratch (box voxel object, removal, addition, etc.).
2. Assigning (if needed) various materials to specific regions of the voxel object to define a material distribution. This is done by carefully drawing the regions where materials are to assign. If the voxel object is homogenous, this step is not necessary.
3. Setting boundary conditions and possible loads.
4. Selecting the relevant stimulus (electric field, magnetic field, heat, light, moisture) with reasonable amplitudes if the voxel object has voxels made of smart materials.
5. Running the simulation.

Case studies

To illustrate the capabilities of the proposed tool, a few examples have been run.

Modeling and simulation of known material distributions

In the following examples, various VOs with prescribed material distributions are simulated.

Bimaterial beam

This case study showcases a beam made of two conventional materials: one of them, softer than the other, is assigned to the beam's middle section. It is loaded upward at one of its ends and fixed at the other. As one could expect, the simulation shows that it bends where the material is softer. The VoxSmart definition along with the visualization of the simulation's result is shown in Fig. 12.6.

Hydrogel actuator

In this example, an actuator made of inert material and hydrogel is simulated (Fig. 12.7). It is anchored at one of its ends and the material distribution is such that there are sections alternating between the inert material and the hydrogel. As the temperature increases, the hydrogel sections shrink and the actuator bends upward. As shown in Fig. 12.8, the tool is able to capture this behavior.

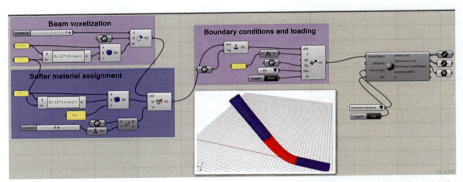

Fig. 12.6 VoxSmart modeling of a loaded bimaterial beam with a softer material in the middle. No Permission Required.

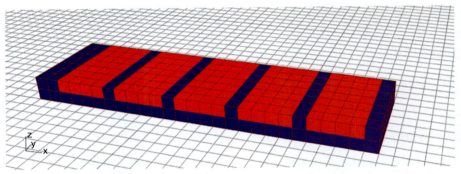

Fig. 12.7 Hydrogel actuator in its rest state. No Permission Required.

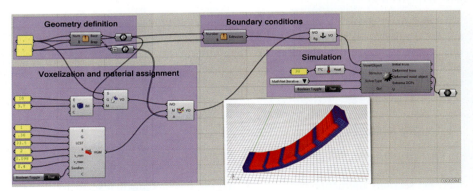

Fig. 12.8 VoxSmart model of the hydrogel actuator. No Permission Required.

Magnetostrictive actuator

Magnetostrictive materials elongate (positive magnetostriction) or shrink (negative magnetostriction) along an applied magnetic field. There are also slight dimensional changes in directions transverse to the applied field; however, these are very small compared to the strain developed along the field. In this example, two different material distributions combining a magnetostrictive material (exhibiting positive magnetostriction) and inert material are explored (Fig. 12.9). The magnetic field is applied along the X-axis. The deformations that such a field would induce in the two distributions are presented in Figs. 12.10 and 12.11. Though less intuitive, the deformation found in distribution 1 is nevertheless consistent with the magnetostrictive material behavior: the top and front views in Fig. 12.9 clearly indicate an overall elongation of the part in the X-direction. The bending motion toward the bottom could be explained by the fact that sections of the actuator parallel to the XZ plane do actually form bilayers with their top sections elongating (along X) while the bottom ones tend to resist that motion thus a bending motion.

Material distribution generation

While the previous examples run a simulation on a specified MD (Dimassi et al., 2021), in this subsection, the behavior-driven material distribution generation capabilities of VoxSmart are illustrated. The VoxSmart component mainly responsible for finding material distributions is shown in Fig. 12.12 and described in detail in the Appendix.

Attempt to retrieve a known distribution

In this example, a known MD, made of hydrogel and inert material, is first simulated with heat as a stimulus. The DOFs resulting from this simulation are then used as target DOFs in the *distributionComputation* component. The goal is to gauge the ability

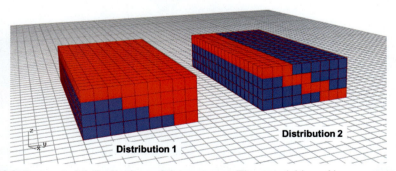

Fig. 12.9 Two material distributions of the same part. The material in *red* is magnetostrictive material, whereas the one in *blue* is an unresponsive (inert) material.
No Permission Required.

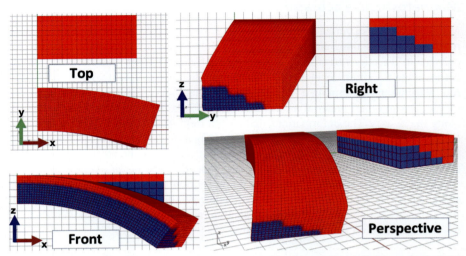

Fig. 12.10 Views of distribution 1 in its initial state and its deformed state (with magnetic field applied along the X-axis).
No Permission Required.

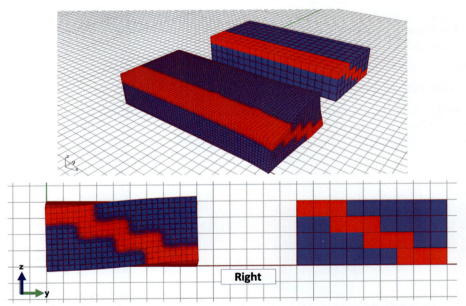

Fig. 12.11 Views of distribution 2 in its initial state and with magnetic field applied along the X-axis.
No Permission Required.

Multimaterial 4D printing simulation using a grasshopper plugin 339

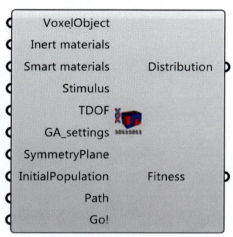

Fig. 12.12 The *distributionComputation* component.
No Permission Required.

of VoxSmart to retrieve a known material distribution of an object. The VoxSmart definition of this example is shown in Fig. 12.13.

The result of the search is shown in Fig. 12.14. The retrieved material distribution is not quite equal to the one expected, however, as shown in the different views in Fig. 12.14, the deformed shapes of both distributions are pretty close. An observation suggests the idea that the solution to the material distribution computation problem may not be unique; in other words, two different material distributions may have the same behavior in a way similar to how two twins can have the same traits but different DNAs (Jonsson et al., 2021; Fig. 12.14).

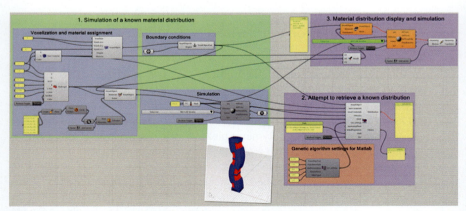

Fig. 12.13 VoxSmart definition for retrieving a known distribution whose deformation upon exposure to heat is shown in the Rhinoceros viewport.
No Permission Required.

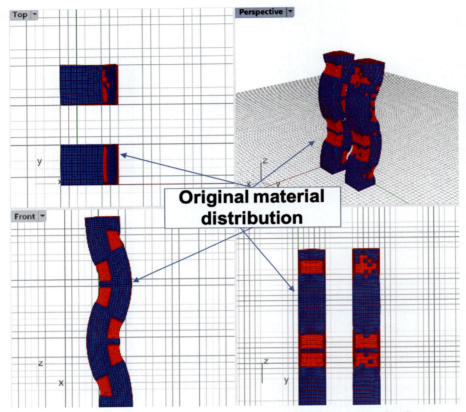

Fig. 12.14 Original and found material distributions in their deformed (stimulated) states.

Distribution computation with enforced symmetry and initial population

A field that is likely to benefit from 4D printing is the one of soft robotics, in which soft compliant actuators and/or soft bodies are used for tasks such as locomotion or grasping fragile objects (Boddeti et al., 2021; Scalet, 2020; Wang, Nurzaman, & Lida, 2017). In this example, a hypothetical stingray-like soft robot is designed. Two materials are considered: a polymeric piezoelectric material and a conventional material. The poling direction of the piezoelectric material is along the robot's transverse direction, which essentially means that most of its deformation is elongated along that direction with a field applied in the same direction. The problem is formulated as a sought deformation from a source and a target shape corresponding to a flat and bent wings configuration, respectively, as shown in Fig. 12.15. To obtain the target shape, source one was bent about the longitudinal direction.

To solve this problem, symmetry was enforced in the distribution, which has the consequence of halving the search space. Furthermore, a few MD candidates were

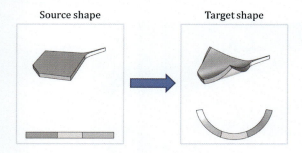

Fig. 12.15 Distribution computation problem. No Permission Required.

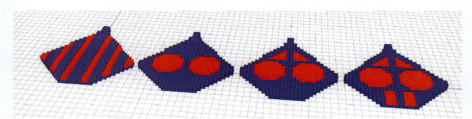

Fig. 12.16 Material distributions included in the initial population. No Permission Required.

provided as an initial population to the *distributionComputation* component. These candidates are shown in Fig. 12.16.

The found distribution is the one shown in Fig. 12.17. Its deformation upon exposure to the electric field is quite similar to the sought deformation as shown in Fig. 12.18.

Conclusion and future work

Multimaterial 4D printing is a strategic scientific path (i) to elaborate objects able to evolve in terms of shape and properties by leveraging a reasonable amount of active

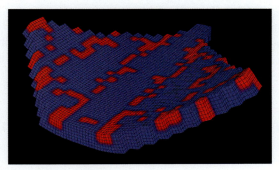

Fig. 12.17 Computed material distribution. No Permission Required.

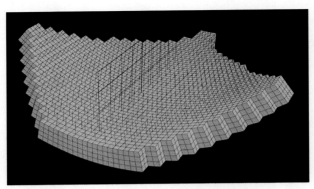

Fig. 12.18 Voxelized sought deformation with enforced symmetry.
No Permission Required.

materials and (ii) to propose heterogeneous structures embedding both elastic and rigid properties. This research field brings freedom in design and simulation but also introduces combinatorial complexities to be tackled. The developed VoxSmart tool within the Rhinoceros 3D/Grasshopper environment allows the designer to play on many parameters regarding active materials and stimuli by relying on a voxel-based modeling strategy. It offers great flexibility in the implementation of material distributions through the genetic algorithm-based computation and the simulation based on the Euler-Bernoulli beams.

In its current shape, the proposed tool requires additional development in terms of simulation setup. The stimulus (and its value) is currently considered an environment variable globally applied and uniformly sensed by each voxel of the model. While such assumption is valid for the electric or magnetic field, for heat and light, it is questionable. Indeed, heat propagates and light gets absorbed as it propagates through a medium. The proposed computational capacities remain, however, useful in embodiment design. Nevertheless, the framework may be extended to model all the stimuli fields in the simulated object more accurately.

Future research will focus on multiscale voxel-based modeling to identify distributions and new shape-changing behaviors that will be coupled to deep neural networks. Additional developments are also underway to connect the tool to a knowledge base structured by ontology in order to capitalize the calculated distributions, interpret them, and make recommendations for future projects as initiated in Dimassi et al. (2021).

Appendix: The *distribution Computation* component

This is the component that runs the material distribution computation. It works, among others, with the MATLAB genetic algorithm toolbox. Its inputs and outputs are described in Tables 12.1 and 12.2, respectively.

Table 12.1 Inputs.

VoxSmart object	Explanation
VoxelObject (generic parameter: VoxSmart. VoxelObject)	Voxel object in homogeneous form with all boundary conditions set
Inert materials (generic parameter: VoxSmart. Material)	List of inert materials. The order in this list does matter. First material is labeled "0," second is labeled "1," and so on
Smart materials (generic parameter: VoxSmart. Material)	List of smart materials of the same type. The order in this list does matter. The ordering continues on the one of the inert materials list. That means, if there are 2 inert materials (material "0" and material "1"), then the first smart material will be labeled "2," the second is labeled "3," and so on
Stimulus (generic parameter: VoxSmart. Stimulus)	Stimulus triggering the smart materials
TDOF (Number parameter)	Required degrees of freedom to achieve the target shape
GA_settings (Number parameter)	List of options to control the genetic algorithm. This list should be provided by the *GA_settings* component
SymmetryPlane (Plane parameter)	Plane of symmetry that must be parallel to XY, XZ, or YZ. This input is optional. If it is provided, then the search space is halved: instead of looking for a material for each voxel, the component will only look for a material for half the voxels. Providing this input assumes, then that the voxel object (which may slightly differ from the geometry it may have been voxelized from) is actually symmetrical with respect to the provided plane. If that symmetry is not verified, then the component will throw an error
InitialPopulation (Integer parameter)	A tree (list of lists) of integers indicating material distributions that the user suspects to be good candidates. This input is optional. If provided, then the candidates will be included in the first population
Path (Text parameter)	Path to a .txt file indicating where to store the result. This input is optional. It must be provided in the form: *path\filename.txt*. The genetic algorithm is likely to take hours before it stops itself or before the user decides to stop it. After completion, the results are available to use in the output ports. And, one may just copy-paste the found distribution somewhere. However, changing a data stream on the canvas that ultimately goes into the component relaunches the computation. A situation that may result in a (huge) waste of time, if the result has not been copy-pasted somewhere. This storage feature is meant to avoid such waste. In case no path is provided, the results are automatically saved in a directory like "C:\Program Files \Rhinoceros 6 (64-bit)\System" under the name "yyyyMMdd_HHmmVoxSmartDistribution.txt"
Go! (Boolean parameter)	A Boolean to launch the algorithm

Table 12.2 Outputs.

VoxSmart object	Explanation
Distribution (Integer)	List of integers defining the best found distribution.
Fitness (Number parameter)	Fitness of the distribution. It is defined by: $\|U_{target} - U(distribution)\|^2$ The U vectors representing the degrees of freedom.

No Permission Required.

References

André, J. C., Méhauté, L., & Witte, D. (1984). *Dispositif pour réaliser un modèle de pièce industrielle.*

Bakarich, S. E., Gorkin, R., Panhuis, M. I. H., & Spinks, G. M. (2015). 4D printing with mechanically robust, thermally actuating hydrogels. *Macromolecular Rapid Communications, 36* (12), 1211–1217. https://doi.org/10.1002/marc.201500079.

Boddeti, N., Van Truong, T., Joseph, V. S., Stalin, T., Calais, T., Lee, S. Y., et al. (2021). Optimal soft composites for under-actuated soft robots. *Advanced Materials Technologies.* https://doi.org/10.1002/admt.202100361.

Chen, D., & Zheng, X. (2018). Multi-material additive manufacturing of metamaterials with giant, tailorable negative Poisson's ratios. *Scientific Reports, 8*(1). https://doi.org/10.1038/s41598-018-26980-7.

Dimassi, S., Demoly, F., Cruz, C., Qi, H. J., Kim, K. Y., André, J. C., et al. (2021). An ontology-based framework to formalize and represent 4D printing knowledge in design. *Computers in Industry, 126.* https://doi.org/10.1016/j.compind.2020.103374.

Ding, Z., Weeger, O., Qi, H. J., & Dunn, M. L. (2018). 4D rods: 3D structures via programmable 1D composite rods. *Materials and Design, 137,* 256–265. https://doi.org/10.1016/j.matdes.2017.10.004.

End User (Aerospace, Automotive, Clothing, Construction, Defense, Healthcare & Utility) & Geography—Global Trends & Forecasts to. (2015). *4D printing market by material (programmable carbon fiber, programmable wood—Custom printed wood grain, programmable textiles).*

Ge, Q., Qi, H. J., & Dunn, M. L. (2013). Active materials by four dimensions printing. *Applied Physics Letters, 103.*

Hamel, C. M., Roach, D. J., Long, K. N., Demoly, F., Dunn, M. L., & Qi, H. J. (2019). Machine-learning based design of active composite structures for 4D printing. *Smart Materials and Structures, 28*(6). https://doi.org/10.1088/1361-665X/ab1439.

Han, D., Lu, Z., Chester, S. A., & Lee, H. (2018). Micro 3D printing of a temperature-responsive hydrogel using projection micro-stereolithography. *Scientific Reports, 8*(1). https://doi.org/10.1038/s41598-018-20385-2.

Hiller, J., & Lipson, H. (2010). Tunable digital material properties for 3D voxel printers. *Rapid Prototyping Journal, 16*(4), 241–247. https://doi.org/10.1108/13552541011049252.

Hull, C. (1984). Method for production of three-dimensional objects by stereo-lithography. US Patent, 5.

Jacobson, A., Deng, Z., Kavan, L., & Lewis, J. (2014). Skinning: Real-time shape deformation. In *ACM SIGGRAPH.*

Jonsson, H., Magnusdottir, E., Eggertsson, H. P., Stefansson, O. A., Arnadottir, G. A., Eiriksson, O., et al. (2021). Differences between germline genomes of monozygotic twins. *Nature Genetics*, *53*(1), 27–34. https://doi.org/10.1038/s41588-020-00755-1.

Kaufman, A., Cohen, D., & Yagel, R. (1993). Volume graphics. *Computer*, *26*(7), 51–64. https://doi.org/10.1109/MC.1993.274942.

Meier, H., Zarnetta, R., & Frenzel, C. H. J. (2009). Selective laser melting of NiTi shape memory components. In *Innovative developments in design and manufacturing* (6 pp.). CRC Press.

Momeni, F., & Ni, J. (2020). Laws of 4D printing. *Engineering*, *6*, 1035–1055.

Okereke, M., & Keates, S. (2018). Direct stiffness method. In *Springer tracts in mechanical engineering* (pp. 47–106). Springer International Publishing. https://doi.org/10.1007/978-3-319-67125-3_3 (Issue 9783319671246).

Raviv, D., Zhao, W., McKnelly, C., Papadopoulou, A., Kadambi, A., Shi, B., et al. (2014). Active printed materials for complex self-evolving deformations. *Scientific Reports*, *4*. https://doi.org/10.1038/srep07422.

Rutten, D.. (n.d.). Grasshopper, www.grasshopper3d.com.

Scalet, G. (2020). Two-way and multiple-way shape memory polymers for soft robotics: An overview. *Actuators*, *9*(1), 10. https://doi.org/10.3390/act9010010.

Sossou, G., Demoly, F., Belkebir, H., Qi, H. J., Gomes, S., & Montavon, G. (2019). Design for 4D printing: A voxel-based modeling and simulation of smart materials. *Materials and Design*, *175*. https://doi.org/10.1016/j.matdes.2019.107798.

Sun, Y. C., Wan, Y., Nam, R., Chu, M., & Naguib, H. E. (2019). 4D-printed hybrids with localized shape memory behaviour: Implementation in a functionally graded structure. *Scientific Reports*, *9*(1). https://doi.org/10.1038/s41598-019-55298-1.

Sydney Gladman, A., Matsumoto, E. A., Nuzzo, R. G., Mahadevan, L., & Lewis, J. A. (2016). Biomimetic 4D printing. *Nature Materials*, 413–418. https://doi.org/10.1038/nmat4544.

Wang, L., Nurzaman, S., & Lida, F. (2017). Soft-material robotics. *Foundations and Trends in Robotics*, *5*, 191–259.

Westbrook, K. K., & Qi, H. J. (2008). Actuator designs using environmentally responsive hydrogels. *Journal of Intelligent Material Systems and Structures*, *19*(5), 597–607. https://doi.org/10.1177/1045389X07077856.

Yu, K., Dunn, M. L., & Qi, H. J. (2015). Digital manufacture of shape changing components. *Extreme Mechanics Letters*, *4*, 9–17. https://doi.org/10.1016/j.eml.2015.07.005.

Zarek, M., Layani, M., Cooperstein, I., Sachyani, E., Cohn, D., & Magdassi, S. (2016). 3D printing of shape memory polymers for flexible electronic devices. *Advanced Materials*, *28*(22), 4449–4454. https://doi.org/10.1002/adma.201503132.

Zhang, H., Guo, X., Wu, J., Fang, D., & Zhang, Y. (2018). Soft mechanical metamaterials with unusual swelling behavior and tunable stress-strain curves. *Science Advances*, *4*(6), eaar8535. https://doi.org/10.1126/sciadv.aar8535.

Zhao, T., Yu, R., Li, X., Cheng, B., Zhang, Y., Yang, X., et al. (2018). 4D printing of shape memory polyurethane via stereolithography. *European Polymer Journal*, *101*, 120–126. https://doi.org/10.1016/j.eurpolymj.2018.02.021.

Origami-inspired 4D tunable RF and wireless structures and modules

13

Yepu Cui, Syed A. Nauroze, and Manos M. Tentzeris
School of Electrical and Computer Engineering, Georgia Institute of Technology, Atlanta, GA, United States

Introduction

Modern mobile and wireless devices require multiple communication modules and sensors that allow them to communicate with other devices as well as collect information about the environment around them for applications such as smart cities, smart skins, 5G, Internet of things, and quality of life. Typically, these modules require packaging a number of multiband RF components to reduce overall size and cost while improving system efficiency. The number of RF components in a module can be significantly reduced by realizing tunable multilayer RF structures (such as antennas, filters, matching networks) that can change their electromagnetic (EM) behavior in response to external stimuli. Such structures are very hard to realize using traditional subtractive manufacturing technologies (SMTs) such as photolithography, milling, and chemical etching that are widely used to fabricate modern wireless and communication devices. Some other disadvantages of these technologies include high cost, long lead time, and production of harmful chemicals waste.

The aforementioned disadvantages of the SMTs can be addressed by using additive manufacturing technologies (AMTs) such as inkjet printing and 3D printing. These technologies use layer-by-layer deposition of conductive, dielectric, or semiconductive material to realize complex 3D/4D multilayer flexible RF structures with lower cost, high reliability, and repeatability as compared to SMTs (Ramasubramaniam, Chen, & Liu, 2003; Wong, Chabinyc, Ng, & Salleo, 2009; Yang, Rida, Vyas, & Tentzeris, 2007). The current state-of-the-art inkjet and 3D printers can achieve 1–20 μm and 10–50 μm feature size, respectively. That is why they are gaining widespread attraction in industry and academia to realize the next generation of RF and mm-wave communication modules and systems. AMTs have very little material wastage and do not use harmful chemicals making them extremely environmentally friendly as compared to traditional SMTs. Moreover, these technologies allow the designer to tap into the 3D (and in some cases 4D) realm to design highly compact and efficient RF systems—a feature that is impossible to achieve using SMTs.

Smart Materials in Additive Manufacturing, Volume 2: 4D Printing Mechanics, Modeling, and Advanced Engineering Applications
https://doi.org/10.1016/B978-0-323-95430-3.00013-0
Copyright © 2022 Elsevier Inc. All rights reserved.

One of the most popular and widely used techniques to realize tunable RF structures is by using electronic components (such as diodes, varactors, and micro-electromechanical systems) to change the electromagnetic properties of the resonant structures. These structures can achieve high tuning speed but they suffer from a high failure rate and limited tunability range, and become harder to realize as their overall size increases. Other techniques include using ferrite substrates and integrated micro-fluidic channels that change the RF structures' electromagnetic response by varying the substrate's effective permittivity and selectively loading the resonant structures, respectively. However, they require high operating voltages and are hard to control. That is why mechanically tunable RF structures are gaining popularity due to their ability to realize wideband (continuous range) tunability, quality factor, linearity, and superior power handling capabilities. But their bulky size, weight, fabrication complexity, and low switching speed limit their use in most modern-day applications.

Origami-inspired RF structures have recently gained interest in the scientific community due to their ability to achieve wideband continuous range tunability by simply changing their physical shape in response to external stimuli. These structures can be completely folded into a small volume allowing them to be stowed during transportation and then fully deployed at their destination making them attractive for various terrestrial, bio-medical, and outer-space applications. These mechanically tunable 3D/4D RF structures offer wide-range continuous range tunability, ease of fabrication, and high-power handling capability. In contrast to typical 3D RF structures that require complex metallization techniques to fabricate conductive traces, these structures are realized by first inkjet-printing conductive traces on a flat substrate that is folded into desired reconfigurable 3D/4D structure. Hence, they are extremely lightweight, robust, and can achieve unprecedented mechanical strength.

While origami-inspired RF structures present a promising methodology to realize novel communication modules and systems for next-generation mm-wave systems, they present a number of novel challenges that need to be addressed for a robust design. These include the realization of truly flexible conductive traces, multilayer configuration with interconnects (that maintain high conductivity during folding or bending process), folding automation techniques (especially for thick substrates), mathematical modeling of origami structures (including its kinematics and design), and their relationship to the EM behavior of the RF structures. In this work, inkjet-printing technology is used to fabricate highly flexible conductive traces as it facilitates rapid fabrication of complex multilayer multimaterial 2D/3D flexible RF structures on a wide range of substrates.

The origami-inspired RF structures presented in the work are first-of-their-kind that truly show a direct relationship between origami kinematics and RF characteristics for the first time. This is realized by using Miura-Ori tessellation, which is the most well-studied and widely used origami structure in a wide range of engineering disciplines and allows shrinkage in both axes (Boatti, Vasios, & Bertoldi, 2017; Filipov, Tachi, & Paulino, 2015; Miura, 1985; Nauroze, 2019; Schenk, Guest, & McShane, 2014; Silverberg et al., 2014; Wei, Guo, Dudte, Liang, & Mahadevan, 2013). Moreover, it also introduced effective mechanisms to address the aforementioned challenges to realize origami-inspired RF structures that can be used in various terrestrial and outer-space applications. Lastly, due to the fundamental nature of the origami and RF structures used in this work, the results can be easily applied to analyze (and realize) a wide variety of origami-inspired RF structures.

Inkjet-printing technologies

The recent advancements in inkjet-printing technologies have made it possible to realize RF structures with 20 μm feature size a reality using commercially available piezo-based desktop-sized inkjet printers. Smaller feature size (<1 μm) is possible with electro-hydrodynamic-based inkjet printer. However, for this study, all proof-of-concept RF structures are realized using Dimatix DMP-2831 inkjet printer as shown in Figs. 13.1 and 13.2. It consists of a printer head and a movable platform that can be

Fig. 13.1 Lab-size inkjet printer (DMP-2831).
From Dimatix Material Printer DMP-2850. https://www.fujifilm.com/us/en/business/inkjet-solutions/inkjet-technology-integration/dmp-2850.

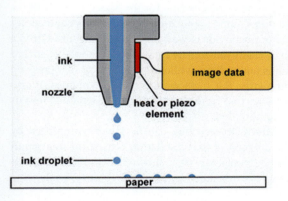

Fig. 13.2 Schematic of a typical inkjet printer head.
From http://www.dp3project.org/technologies/digital-printing/inkjet.

electronically controlled by a computer. The printer head typically consists of a cartridge that stores (conductive/dielectric or semiconductive) ink and an array of 16 nozzles that are individually connected to piezoelectric elements. During the printing process, each nozzle can deposit 1–10 pL droplet of the ink onto the substrate that is placed on the movable platform. The printer head and movable platform move in a predefined manner to realize the desired pattern on the substrate using the computer.

A key advantage of the inkjet printer includes the availability of a wide variety of inks that can be used to fabricate complex multilayer multimaterial RF structures. Typically, the inks consist of dielectric, conductive, or active materials dispersed in ethanol or water-based solvent. The choice of solvent is generally determined by the type of dispersed material, target substrate, print head jetting ability, sintering conditions, etc. These properties give inkjet-printing technology a unique advantage to realize 2D/3D high-resolution low-cost multilayer flexible proof-of-concept RF structures on a variety of substrates that can be easily scaled to large numbers (Dimatix Material Printer DMP-2850, 2016). Table 13.1 outlines key processing parameters for some commonly used inks.

An efficient implementation of multilayer inkjet-printed RF structures requires proper pattern alignment and evaluation of ink wettability on the desired substrate. The former can be achieved by using the build-in fiducial system and alignment markers or patterns on the target substrate, while the latter requires matching surface energy and tension of the ink and substrate, respectively. Typically for efficient design, this difference is kept around 10 mN/m and can be customized by exposing the substrate to UV-ozone and plasma treatment (Tehrani, Mariotti, Cook, Roselli, & Tentzeris, 2016).

Miura-Ori tessellation

Miura-Ori tessellation is one of the most fundamental origami structures that was originally developed by Japanese astrophysicist Koryo Miura. One of its first applications involved the implementation of deployable solar panels for the Space Flyer Unit—a Japanese spacecraft in 1996. Since it has found applications in numerous engineering disciplines, its kinematic and structure behavior is well documented. Miura-Ori is a developable surface and hence can be easily realized by folding a flat substrate/surface along foldlines in form of mountains and valleys without stretching or compressing the sheet thereby facilitating folding large flat surfaces into smaller volumes. It also exhibits "shape-memory" characteristics that facilitate re-folding the structure into its original compact configuration once unfolded.

The kinematic analysis of the Miura-Ori structure can be greatly simplified by exploiting its repetitive nature. The unit cell of the tessellation consists of four parallelograms (each with length a, b and internal angle α) as shown in Fig. 13.3. The folding behavior of the structure is completely depicted by dihedral (folding) angle theta

Table 13.1 Processing parameters of different inkjet-printed materials.

Material	Standard drop spacing (µm)	Per-layer thickness (nm)	Per-layer thickness (nm)	UV cross-linking	Sintering	Application
SNP	20	500–800	100	No	<200°C	Conductive pattern
SU-8	20	4000–6000	100	Yes	No	Thick insulators/printed RF structures
PVP	25	300–500	150	No	No	Thin dielectric films
CNT	25	10–500	150	No	No	Transparent conductors and sensors

Fig. 13.3 Unit cell of Miura-Ori tessellation.
From Nauroze, S. A. (2019). *Additively manufactured origami-inspired "4D" RF structures with on-demand continuous-range tunability.* Georgia Institute of Technology. http://hdl.handle.net/1853/62330.

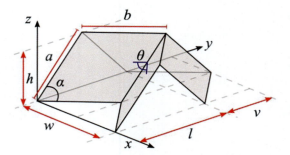

(θ) or equivalently lengths w or l that are related to each other by Nauroze, Novelino, Tentzeris, and Paulino (2017):

$$l = 2a\zeta, h = \frac{2a\zeta\alpha\cos\theta}{2}, \quad \omega = 2b\xi, \nu = b(1-\xi^2)^{1/2}$$

$$\text{where } \zeta = \cos\alpha(a-\xi^2)^{-1/2}, \quad \xi = \sin\alpha\sin\theta/2 \tag{13.1}$$

where h and v give the Miura cell's height and expansion during the folding/unfolding process. This work chooses frequency selective surfaces (FSSs) as a prime candidate for origami-based RF structure since it exhibits many structural similarities with the Miura-Ori tessellation, hence facilitating the realization of the fundamental relationship between the kinematics of origami structure and its integrated RF structure. Note that the principles established through this study could be extended to a wide variety of RF structures due to their fundamental nature.

Frequency selective surfaces

FSSs consist of a 2D array of 2D-/3D-shaped resonators uniformly spaced on a thin substrate. This unique configuration allows them to behave like spatial filters that allow certain electromagnetic waves to pass through it depending on the size, shape, and interelement spacing. The resonant elements can be generally categorized into N-pole, closed-shape, and patch (Munk, 2000). Since substrate thickness is typically very thin as compared to operating wavelength ($\ll \lambda/10$) (Munk, 2000), its effect can be ignored. However, for thicker substrates, the substrate effect results in shift in resonant frequency by a factor of (ε_r+1)/2. Some of their applications include smart skins, absorbers, radome design, EMI filters, and antenna radar cross-section reduction (Nauroze et al., 2017). Due to their periodic nature, their EM response can be completely characterized by applying Floquet periodic boundary conditions on its unit cell. The periodic boundary conditions ensure that the phase of currents in resonant elements matches the incident fields. It can be easily observed that Miura-Ori tessellation and FSSs have the same analysis methodology making them ideal candidates to establish a fundamental relationship between kinematics and electromagnetic behavior of origami-inspired RF structures. Moreover, the results obtained in the study can be easily extended to realize complex structures such as reflect/transmit arrays and radomes.

FSS structures typically require tuning elements such as diodes, MEMS, varactors, microfluidics, or mechanically movable parts to achieve tunability (Jung, Lee, Li, & De Flaviis, 2006; Sung, Jang, & Kim, 2004; White & Rebeiz, 2009; Wong et al., 2009). Among them, electronically tunable techniques are most widely used due to their high switching speed. However, their fabrication complexity, cost, tuning range, power requirements, and nonlinear EM behavior increase with size, limiting their use in most modern-day applications. In contrast, mechanically tunable FSS structures feature wideband continuous range tunability range, high power handling capability, and linear EM response but due to their bulky size, slow tuning/switching speed, and fabrication complexity, they have only found use in limited applications. The afore-mentioned problems can be addressed by using origami-inspired FSS structures that feature easy-to-fabricate 3D/4D RF structures that can realize wideband continuous range tunability in response to the change in their external environment. A detailed comparison of various state-of-the-art tuning mechanisms for FSS structures is given in Table 13.2.

Origami-inspired inkjet-printed FSS structures

In this work, a dipole-based origami-inspired FSS structure is designed and analyzed. Dipoles are one of the most fundamental resonant structures; therefore, the results obtained for the proposed design can be easily extended to complex RF structures without loss of generality. The unit cell of the dipole-based Miura-FSS structure is shown in Figs. 13.4 and 13.5 where the dipoles are inkjet-printed across the mountain foldlines so that their effective length and interelement decrease proportionally as it is folded, hence changing its resonant frequency to higher values. The EM behavior of Miura-FSS can be fully described by a change in its folding angle. Moreover, it also presents a robust proof-of-concept methodology to realize highly flexible conductive traces that can maintain good conductivity during the bending/folding process. Two slits are introduced along the dipole length to implement a "bridge-like" structure that allows the dipole to fold in a curve-like fashion along the foldlines during the bending/folding process instead of a sharp bend. These "bridge-like" structures improve the flexibility of the conductive traces and avoid breakage during the folding process while maintaining high conductivity (Nauroze et al., 2017).

Fabrication process

A step-by-step fabrication process of the proposed dipole-based Miura-FSS is depicted in Fig. 13.6. In this work, cellulose-based substrate is used because of its wide use in the implementation of origami structures and ease of use. First, the Miura pattern is perforated on the 110-μm-thick cellulose paper, and dipoles were inkjet-printed

Table 13.2 Advantages and disadvantages of different tunability techniques for FSS structures.

Type		Tuning mechanism	Advantages	Disadvantages
Active	Discrete state	PIN diode	Fast tuning High frequency operation High isolation	Complex biasing network Requires RF/DC choke Nonlinear behavior High, continuous power consumption Only two state tunability
		MEMS switches	Compact size Diversity of input source Easy to integrate	High fabrication cost and complexity Narrow band tunability (~10%) Nonuniform EM response (for different states and AoI)
	Continuous state	Varactor	Low cost Up to 60% tunability Low power consumption	Complex biasing network required High bias voltage required Poor AoI performance Performance sensitive to surface mount component (degrades with size)
		Ferrite substrates	Wide-band tunability (50%–100%) High dielectric constant (thin layer ~10 μm)	Complex conductivity control High H-field required Narrow bandwidth
Microfluidics		Liquid (metal and dielectric)	High power handling Wideband tunability (~200%) Multifunctional FSS possible Flexible and self-healing Diversity in choice of liquids	Multilayer configuration is difficult Not suitable for band-pass FSS Difficult to fabricate Oxidation of GaIn (Ltd. reusability) Requires pumps—high setup time
Mechanical		Shape change	High power handling Wideband (continuous-range) tunability Linear and high Q factor	Heavy and bulky Complex fabrication Slow tuning speed

Origami-inspired 4D tunable RF and wireless structures and modules 355

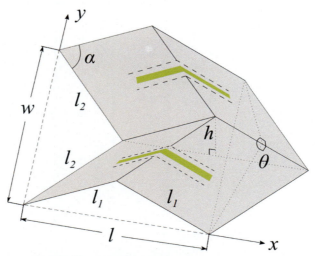

Fig. 13.4 Unit cell of Miura-FSS, folded configuration.
From Nauroze, S. A., Novelino, L., Tentzeris, M. M., & Paulino, G. H. (2017). Inkjet-printed "4D" tunable spatial filters using on-demand foldable surfaces. In: *2017 IEEE MTT-S international microwave symposium (IMS)* (pp. 1575–1578). https://doi.org/10.1109/MWSYM.2017.8058932.

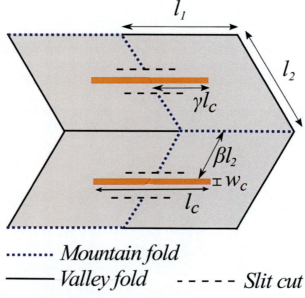

Fig. 13.5 Unit cell of Miura-FSS, flat ($\theta = 180$ degrees).
From Nauroze, S. A., Novelino, L., Tentzeris, M. M., & Paulino, G. H. (2017). Inkjet-printed "4D" tunable spatial filters using on-demand foldable surfaces. In: *2017 IEEE MTT-S international microwave symposium (IMS)* (pp. 1575–1578). https://doi.org/10.1109/MWSYM.2017.8058932.

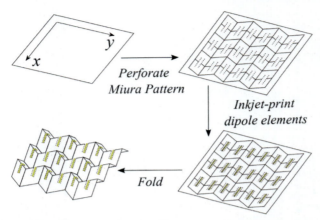

Fig. 13.6 The fabrication process of Miura-FSS.
From Nauroze, S. A., Novelino, L., Tentzeris, M. M., & Paulino, G. H. (2017). Inkjet-printed "4D" tunable spatial filters using on-demand foldable surfaces. In: *2017 IEEE MTT-S international microwave symposium (IMS)* (pp. 1575–1578). https://doi.org/10.1109/MWSYM.2017.8058932.

across the foldlines using 10 layers of silver nanoparticle (SNP) ink and sintered for 2.5 h at 150°C in an oven. The ink is composed of SNPs in an ethanol-based solvent that is absorbed by the paper substrate during the inkjet-printing process due to its hydrophilic nature. During the sintering process, the solvent is evaporated and the SNP coalesces to form highly conductive traces that are embedded into the substrate itself making them very flexible (Nauroze et al., 2017). The improvement in flexibility of the conductive traces due to the "bridge-like" structures can be accessed by observing their folding behavior along the foldlines shown in Fig. 13.7. Finally, the Miura-FSS pattern was cut out and folded manually as shown in Fig. 13.8; however, this process can be easily automated by using hydrofolding techniques, heat-sensitive substrates, or 3D-printing technologies (Nauroze et al., 2017).

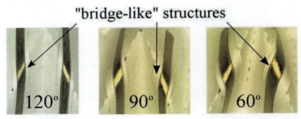

Fig. 13.7 Micrograph of dipole element with different folding angle θ.
From Nauroze, S. A., Novelino, L., Tentzeris, M. M., & Paulino, G. H. (2017). Inkjet-printed "4D" tunable spatial filters using on-demand foldable surfaces. In: *2017 IEEE MTT-S international microwave symposium (IMS)* (pp. 1575–1578). https://doi.org/10.1109/MWSYM.2017.8058932.

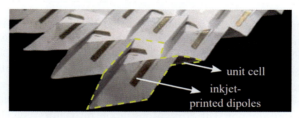

Fig. 13.8 Close-up picture of fabricated Miura-FSS.
From Nauroze, S. A., Novelino, L., Tentzeris, M. M., & Paulino, G. H. (2017). Inkjet-printed "4D" tunable spatial filters using on-demand foldable surfaces. In: *2017 IEEE MTT-S international microwave symposium (IMS)* (pp. 1575–1578). https://doi.org/10.1109/MWSYM.2017.8058932.

Results and discussions

The single-layer Miura-FSS structure was designed and simulated on HFSS. In order to reduce computation time and complexity, only the unit cell with master/slave periodic boundary conditions was used that was excited by a Floquet port (Boatti et al., 2017; Filipov et al., 2015; Miura, 1985; Nauroze, 2019; Schenk et al., 2014; Silverberg et al., 2014; Wei et al., 2013). The negative in-plane Poisson's ratio of the Miura (Nauroze, 2019; Wei et al., 2013) facilitated a symmetrical and proportional variation in the effective electrical length and spacing between the dipole elements during the folding/unfolding process. This allows an on-the-fly continuous-range change in frequency response without the use of any electronic components. The simulation results were verified using the measurement setup shown in Fig. 13.9 consisting of two wideband horn antennas placed in the line of sight of each other while the fabricated Miura-FSS is inserted in the middle. The space between the horn and the Miura-FSS can be easily accounted for by careful calibration at the vector network analyzer. The structure was held to three distinct folding angles ($\theta = 60$, 90, and 120 degrees) by using

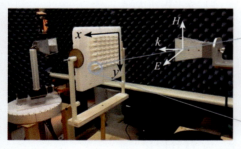
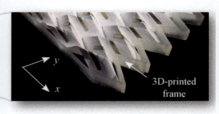

Fig. 13.9 Measurement setup with a close-up showing the 3D-printed frame to uniformly fold Miura at a given folding angle (frame for $\theta = 90$ is shown here).
From Nauroze, S. A., Novelino, L., Tentzeris, M. M., & Paulino, G. H. (2017). Inkjet-printed "4D" tunable spatial filters using on-demand foldable surfaces. In: *2017 IEEE MTT-S international microwave symposium (IMS)* (pp. 1575–1578). https://doi.org/10.1109/MWSYM.2017.8058932.

3D-printed frames. This ensures that each unit cell in the Miura structure is kept at the same folding angle value. The simulated and measured insertion loss (S_{21}) of the fabricated Miura-FSS with respect to different folding angles is shown in Fig. 13.10 that clearly indicates a systematic variation of resonant frequency with folding.

The substrate effects can be ignored in this work since it is very thin as compared to the operational wavelength. This can be verified by a simple observation; a completely flat Miura-FSS configuration with 20 mm-long dipole resonators typically resonates at 7.5 GHz in free space. The slight shift in frequency in our results is primarily due to ink spreading within the substrate, increasing the dipole length by 0.3 mm on each side, which corresponds to the observed resonance frequency of 7.2 GHz. The proposed design also exhibits a wide-range angle of incidence (AoI) rejection and its bandwidth increases for higher values of AoI as shown in Fig. 13.11. The latter is an inherent property of dipole-based FSS structures whose bandwidth varies by 1/cos(AoI) in TE-mode (Munk, 2000).

Typically, FSS structures are designed to operate in "no grating lobe" region. This is achieved by limiting the interelement periodicity less than first high-order Floquet harmonic wavelength $\lambda_g^{\varepsilon_r} = D_l \left(\sqrt{\varepsilon_r} + sin\left(AoI\right) \right)$, where ε_r represents the effective dielectric constant of the substrate and is the interelement distance along the x-axis.

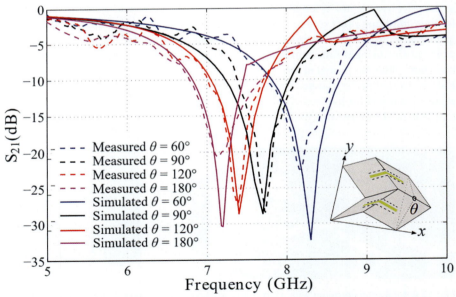

Fig. 13.10 Frequency response of single-layer Miura-FSS with different values of folding angle θ.
From Nauroze, S. A., Novelino, L., Tentzeris, M. M., & Paulino, G. H. (2017). Inkjet-printed "4D" tunable spatial filters using on-demand foldable surfaces. In: *2017 IEEE MTT-S international microwave symposium (IMS)* (pp. 1575–1578). https://doi.org/10.1109/MWSYM.2017.8058932.

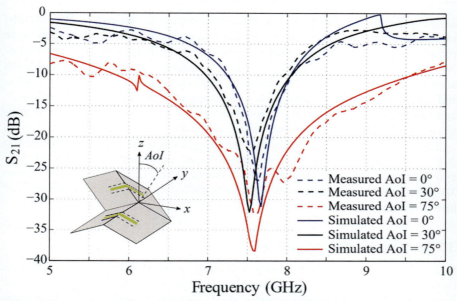

Fig. 13.11 Frequency response of single-layer Miura-FSS with different values of AoI at $\theta = 90$ degrees.
From Nauroze, S. A., Novelino, L., Tentzeris, M. M., & Paulino, G. H. (2017). Inkjet-printed "4D" tunable spatial filters using on-demand foldable surfaces. In: *2017 IEEE MTT-S international microwave symposium (IMS)* (pp. 1575–1578). https://doi.org/10.1109/MWSYM.2017.8058932.

The resonant wavelength of the proposed structure is much larger than λ_g; hence, only fundamental mode dominates and higher-order modes are evanescent that decay exponentially further away from the FSS structure. The sudden null in frequency response designates onset of the grating lobe region ($D_l = \lambda_g$ at AoI = 0 degrees). It is interesting to note here that unlike traditional FSS structures, the grating lobe region of proposed Miura-FSS structure varies with folding angle and hence no further adjustment is required for error correction.

Fabrication process of 4D-printed origami-inspired RF structures

The traditional fabrication process of origami-inspired RF structures uses paper as substrate and involves manual cutting and folding, limiting the accuracy, scalability, durability for real-life mm-wave applications. As a potential solution to this problem, 4D printing technology with 3D-printing and inkjet-printing techniques can be utilized to deposit both dielectric and conductive materials in a 3D fashion, where complex origami-inspired designs can be realized directly. 3D printing allows the rapid design and fabrication of free-form 3D objects with ease. 3D-printed flexible substrates eliminate the requirement of manual folding and cutting which enables more complicated design elements such as slots and round holes (Babaei, Ramos, Lu, Webster, & Matusik, 2017). Meanwhile, the inkjet-printing technique allows the selective deposition of a wide variety of materials, including metals, dielectrics, and nanomaterials. Fully automated high-accuracy additive manufacturing methodologies can easily scale up in production, opening the potential for miniaturized mm-wave origami-inspired RF devices (Hester et al., 2015).

This section will describe a hybrid 4D-printing process (Cui, Nauroze, & Tentzeris, 2019) that combines 3D-printing and inkjet-printing techniques using Miura-Ori FSS as an example. The fabrication process as shown in Fig. 13.12 includes three steps: 3D-print the substrate, inkjet-print SU-8 dielectric buffer layer, and inkjet-print silver nanoparticle conductive layer.

3D-printing of the substrate

Stereolithography (SLA) 3D-printing technology is an ideal candidate to fabricate flexible substrates with complex origami-inspired structures because of its balanced cost, material availability, and accuracy. SLA 3D-printing method uses a laser, projector (DLP), or liquid crystal display (LCD) screen as a light source to cure the light-sensitive resin layer-by-layer to form a 3D geometry. Compared with the fused deposition modeling (FDM), 3D-printing method which heats the thermoplastic filament to its melting point and then extrudes the filament to create a 3D object, the SLA method has better accuracy and consistency, thereby making it superior for mm-wave applications (Bahr, Tehrani, & Tentzeris, 2018). Commercialized SLA 3D printing systems have a wide range of available materials with unique mechanical and RF properties. In the proposed process, we use flexible materials FLGR02 and FL8001 with Formlabs Form2 and Form3 3D-printing systems (Formlabs 3D Printers, 2021). FLGR02 is a

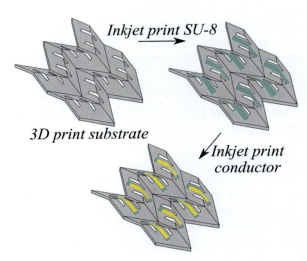

Fig. 13.12 The fabrication process for 4D-printed origami structures.
From Cui, Y., Nauroze, S. A., & Tentzeris, M. M. (2019). Novel 3D-printed reconfigurable origami frequency selective surfaces with flexible inkjet-printed conductor traces. In: *2019 IEEE MTT-S international microwave symposium (IMS)* (pp. 1367–1370). https://doi.org/10.1109/MWSYM.2019.8700994.

dark, "rubber-like" material with a tensile strength of 7.7–8.5 MPa, shore hardness of 80–85A, and tear strength of 10.6 kN/m. FL8001 is a transparent elastomer with a tensile strength of 8.9 MPa, shore hardness of 80A, and tear strength of 24 kN/m. To extract the RF dielectric constant and loss tangent properties, we utilized Nicolson-Ross-Weir (NRW) methodology with A-INFO 28CLKA2 waveguide calibration kit to measure characterize the material from 26 GHz to 40 GHz. The measurement results for FL8001 are shown in Fig. 13.13. The substrate is printed with a 50 μm layer thickness followed by a wash-and-cure postprocess to have a smoother surface, reduced chance of failure, more consistency, and less electromagnetic losses. The wash process takes 15 min with 99% isopropyl alcohol to remove extra resin left on the surface without damaging the substrate. Then, the sample takes a 60-min curing process under 405 nm LED UV lights in a 60°C heated environment to improve the cross-linking of the polymer.

Inkjet-print SU-8 buffer layer

As SLA 3D-printing techniques cure the resin in a layer-by-layer manner, the surface of the printed geometry will have periodic micro-structures with "hills" and "valleys" perpendicular to the printing plane as shown in Fig. 13.14. To characterize the roughness of the surface, we measure the surface profile with a KLA Tencor Alpha-step D-100 stylus profilometer. The measurement result shows 42.51 μm root mean square (RMS) perpendicular to the printing plane and 26.51 μm RMS to the printing plane.

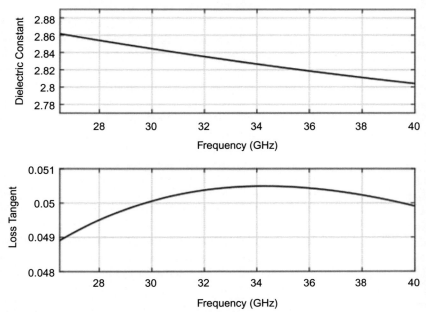

Fig. 13.13 Material characterization results for FL8001.
No Permission Required.

Fig. 13.14 The surface of an SLA 3D-printed sample with Formlabs FLGR02 resin. No Permission Required.

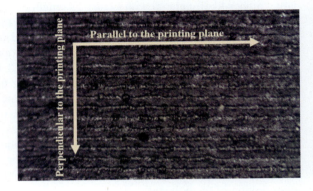

Therefore, the 3D-printed surface is too rough for one layer of inkjet-printed silver nanoparticle ink (0.8 μm). We will reduce the surface roughness by inkjet-printing a few layers of MicroChem SU-8 dielectric ink as a buffer layer to solve this issue. To minimize the thickness of the SU-8 layer to maintain the best substrate flexibility, we measured the surface profile with different layers of SU-8 perpendicular to the printing plane with the surface profilometer. From the measurement results shown in Fig. 13.15, after four layers of inkjet-printed SU-8, the surface roughness reduced to 1.59 μm (RMS), which will be smooth enough to print silver nanoparticles inks.

Inkjet-print the conductive layer

The conductive layer will be inkjet-printed with SunChemical EMD5730 SNP ink. Before printing the ink on top of SU-8, we utilize 90 s of ultraviolet (UV)-ozone treatment to improve the wettability and adhesion without losing too much resolution. With the UV-ozone surface treatment, the contact angle of the SNP ink decreased from 46 degrees to 29 degrees as shown in Fig. 13.16, which is the optimal contact angle for the inkjet printing.

SNP inks generally require a high sintering temperature (above 180°C) to ensure good conductivity and adhesion. High temperature damages most of the 3D-printed substrates; hence, placing FLGR02 substrate at 150°C for 30 min reduces the flexibility and elasticity dramatically, causing the substrate to break after just a few folds. Additionally, the substrate deforms when the temperature suddenly gets changed, causing cracks on the inkjet-printed conductive traces as shown in Fig. 13.17. To solve this problem, we utilized a low-temperature gradient sintering process. After SNP printing, the sample was placed on a hot plate ramping from 25°C to 90°C with a 150°C/h temperature ramp and hold at 90°C for 30 min to dry the pattern completely. Then, the temperature was increased from 90°C to 120°C with a 150°C/h ramp and hold at 120°C for 30 min to sinter the pattern without breaking the substrate. Finally, the hot plate was turned off to let the sample slowly cool back to room temperature to avoid deformation caused by sudden temperature change. With this low-temperature gradient sintering process, the conductor quality was greatly improved and the realized conductor trace showed a conformal and smooth surface on a curved 3D substrate as shown in Fig. 13.18.

Origami-inspired 4D tunable RF and wireless structures and modules 363

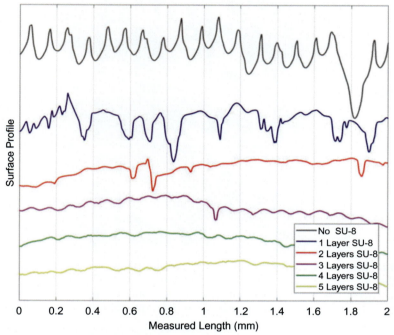

Fig. 13.15 Surface profile perpendicular to the printing line.
No Permission Required.

Fig. 13.16 The contact angle of silver nanoparticle ink before and after UV-ozone treatment..
No Permission Required.

4D-printed origami-inspired frequency selective surfaces

Origami-inspired FSS substrate can eliminate the need for active components, and enables unprecedented capabilities for deployability, power handling, and continuous-range tunability. With the hybrid 4D-printing process that combines 3D-printing and inkjet-printing technologies, we can realize origami-inspired frequency

Fig. 13.17 Inkjet-printed conductive traces with normal sintering process.
From Cui, Y., Nauroze, S. A., & Tentzeris, M. M. (2019). Novel 3D-printed reconfigurable origami frequency selective surfaces with flexible inkjet-printed conductor traces. In: *2019 IEEE MTT-S international microwave symposium (IMS)* (pp. 1367–1370). https://doi.org/10.1109/MWSYM.2019.8700994.

Fig. 13.18 Inkjet-printed conductive traces with low-temperature gradient sintering process.
From Cui, Y., Nauroze, S. A., & Tentzeris, M. M. (2019). Novel 3D-printed reconfigurable origami frequency selective surfaces with flexible inkjet-printed conductor traces. In: *2019 IEEE MTT-S international microwave symposium (IMS)* (pp. 1367–1370). https://doi.org/10.1109/MWSYM.2019.8700994.

Fig. 13.19 Prototype of a single-layer FSS with 8 × 10 Miura-Ori elements.
From Cui, Y., Nauroze, S. A., & Tentzeris, M. M. (2019). Novel 3D-printed reconfigurable origami frequency selective surfaces with flexible inkjet-printed conductor traces. In: *2019 IEEE MTT-S international microwave symposium (IMS)* (pp. 1367–1370). https://doi.org/10.1109/MWSYM.2019.8700994.

selective surfaces that operate at mm-wave frequencies with small feature sizes. Fig. 13.19 shows the prototype of a hybrid 4D-printed single-layer FSS with 8 × 10 Miura-Ori element. The design of a single unit cell is shown in Fig. 13.20. The 3D-printed Miura-Ori structure was realized using Formlabs FLGR02 flexible material as substrate, with inkjet-printed dipole-shaped traces as conductive pattern on top. To reduce the stress applied to the substrate during folding or unfolding process, we added stress-release holes with 0.4 mm radius to the intersection of foldlines. To reduce the bending force applied to the conductor traces, we added stress-release slots by the edge of each conductor to form a "bridge-like" structure. The single-layer Miura-FSS was designed and simulated in Ansys HFSS with master and slave boundary conditions along with Floquet port excitations. The fabricated prototype was measured with Anritsu vector network analyzer (VNA) with two A-INFO LB-180400-20-C-KF horn antennas.

The results were simulated in HFSS with master and slave Floquet boundary conditions. The sample was mounted on a 3D-printed holder and measured with two horn antennas. The simulated and measured frequency response with different folding angles is shown in Fig. 13.21. The folding angle decreases during compression resulting in resonant frequency shift to higher values. This is because the equivalent electrical length of the conductive traces will decrease due to the folding. The measured frequency shift matched well with the simulation results. However, there are some mismatches in the amplitude of the insertion loss. This is because when we fold the structure from $\theta = 110$ degrees to $\theta = 60$ degrees, the size of the Miura-FSS decreases from 50×60 mm to 50×7 mm, the smaller sample cannot cover the entire

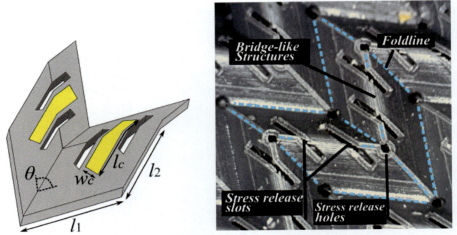

Fig. 13.20 Unit cell design of the hybrid 4D-printed single-layer Miura-Ori FSS. From Cui, Y., Nauroze, S. A., & Tentzeris, M. M. (2019). Novel 3D-printed reconfigurable origami frequency selective surfaces with flexible inkjet-printed conductor traces. In: *2019 IEEE MTT-S international microwave symposium (IMS)* (pp. 1367–1370). https://doi.org/10.1109/MWSYM.2019.8700994.

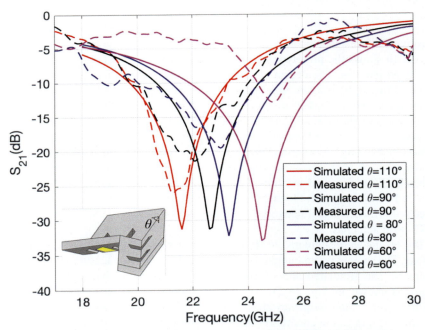

Fig. 13.21 Simulated and measured frequency response with different folding angles. From Cui, Y., Nauroze, S. A., & Tentzeris, M. M. (2019). Novel 3D-printed reconfigurable origami frequency selective surfaces with flexible inkjet-printed conductor traces. In: *2019 IEEE MTT-S international microwave symposium (IMS)* (pp. 1367–1370). https://doi.org/10.1109/MWSYM.2019.8700994.

antenna luminance anymore, causing leakage that reduces the insertion loss. This issue can be improved by implementing multilayer structures. The simulated and measured frequency response with different angles of incidence (AoI) is shown in Fig. 13.22. The single-layer Miura-FSS shows good AoI rejection from 0 degrees to 60 degrees with boosted bandwidth from 14% to 32%.

To improve the insertion loss performance of a single-layer Miura-FSS, we designed a mirror-stacked multilayer Miura-FSS structure (Cui, Nauroze, Bahr, & Tentzeris, 2021). The prototype shown in Fig. 13.23 has a 3D-printed single-piece substrate with two Miura elements being stacked. The conductive traces were printed on both sides. So the density of the conductor is doubled without increasing the size. The measurement result of this multilayer Miura-FSS is shown in Fig. 13.24. The resonant frequency can be tuned from 22.4 GHz to 26.1 GHz by changing folding angle θ from 110 degrees to 60 degrees. Compared to the single-layer structure in Fig. 13.21, the multilayer configuration showed 12 dB better insertion loss performance for lower theta values due to the improved conductor density. Thus, the hybrid 4D printing process with 3D-printing and inkjet-printing technologies has shown the ability to prototype complex multilayer 3D FSSs that are hard to realize with traditional fabrication processes.

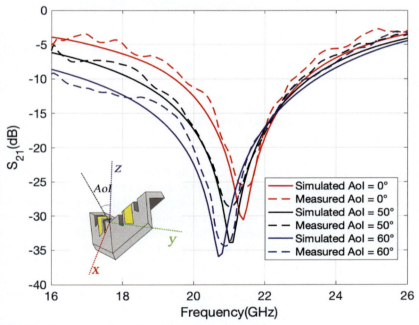

Fig. 13.22 Simulated and measured frequency response with different AoI.
From Cui, Y., Nauroze, S. A., & Tentzeris, M. M. (2019). Novel 3D-printed reconfigurable origami frequency selective surfaces with flexible inkjet-printed conductor traces. In: *2019 IEEE MTT-S international microwave symposium (IMS)* (pp. 1367–1370). https://doi.org/10.1109/MWSYM.2019.8700994.

Fig. 13.23 Prototype of a multilayer FSS with 8 × 10 mirror-stacked Miura-Ori elements. From Cui, Y., Nauroze, S., Bahr, R., & Tentzeris, M. (2021). A novel additively 4D printed origami-inspired tunable multi-layer frequency selective surface for mm-wave IoT, RFID, WSN, 5G, and smart city applications. In: *2021 IEEE/MTT-S international microwave symposium (IMS)*.

4D-printed chipless RFID pressure sensors for WSN applications

Wireless sensor network (WSN) is an emerging technology that utilizes wirelessly connected sensor nodes to collect and monitor the surrounding environment. One of the key challenges of realizing WSN is the lack of sustainable power supplies to enable consistent data readings over a long period of time without maintenance (Stankovic, 2008). A promising solution is to use passive sensors along with chipless radio frequency identification (RFID) technologies to eliminate the need for a power source. Chipless RFID can be realized with multiple resonators, microfluidic channels, variant modulations, etc. (Cui, Nauroze, et al., 2019). However, it is challenging to design passive sensors especially pressure sensors that can cooperate with chipless RFID transmitters.

Pressure sensor plays an essential role in detecting force, strain, and displacement. Among various sensors, conventional pressure sensors use capacitors or piezoelectric that requires biasing and modulation circuits with a power source. 4D-printed electromagnetic (EM)-based pressure sensors with origami-inspired elements such as hinges and meshes can be an ideal solution to this challenge. Origami-inspired structures can provide consistent and predictable movement related to real-world physical properties

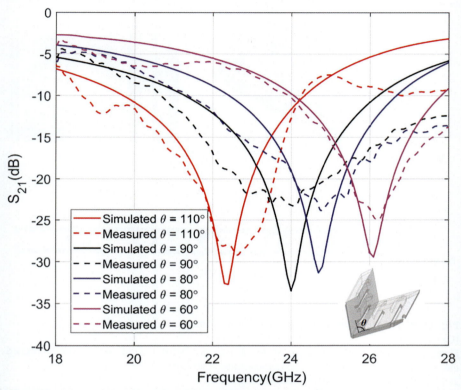

Fig. 13.24 Simulated and measured frequency response with different folding angles. From Cui, Y., Nauroze, S., Bahr, R., & Tentzeris, M. (2021). A novel additively 4D printed origami-inspired tunable multi-layer frequency selective surface for mm-wave IoT, RFID, WSN, 5G, and smart city applications. In: *2021 IEEE/MTT-S international microwave symposium (IMS)*.

such as force, strain, pressure, and displacement. Passive electromagnetic components such as resonators and frequency selective surfaces (FSSs) can relate the physical movement to resonant frequencies and transmit the data via chipless RFID techniques. With 4D hybrid printing technology, we can realize origami-inspired EM-based sensors with simplified design, low fabrication cost, robust structure, and no need for an external power source.

4D-printed planar pressure sensor using metamaterial absorber

Fig. 13.25 shows the design of the 4D-printed planar pressure sensor using a metamaterial absorber (Jeong, Cui, Tentzeris, & Lim, 2020). The one-piece 3D-printed flexible substrate has a "sandwich" structure with a planar sensing area on top, a ground plane on the bottom, and a series of origami-inspired tunable hinge

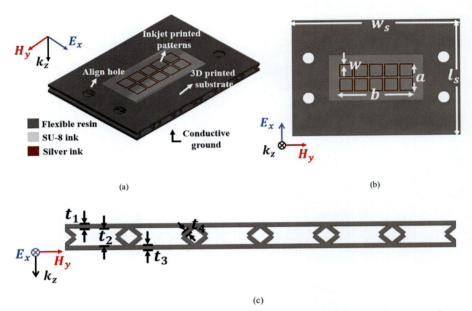

Fig. 13.25 The electromagnetic pressure sensor geometry: (A) Perspective, (B) top, and (C) side view.
From Jeong, H., Cui, Y., Tentzeris, M. M., & Lim, S. (2020). Hybrid (3D and inkjet) printed electromagnetic pressure sensor using metamaterial absorber. *Additive Manufacturing, 35*, 101405. https://doi.org/10.1016/j.addma.2020.101405.

structures in the middle. The planar sensing area consists of an array of inkjet-printed metamaterial waveguide elements sitting on inkjet-printed SU-8 dielectric buffer layer. When applying pressure on top of the sensing area, the origami-inspired hinge will deform, changing the metamaterial waveguide's height and shifting the resonate frequency.

The mechanical performance of this sensor was simulated in COMSOL Multiphysics by applying forces on top of the metamaterial structure. The material properties were set as 5.6 MPa Young's modulus and 0.45 Poisson's ratio. As shown in Fig. 13.26, 45 N compressive force is required in this design to increase the compression length from 0 mm to 1 mm. The electromagnetic performance sensor was simulated in ANSYS HFSS and measured with INSTRON 5569 compression electromechanical device. The simulated and measured reflection coefficient is shown in Fig. 13.27. In the initial state, the pressure sensor resonates at 5.2 GHz with a −16 dB reflection coefficient. When the height was varied from 5.9 mm to 6.5 mm, the resonant frequency increased from 5.20 GHz to 5.66 GHz, resulting in a 0.2×108 Hz/N sensitivity.

This pressure sensor can be used as passive nodes in WSN and get the readings by measuring the reflected signal. The size of the sensor is expandable (ideally to infinite

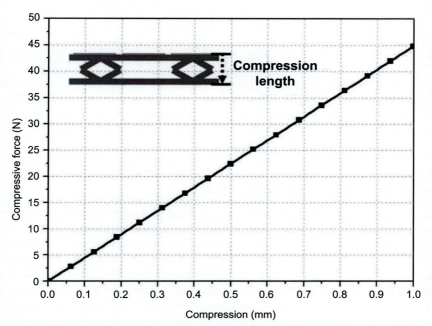

Fig. 13.26 COMSOL simulation results for the proposed EM pressure sensor.
From Jeong, H., Cui, Y., Tentzeris, M. M., & Lim, S. (2020). Hybrid (3D and inkjet) printed electromagnetic pressure sensor using metamaterial absorber. *Additive Manufacturing*, *35*, 101405. https://doi.org/10.1016/j.addma.2020.101405.

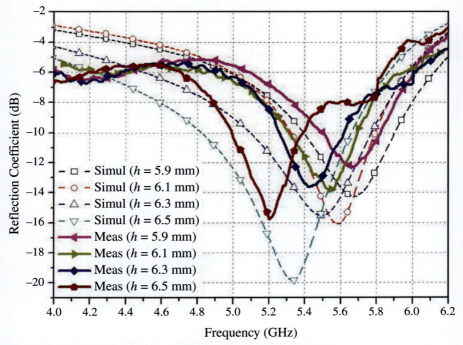

Fig. 13.27 Simulated and measured EM results for the proposed EM pressure sensor.
From Jeong, H., Cui, Y., Tentzeris, M. M., & Lim, S. (2020). Hybrid (3D and inkjet) printed electromagnetic pressure sensor using metamaterial absorber. *Additive Manufacturing*, *35*, 101405. https://doi.org/10.1016/j.addma.2020.101405.

size) to be deployed in a larger area. In addition, by tuning the thickness of the origami-inspired hinge structure, the stiffness of the sensor can be changed. Thus, the sensor sensitivity can be easily tuned to fulfill the needs for more application scenarios.

4D-printed planar pressure sensor using substrate integrated waveguide (SIW) technology

The 4D-printed metamaterial-based planar pressure sensor has shown excellent sensitivity for large area pressure sensing applications. However, it is challenging for metamaterial-based structures to be miniaturized for small area pressure sensing. Therefore, we designed a substrate integrated waveguide (SIW)-based 4D-printed pressure sensor for small contact area applications (Kim, Cui, Tentzeris, & Lim, 2020). The schematic and fabricated sample of the pressure sensor is shown in Fig. 13.28. It consists of inkjet-printed conductive SIW cavity pattern and a 3D-printed flexible substrate. At the center of the substrate, there is a cylindrical cavity which is filled with a 3D-printed mesh structure to tune the stiffness of the cavity. When pressure is applied on top of the SIW cavity resonator, the 3D-printed mesh structure compresses resulting in a change in effective permittivity and resonant frequency of the SIW cavity resonator.

The structure was simulated in COMSOL Multiphysics to evaluate the mechanical performances. With added 0–2.4 kPa pressure, the height change increased from 0 mm to 1.52 mm. The measurement result is shown in Fig. 13.29. As the height change increased from 0.29 mm to 1.5 mm, the resonant shifted from 4.21 GHz to 3.79 GHz, respectively, resulting in a sensitivity of 2.4×10^8 Hz/kPa. The results showed good linearity and sensitivity for a passive pressure sensor design with good frequency response stability over 100 repeats. By changing the density of the mesh structure, the stiffness and sensitivity of sensor can be easily tuned based on the actual

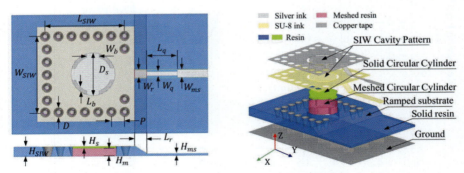

Fig. 13.28 Proposed SIW cavity pressure sensor: Top and perspective view.
From Kim, Y., Cui, Y., Tentzeris, M. M., & Lim, S. (2020). Additively manufactured electromagnetic based planar pressure sensor using substrate integrated waveguide technology. *Additive Manufacturing*, 34, 101225. https://doi.org/10.1016/j.addma.2020.101225.

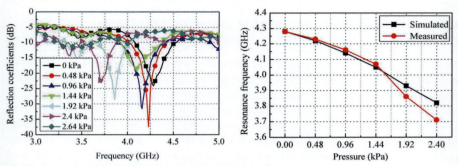

Fig. 13.29 Measurement results for the fabricated SIW pressure sensor.
From Kim, Y., Cui, Y., Tentzeris, M. M., & Lim, S. (2020). Additively manufactured electromagnetic based planar pressure sensor using substrate integrated waveguide technology. *Additive Manufacturing, 34,* 101225. https://doi.org/10.1016/j.addma.2020.101225.

need. Along with the metamaterial-based design, the 4D-printed pressure sensor can find their applicable scenarios for both large area and small area pressure/strain sensing applications.

4D-printed origami-inspired deployable and reconfigurable antennas

With the advancement of 5G and mm-wave technologies, high-gain antennas are required to be installed on small form factor devices such as mobile stations and CubeSats. Traditional high-gain antennas such as parabolic reflectors and corner reflectors are bulky, expensive, and hard to fabricate. Thus, antenna designers must find solutions for deployable high-gain antennas that can be retracted into a small space. Origami-inspired structures can be an excellent option because of their lightweight, low cost, and ease of deployment. Various origami-inspired deployable antennas have been studied over the years with promising performance, deployability, and reconfigurability (Shah & Lim, 2021). However, most of the origami-inspired antennas are fabricated with paper as the substrate. While paper can be an ideal substrate for proof-of-concept demonstrations, its inferior durability makes it difficult to utilize in real-life applications. Moreover, the fabrication process involves manual cutting and folding, making it difficult to realize complicated 3D structures.

With 4D printing technology, fully featured origami-inspired antennas via holes, transmission lines, microfluidic channels can be fabricated with an automated process with exceptional resolution and accuracy. The 3D printing technique provides a wide range of flexible and durable photopolymers to realize the structure of the origami antenna. Meanwhile, the conductive patterns or traces can be realized with inkjet-printing techniques or liquid metals.

4D-printed one-shot deployable dielectric reflectarray antenna for mm-wave applications

Reflectarray is a type of antenna that combines some of the advantages of phased arrays and parabolic reflectors with a planar design, high gain, ease to fabricate, and have no requirement for feeding circuits. Traditional reflectarray antennas usually use microstrip patch elements with various phase delay lines or different element sizes to control phase shift. However, at mm-wave or terahertz (THz) frequencies, conductor losses can be a major performance limitation. To solve this issue, dielectric reflectarrays have attracted considerable attention in recent years as it uses dielectric-based materials to achieve phase delays which not only eliminate high-frequency conductor losses but also improve the bandwidth (Deng, Yang, Xu, & Li, 2016; Jamaluddin et al., 2010; Nayeri et al., 2014; Sun & Leung, 2018; Zhang, 2017). However, most dielectric reflectarrays use solid blocks as unit cells and are fabricated with rigid materials, preventing them from being able to transform to a portable or retractable design. Thus, compared with traditional microstrip reflectarrays, current dielectric reflectarrays feature larger volumes and lack compressibility and deployability. Therefore, applicable scenarios such as satellite and mobile communications are extensively limited.

To solve this challenge, we designed a novel "snapping-like" unit cell structure (Cui, Nauroze, Bahr, & Tentzeris, 2020) as shown in Fig. 13.30 that was inspired by previous research of both a variation of origami involving cut sections called Kirigami (Castle, Sussman, Tanis, & Kamien, 2016) and mechanical two-stage metamaterials (Rafsanjani, Akbarzadeh, & Pasini, 2015). The structure was designed and simulated in Abaqus with Formlabs Flexible 80A material. As shown in Fig. 13.31, when applying pressure on top of the structure, at the 0.85 N/m threshold, the snapping segment will "snap" onto the bearing segment, and prompt the structure to its second stable "retracted" state. As a result, the structure can be retracted effortlessly to a stable state and can be deployed in "one shot" with a burned switch.

The unit cell was designed and simulated in CST Studio Suite 2019 frequency-domain solver with unit cell boundary conditions. As shown in Fig. 13.32, a full 360-degree phase shift can be obtained by varying the element height h from 9.83 mm to 31.10 mm. The proposed dielectric reflectarray configuration is shown in Fig. 13.33. The feeding antenna has a 20-degree offset theta to avoid radiation blockage. The main bean direction is also tilted at 20 degrees to minimize beam squint.

In this work, a 117 mm × 118 mm array with 14 × 16 elements was generated with a Visual Basic Script and simulated with a time-domain solver in CST Studio Suite 2019. The prototype was fabricated with Formlabs Form 2 3D orienting system. A fully deployed sample is shown in Fig. 13.34, and a fully retracted sample is shown in Fig. 13.35. The prototype was held by a 3D-printed frame and can be deployed in "one shot" with a burned switch. The retracted sample demonstrates a 65% volume reduction in comparison with the fully deployed state.

The prototype was measured with two AINFO LB-180400-20-C-KF wideband horn antennas in an anechoic chamber. The measured copolarization radiation pattern on E-plane and H-plane is shown in Fig. 13.36. The prototype has a measured realized

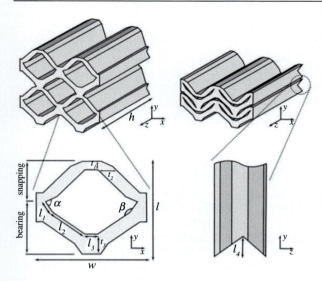

Fig. 13.30 "Snapping-like" unit cell structure design. From Cui, Y., Nauroze, S. A., Bahr, R., & Tentzeris, E. M. (2020). 3D printed one-shot deployable flexible "kirigami" dielectric reflectarray antenna for mm-wave applications. In: *2020 IEEE/MTT-S international microwave symposium (IMS)* (pp. 1164–1167). https://doi.org/10.1109/IMS30576.2020.9224010.

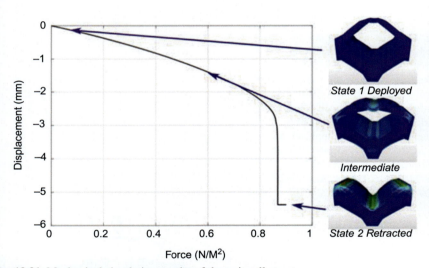

Fig. 13.31 Mechanical simulation results of the unit cell structure. No Permission Required.

gain of 25.5 dBi at 29 GHz with the simulated realized gain of 25.4 dBi. The measured 1 dB bandwidth is 16.34% with a simulation result of 18.75%. The measured side lobe level (SLL) of both E-plane and H-plane is under −15 dB, matching the simulation results. The prototype demonstrated exceptional performance with deplorability, enabling more space-limited communication applications for dielectric reflectarrays.

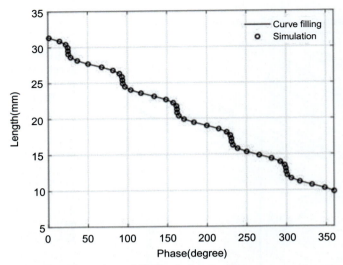

Fig. 13.32 Unit cell length vs phase shift simulation results.
From Cui, Y., Nauroze, S. A., Bahr, R., & Tentzeris, E. M. (2020). 3D printed one-shot deployable flexible "kirigami" dielectric reflectarray antenna for mm-wave applications. In: *2020 IEEE/MTT-S international microwave symposium (IMS)* (pp. 1164–1167). https://doi.org/10.1109/IMS30576.2020.9224010.

Fig. 13.33 Reflectarray design: Feed placement and main beam direction.
From Cui, Y., Nauroze, S. A., Bahr, R., & Tentzeris, E. M. (2020). 3D printed one-shot deployable flexible "kirigami" dielectric reflectarray antenna for mm-wave applications. In: *2020 IEEE/MTT-S international microwave symposium (IMS)* (pp. 1164–1167). https://doi.org/10.1109/IMS30576.2020.9224010.

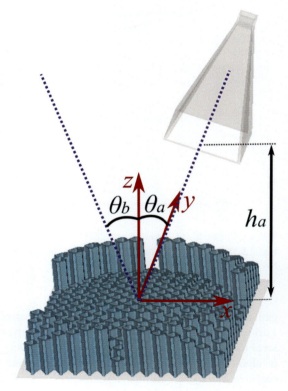

Fig. 13.34 Fabricated sample: Fully deployed sample (front side).
From Cui, Y., Nauroze, S. A., Bahr, R., & Tentzeris, E. M. (2020). 3D printed one-shot deployable flexible "kirigami" dielectric reflectarray antenna for mm-wave applications. In: *2020 IEEE/MTT-S international microwave symposium (IMS)* (pp. 1164–1167). https://doi.org/10.1109/IMS30576.2020.9224010.

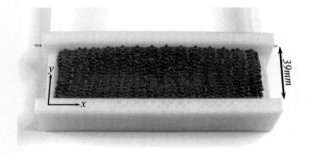

Fig. 13.35 Fabricated sample: Fully retracted sample (backside).
From Cui, Y., Nauroze, S. A., Bahr, R., & Tentzeris, E. M. (2020). 3D printed one-shot deployable flexible "kirigami" dielectric reflectarray antenna for mm-wave applications. In: *2020 IEEE/MTT-S international microwave symposium (IMS)* (pp. 1164–1167). https://doi.org/10.1109/IMS30576.2020.9224010.

Liquid-metal-alloy microfluidic-based 4D-printed reconfigurable origami antennas

To achieve unprecedented reconfigurability on an origami-inspired antenna while avoiding metal breakage during bending, liquid-metal-alloy (LMA) can be utilized as a conductor to be filled into microfluidic-based antenna elements. LMA such as Eutectic Gallium Indium (EGaIn) has good mechanical properties due to its flowable

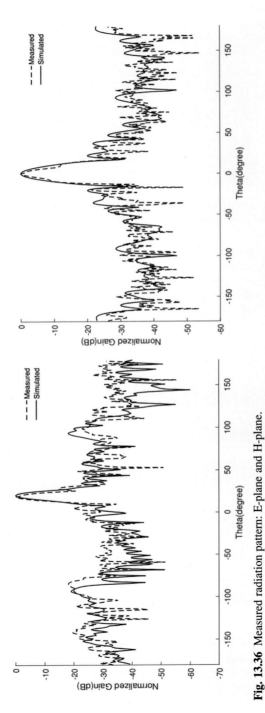

Fig. 13.36 Measured radiation pattern: E-plane and H-plane.
From Cui, Y., Nauroze, S. A., Bahr, R., & Tentzeris, E. M. (2020). 3D printed one-shot deployable flexible "kirigami" dielectric reflectarray antenna for mm-wave applications. In: *2020 IEEE/MTT-S international microwave symposium (IMS)* (pp. 1164–1167). https://doi.org/10.1109/IMS30576.2020.9224010.

and stretchable nature and features good electrical conductivity. Unlike solid materials, stretching liquid metals results literally no failure point and is only limited by the stretchability of the microfluidic channels. Therefore, origami-inspired antennas with LMA microfluidic channels can have a better range of tunability, flexibility, and durability. Fig. 13.37 shows an LMA microfluidic-based bow-tie antenna with a "Chinese fan" origami structure (Su, Bahr, Nauroze, & Tentzeris, 2017).

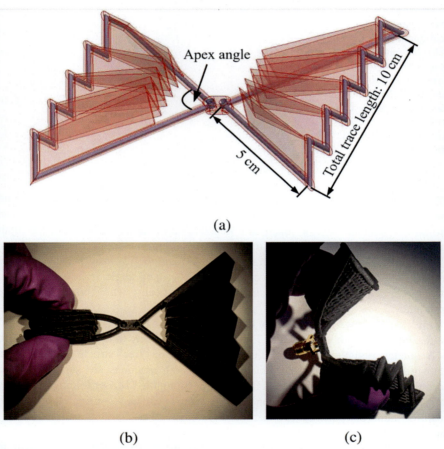

Fig. 13.37 The design and fabricated prototype of the "Chinese fan" bow-tie antenna. From Su, W., Bahr, R., Nauroze, S. A., & Tentzeris, M. M. (2017). Novel 3D-printed "Chinese fan" bow-tie antennas for origami/shape-changing configurations. In: *2017 IEEE international symposium on antennas and propagation & USNC/URSI national radio science meeting* (pp. 1245–1246). https://doi.org/10.1109/APUSNCURSINRSM.2017.8072665.

Fig. 13.38 Simulation results of the "Chinese fan" bow-tie antenna. From Su, W., Bahr, R., Nauroze, S. A., & Tentzeris, M. M. (2017). Novel 3D-printed "Chinese fan" bow-tie antennas for origami/shape-changing configurations. In: *2017 IEEE international symposium on antennas and propagation & USNC/URSI national radio science meeting* (pp. 1245–1246). https://doi.org/10.1109/APUSNCURSINRSM.2017.8072665.

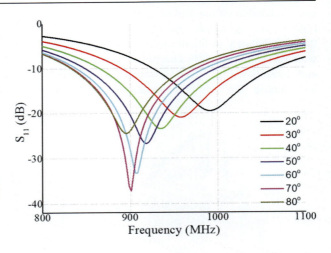

The structure's outer edge has two microfluidic channels that create a typical bow-tie antenna filled with LMA. When folding the "Chinese fan" structure, the bandwidth of the bow-tie antenna will be tuned as it depends on the apex angle. Also, the input impedance will be changed during the folding, resulting in resonant frequency shifts. The simulated return loss is shown in Fig. 13.38, and the resonant frequency shifts from 896 MHz to 992 MHz when compressing the apex angle from 80 degrees to 20 degrees. The bandwidth also increased dramatically when compressing the structure.

Fig. 13.39 shows another LMA microfluidic-based antenna with a novel "tree" configuration (Su, Bahr, et al., 2017) that enables tunability in multiple aspects such as frequency, radiation pattern, and/or polarization. The "tree" structure consists of a 3D zigzag antenna and a 3D helical antenna, both antennas are supported by a voronoized Miura origami support. By filling/unfilling the LMA or folding/unfolding the origami structure, the performance of the antenna can be reconfigured on the fly, enabling numerous potential applications including satellite communications and collapsible/portable radars.

The simulated and measured return loss of the zigzag antennas is shown in Fig. 13.40 (left), demonstrating a good impedance matching for effective radiation across all folding states. The simulated and measured radiation patterns of the zigzag antenna are shown in Fig. 13.41A. When compressing the structure, the radiation pattern transformed from a directional beam to an omnidirectional beam with the half-power beamwidth (HPBW) increased from 28 degrees to 60 degrees. The simulated and measured return loss of the helical antenna is shown in Fig. 13.40 (right) with good impedance matching for different folding states. Similar to the zigzag antenna, when the structure is compressed, the HPBW increased from 60 degrees to 90 degrees.

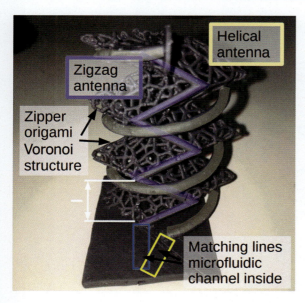

Fig. 13.39 LMA microfluidic-based antenna "tree." From Su, W., Nauroze, S. A., Ryan, B., & Tentzeris, M. M. (2017). Novel 3D printed liquid-metal-alloy microfluidics-based zigzag and helical antennas for origami reconfigurable antenna "trees". In: *2017 IEEE MTT-S international microwave symposium (IMS)* (pp. 1579–1582). https://doi.org/10.1109/MWSYM.2017.8058933.

The 4D-printed origami-inspired antennas with LMA microfluidic channels have shown the ability to change multiple aspects such as frequency, bandwidth, radiation patterns, and polarization to fit various application scenarios such as satellite communications, RFID, mobile communications, and collapsible radars.

Conclusion

With 4D printing technology, origami-inspired high-performance reconfigurable and deployable microwave devices can be realized effortlessly for 5G and mm-wave applications. The efforts of this research focus on the realization of a fully automated 4D hybrid printing process to realize novel origami-inspired FSSs, antennas, and reflectarrays. The research results demonstrated a series of novel origami-inspired FSS that utilizes a Miura-Ori structure that enables more tunable dimensions, wider frequency band, and dual polarization. The hybrid printing process successfully combined 3D-printing with inkjet-printing techniques to realize flexible reconfigurable devices. With the hybrid printing process, the first mm-wave origami-inspired FSS with single- and multilayer configurations was realized. In addition, "Kirigami" inspired structures demonstrated a high-performance reflectarray design with foldability that can reduce the size by 66%. The origami-inspired structure can also find applicable scenarios in RFID, microfluidics, and wireless sensing applications.

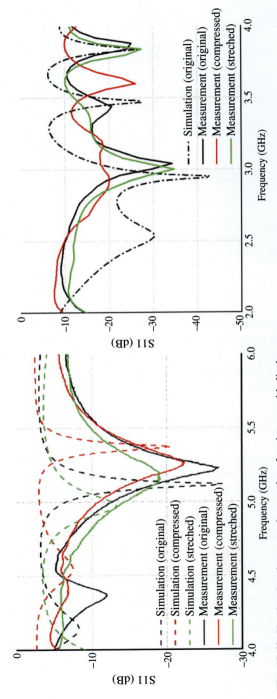

Fig. 13.40 Simulated and measured return loss for zigzag and helical antenna.
From Su, W., Nauroze, S. A., Ryan, B., & Tentzeris, M. M. (2017). Novel 3D printed liquid-metal-alloy microfluidics-based zigzag and helical antennas for origami reconfigurable antenna "trees". In: *2017 IEEE MTT-S international microwave symposium (IMS)* (pp. 1579–1582). https://doi.org/10.1109/MWSYM.2017.8058933.

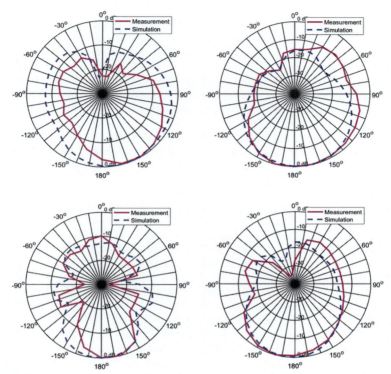

Fig. 13.41 Simulated and measured radiation patterns for the zigzag and helical antennas. From Su, W., Nauroze, S. A., Ryan, B., & Tentzeris, M. M. (2017). Novel 3D printed liquid-metal-alloy microfluidics-based zigzag and helical antennas for origami reconfigurable antenna "trees". In: *2017 IEEE MTT-S international microwave symposium (IMS)* (pp. 1579–1582). https://doi.org/10.1109/MWSYM.2017.8058933.

References

Babaei, V., Ramos, J., Lu, Y., Webster, G., & Matusik, W. (2017). FabSquare: Fabricating photopolymer objects by mold 3D printing and UV curing. *IEEE Computer Graphics and Applications*, *37*(3), 34–42. https://doi.org/10.1109/MCG.2017.37.

Bahr, R., Tehrani, B., & Tentzeris, M. M. (2018). Exploring 3-D printing for new applications: Novel inkjet- and 3-D-printed millimeter-wave components, interconnects, and systems. *IEEE Microwave Magazine*, *19*(1), 57–66. https://doi.org/10.1109/MMM.2017.2759541.

Boatti, E., Vasios, N., & Bertoldi, K. (2017). Origami metamaterials for tunable thermal expansion. *Advanced Materials*, *29*(26), 1700360. https://doi.org/10.1002/adma.201700360.

Castle, T., Sussman, D. M., Tanis, M., & Kamien, R. D. (2016). Additive lattice kirigami. *Science Advances*, *2*(9), e1601258. https://doi.org/10.1126/sciadv.1601258.

Cui, Y., Nauroze, S. A., Bahr, R., & Tentzeris, E. M. (2020). 3D printed one-shot deployable flexible "kirigami" dielectric reflectarray antenna for mm-wave applications. In *2020 IEEE/MTT-S international microwave symposium (IMS)* (pp. 1164–1167). https://doi.org/10.1109/IMS30576.2020.9224010.

Cui, Y., Nauroze, S., Bahr, R., & Tentzeris, M. (2021). A novel additively 4D printed origami-inspired tunable multi-layer frequency selective surface for mm-wave IoT, RFID, WSN, 5G, and smart city applications. In *2021 IEEE/MTT-S international microwave symposium (IMS)*.

Cui, Y., Nauroze, S. A., & Tentzeris, M. M. (2019). Novel 3D-printed reconfigurable origami frequency selective surfaces with flexible inkjet-printed conductor traces. In *2019 IEEE MTT-S international microwave symposium (IMS)* (pp. 1367–1370). https://doi.org/10.1109/MWSYM.2019.8700994.

Deng, R., Yang, F., Xu, S., & Li, M. (2016). Radiation performances of conformal dielectric reflectarray antennas at sub-millimeter waves. In *Vol. 1. 2016 IEEE international conference on microwave and millimeter wave technology (ICMMT)* (pp. 217–219). https://doi.org/10.1109/ICMMT.2016.7761728.

Dimatix Material Printer DMP-2850. (2016). https://www.printingnews.com/digital-inkjet/product/12232165/fujifilm-dimatix-inc-dimatix-material-printer-dmp2850.

Filipov, E. T., Tachi, T., & Paulino, G. H. (2015). Origami tubes assembled into stiff, yet reconfigurable structures and metamaterials. *Proceedings. National Academy of Sciences. United States of America, 112*(40), 12321. https://doi.org/10.1073/pnas.1509465112.

Formlabs 3D Printers. (2021). https://formlabs.com/3d-printers/ (Original work published 2021).

Hester, J. G., Kim, S., Bito, J., Le, T., Kimionis, J., Revier, D., et al. (2015). Additively manufactured nanotechnology and origami-enabled flexible microwave electronics. *Proceedings of the IEEE, 103*(4), 583–606. https://doi.org/10.1109/JPROC.2015.2405545.

Jamaluddin, M. H., Sauleau, R., Castel, X., Benzerga, R., Coq, L., Gillard, R., et al. (2010). Design, fabrication and characterization of a dielectric resonator antenna reflectarray in Ka-band. *Progress In Electromagnetics Research B, 25*, 261–275. https://doi.org/10.2528/PIERB10071306.

Jeong, H., Cui, Y., Tentzeris, M. M., & Lim, S. (2020). Hybrid (3D and inkjet) printed electromagnetic pressure sensor using metamaterial absorber. *Additive Manufacturing, 35*, 101405. https://doi.org/10.1016/j.addma.2020.101405.

Jung, C. W., Lee, M.-J., Li, G. P., & De Flaviis, F. (2006). Reconfigurable scan-beam single-arm spiral antenna integrated with RF-MEMS switches. *IEEE Transactions on Antennas and Propagation, 54*(2), 455–463. https://doi.org/10.1109/TAP.2005.863407.

Kim, Y., Cui, Y., Tentzeris, M. M., & Lim, S. (2020). Additively manufactured electromagnetic based planar pressure sensor using substrate integrated waveguide technology. *Additive Manufacturing, 34*, 101225. https://doi.org/10.1016/j.addma.2020.101225.

Miura, K. (1985). *Method of packaging and deployment of large membranes in space*. Institute of Space and Astronautical Sciences.

Munk, B. A. (2000). *Frequency selective surfaces theory and design*.

Nauroze, S. A. (2019). *Additively manufactured origami-inspired "4D" RF structures with on-demand continuous-range tunability*. Georgia Institute of Technology. http://hdl.handle.net/1853/62330.

Nauroze, S. A., Novelino, L., Tentzeris, M. M., & Paulino, G. H. (2017). Inkjet-printed "4D" tunable spatial filters using on-demand foldable surfaces. In *2017 IEEE MTT-S international microwave symposium (IMS)* (pp. 1575–1578). https://doi.org/10.1109/MWSYM.2017.8058932.

Nayeri, P., Liang, M., Sabory-Garcia, R. A., Tuo, M., Yang, F., Gehm, M., et al. (2014). 3D printed dielectric reflectarrays: Low-cost high-gain antennas at sub-millimeter waves. *IEEE Transactions on Antennas and Propagation, 62*(4), 2000–2008. https://doi.org/10.1109/TAP.2014.2303195.

Rafsanjani, A., Akbarzadeh, A., & Pasini, D. (2015). Snapping mechanical metamaterials under tension. *Advanced Materials*, *27*(39), 5931–5935. https://doi.org/10.1002/adma.201502809.

Ramasubramaniam, R., Chen, J., & Liu, H. (2003). Homogeneous carbon nanotube/polymer composites for electrical applications. *Applied Physics Letters*, *83*(14), 2928–2930. https://doi.org/10.1063/1.1616976.

Schenk, M., Guest, S. D., & McShane, G. J. (2014). Novel stacked folded cores for blast-resistant sandwich beams. *International Journal of Solids and Structures*, *51*(25), 4196–4214. https://doi.org/10.1016/j.ijsolstr.2014.07.027.

Shah, S. I. H., & Lim, S. (2021). Review on recent origami inspired antennas from microwave to terahertz regime. *Materials & Design*, *198*, 109345. https://doi.org/10.1016/j.matdes.2020.109345.

Silverberg, J. L., Evans, A. A., McLeod, L., Hayward, R. C., Hull, T., Santangelo, C. D., et al. (2014). Using origami design principles to fold reprogrammable mechanical metamaterials. *Science*, *345*(6197), 647. https://doi.org/10.1126/science.1252876.

Stankovic, J. A. (2008). Wireless sensor networks. *Computer*, *41*(10), 92–95. https://doi.org/10.1109/MC.2008.441.

Su, W., Bahr, R., Nauroze, S. A., & Tentzeris, M. M. (2017). Novel 3D-printed "Chinese fan" bow-tie antennas for origami/shape-changing configurations. In *2017 IEEE international symposium on antennas and propagation & USNC/URSI national radio science meeting* (pp. 1245–1246). https://doi.org/10.1109/APUSNCURSINRSM.2017.8072665.

Sun, Y., & Leung, K. W. (2018). Millimeter-wave substrate-based dielectric reflectarray. *IEEE Antennas and Wireless Propagation Letters*, *17*(12), 2329–2333. https://doi.org/10.1109/LAWP.2018.2874082.

Sung, Y. J., Jang, T. U., & Kim, Y. S. (2004). A reconfigurable microstrip antenna for switchable polarization. *IEEE Microwave and Wireless Components Letters*, *14*(11), 534–536. https://doi.org/10.1109/LMWC.2004.837061.

Tehrani, B. K., Mariotti, C., Cook, B. S., Roselli, L., & Tentzeris, M. M. (2016). Development, characterization, and processing of thin and thick inkjet-printed dielectric films. *Organic Electronics*, *29*, 135–141. https://doi.org/10.1016/j.orgel.2015.11.022.

Wei, Z. Y., Guo, Z. V., Dudte, L., Liang, H. Y., & Mahadevan, L. (2013). Geometric mechanics of periodic pleated origami. *Physical Review Letters*, *110*(21), 215501. https://doi.org/10.1103/PhysRevLett.110.215501.

White, C. R., & Rebeiz, G. M. (2009). Single- and dual-polarized tunable slot-ring antennas. *IEEE Transactions on Antennas and Propagation*, *57*(1), 19–26. https://doi.org/10.1109/TAP.2008.2009664.

Wong, W. S., Chabinyc, M. L., Ng, T.-N., & Salleo, A. (2009). Materials and novel patterning methods for flexible electronics. In *Vol. 11. Flexible electronics. Electronic materials: Science & Technology* (pp. 143–181). Springer US. https://doi.org/10.1007/978-0-387-74363-9_6.

Yang, L., Rida, A., Vyas, R., & Tentzeris, M. M. (2007). RFID tag and RF structures on a paper substrate using inkjet-printing technology. *IEEE Transactions on Microwave Theory and Techniques*, *55*(12), 2894–2901. https://doi.org/10.1109/TMTT.2007.909886.

Zhang, S. (2017). Three-dimensional printed millimetre wave dielectric resonator reflectarray. *IET Microwaves, Antennas and Propagation*, *11*. https://doi.org/10.1049/iet-map.2017.0278.

Shape-reversible 4D printing aided by shape memory alloys

Saeed Akbari[a,*], Amir Hosein Sakhaei[b,*], Sahil Panjwani[c,*], Kavin Kowsari[c], and Qi Ge[c,d]
[a]RISE Research Institutes of Sweden, Mölndal, Sweden
[b]School of Engineering and Digital Arts, University of Kent, Canterbury, United Kingdom
[c]Digital Manufacturing and Design Center, Singapore University of Technology and Design, Singapore, Singapore
[d]Department of Mechanical and Energy Engineering, Southern University of Science and Technology, Shenzhen, China

Introduction

If a certain thermomechanical load is applied to the shape memory alloys (SMAs), they can deform into a secondary elongated shape and then restore their original length. At a microstructural level, this is a result of phase transformations because at low temperatures, the SMA is in the martensite phase, while at high temperatures, it is in the austenite phase. This category of smart materials has been extensively exploited over the past 40 years to fabricate shape memory active structures with various applications in robotics, aerospace, biomedical engineering, etc. (Ashir, Nocke, Theiss, & Cherif, 2017; Barbarino, Flores, Ajaj, & Dayyani, & Friswell, 2014; Bruck, Moore, & Valentine, 2002; Holschuh, Obropta, & Newman, 2015; Icardi & Ferrero, 2009; Lagoudas & Tadjbakhsh, 1992; Wang, Hang, Wang, Li, & Du, 2008).

A major drawback of the SMAs is that their maximum recoverable strain is only 4%–8%. This limits their use in applications that require large deformations (Rodrigue, Wang, Han, Kim, & Ahn, 2017). To address this issue, the SMA wires can be integrated into an elastomeric soft matrix in order to transform the SMA small deformations into deformations of at least an order of magnitude larger. This principle can be used in fabricating bending, twisting, and coupled bending-twisting actuators. For instance, to create bending deformations inspired by this methodology, the SMA wires can be inserted eccentrically into a low-stiffness elastomeric beam. After actuation, the SMA wire axial contraction is transformed into a large bending deformation (Ashir et al., 2017; Barbarino et al., 2014; Bruck et al., 2002; Holschuh et al., 2015; Icardi & Ferrero, 2009; Lagoudas & Tadjbakhsh, 1992; Wang et al., 2008).

The common practice to fabricate the SMA-based soft actuators is to pour uncured liquid polymers into molds, fix the SMA wire at both ends of the molds, and then cure the molded polymer at elevated temperatures. These steps have to

---

[*] These authors contributed equally to this work.

be done manually, and for each new set of design parameters, a custom-designed mold is needed (Ghosh, Rao, & Srinivasa, 2013; Hugo, Wei, Dong-Ryul, & Sung-Hoon, 2017; Lelieveld, Jansen, & Teuffel, 2016; Rodrigue, Wang, Han, Quan, & Ahn, 2016; Song et al., 2016; Song, Chen, & Naguib, 2016; Taya, Liang, Namli, Tamagawa, & Howie, 2013; Wang, Rodrigue, Kim, Han, & Ahn, 2016). This may impede a repeatable fabrication process. In addition, the SMA soft actuators fabricated by this method do not have the capability to maintain deformed configuration. For that, the SMA wire needs to be constantly maintained in the actuated state. This, however, requires more energy consumption. Some efforts have been made to solve this issue by integrating phase-changeable materials, for example, fusible alloys (Schubert & Floreano, 2013; Wang, Rodrigue, & Ahn, 2015) and polymers (McEvoy & Correll, 2015; Wang & Ahn, 2017), into the actuator. The stiffness of this class of materials changes significantly with temperature, so they can be embedded into the actuator structure to control its stiffness. Moreover, if a phase-changeable material has the shape recovery characteristics, for example, SMPs, the actuator can exhibit reversible deformation as well (Taya et al., 2013; Tobushi, Hayashi, Sugimoto, & Date, 2009; Wang & Ahn, 2017). It should be noted that one problem with this stiffness-modulation technique is the accurate placement of the phase-changeable segments in the actuator structure.

To address these problems, we need a more flexible fabrication process that does not need cumbersome molding and assembly processes. In this chapter, we propose the use of inkjet four-dimensional (4D) printing as an alternative approach to fabricate multimaterial soft SMA actuators. The 4D-printed composite actuators demonstrate three features simultaneously: variable stiffness, shape retention, and shape restoration (Zolfagharian et al., 2020; Zolfagharian, Kaynak, & Kouzani, 2020). Multimaterial inkjet 4D printing is able to print different digital materials with various mechanical properties in a voxelized domain (Ge, Dunn, Qi, & Dunn, 2014; Ge, Qi, & Dunn, 2013; Mao et al., 2015, 2016). The SMA-printed actuators whose matrix was from a low-stiffness elastomer were demonstrated in our earlier work (Akbari et al., 2019). The actuators presented in that study did not have shape retention and recovery capabilities. To add these features, in this chapter, we show how multimaterial inkjet 4D printing can be used to create SMA-SMP soft composite actuators with variable stiffness. The hinges of the printed bending actuators consist of stiff SMP substructures and a soft elastomeric layer. The temperature of the SMP segments, which have different thicknesses, is controlled using resistive wires to modulate their stiffness and realize shape retention and recovery. While the SMP substructures are at higher temperatures (low stiffness state), the SMA wire actuation creates a large bending deformation. The composite actuator can then recover its original shape due to the SMP shape recovery effect. This chapter is mainly an extension of our earlier work (Akbari, Sakhaei, Panjwani, Kowsari, & Ge, 2021), with a detailed elaboration on the advanced material models needed for numerical simulation of SMAs and SMPs. In addition, we discussed the limitations of the presented fabrication method and made a few suggestions to advance this technology further.

Materials and methods

Design of actuators

We designed two different multimaterial 4D-printed actuators capable of bending deformation, as shown in Fig. 14.1. The thickness, width, and effective length of each actuator were 3, 15, and 100 mm, respectively. Also, the hinge length was 20 mm. Joule heating using embedded resistive wires was used to control the temperature of the SMP segments. The SMP has two functions: to maintain the actuator deformed shape at low temperatures ($T < T_g$) and to help the actuator recover its as-printed shape at higher temperatures ($T > T_g$). Fig. 14.1 shows that the difference between the two actuator designs, referred to as Actuator-1 and Actuator-2, lies in the amount of the elastomeric soft material printed in the hinge area. This causes Actuator-1 to have a significantly lower bending stiffness than Actuator-2. The temperature of each hinge is controlled separately using resistive wires. Prestrained SMA wires are inserted eccentrically to the bending actuator and heated via Joule heating to create the actuation bending force. We also changed the SMP layer thickness in each hinge to study its effect on the actuation performance.

Accurate control of the actuation of the SMA and resistive wires is of critical importance in actuator performance. The timing sequence for applying current to the SMA and resistive wires is shown in Fig. 14.2. First, the temperature of the hinge SMP segments is increased using resistive heating by applying a current of value I_1 from the time zero until the time t_1 to transform them to a low-stiffness rubbery state. Then, the resistive wire current is switched off, and the SMA wire is activated by applying a current of value I_2 from t_1 to t_2 to produce a large bending deformation. The SMA wire at this stage maintains its actuated state until the SMP temperature decreases to a high-stiffness glassy state. After the SMA wire current is disconnected, the actuator maintains the actuated bent shape. This results from the SMP high bending stiffness. Finally, applying a current of value I_1 again to the resistive wire raises the SMP temperature and leads to the actuator shape recovery.

Experimental procedure

The multimaterial 4D printing capability of Stratasys inkjet 3D printer was used to fabricate the body of the bending actuators. The main feature of this printer is its ability to deposit two base polymer droplets, referred to as Agilus30 and VeroClear, on a voxelized domain. VeroClear stiffness is an order of magnitude higher than Agilus30. For this reason, it was used to print the stiff SMP segments. The printer can also print different combinations of these two base materials to generate a broad range of mechanical properties within the printed structure. We performed dynamic mechanical analysis (DMA) tests as well as uniaxial tensile tests to obtain the properties of the printed material. Fig. 14.3A and B demonstrates the effect of temperature on the stress–strain behavior, failure strain, and Young's modulus of VeroClear. Previous research has confirmed excellent shape memory behavior of VeroClear

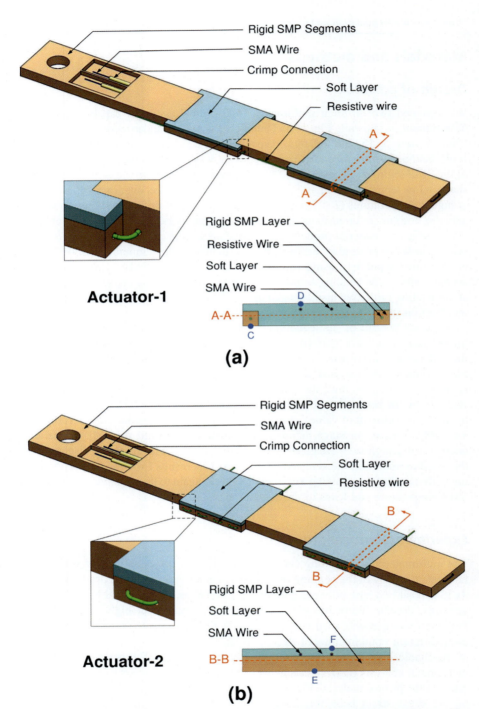

Fig. 14.1 Schematic representation of the actuator designs. Each actuator has two hinges. The SMA wire is embedded into the actuator's entire length, and the resistive wires are embedded into the hinge SMP segments. The resistive wires are inserted (A) longitudinally and (B) transversely into the SMP segments.
No permission required.

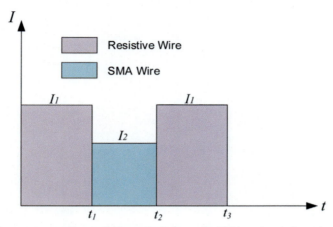

Fig. 14.2 The current profile applied sequentially to the SMA and resistive wires to achieve bending deformation, shape retention, and shape restoration.
No permission required.

(Ge et al., 2013, 2014; Mao et al., 2015, 2016), which enables efficient recovery of the actuator as-printed shape.

Fig. 14.3C and D shows the results of the tensile tests of the digital soft materials obtained at room temperature. It can be noted that from Agilus30 to FLX9950, the tensile modulus goes up while the failure strain decreases. Out of the digital materials tested, FLX9960 ($E = 2.6$ MPa) was selected to print the actuator soft substructures, due to the similarity of its stiffness to polydimethylsiloxane (PDMS, $E = 1.8$ MPa), an elastomer widely used in the conventional fabrication methods of soft actuators (Hugo et al., 2017; Rodrigue et al., 2016; Wang et al., 2016).

The DMA characterization results are illustrated in Fig. 14.4. The T_g of the printed materials was identified as a peak in the Tan δ curve in Fig. 14.4A and plotted in Fig. 14.4B. According to the DMA tests, the T_g of VeroClear and FLX9960 is 58°C and 4°C, respectively. At temperatures higher than T_g, VeroClear is in a rubbery state.

Bending actuators demonstrated here, that is, Actuator-1 and Actuator-2 (Fig. 14.1), were fabricated by the multimaterial inkjet printer. After the samples were printed, the residual support material from the 3D printed parts was cleared using a high-pressure water cleaner (Powerblast, Balco Engineering Ltd., Birmingham, UK). After removing all of the support material, the SMA wire, which was a nickel–titanium alloy (Flexinol, Irvine, CA, USA, properties of Table 14.1) with a diameter of 0.15 mm, was inserted into the actuator printed body and fixed to the two designated points at one end of the actuator using a copper crimp connections (Insulated Crimp Bootlace Ferrule, RS Components, Corby, UK) to stop relative sliding between the soft matrix and the SMA wire.

Moreover, a copper–nickel alloy (Constantan) resistive wire, whose thickness and resistance were 0.20 mm and 15.6 Ω /m, respectively, was inserted into the SMP

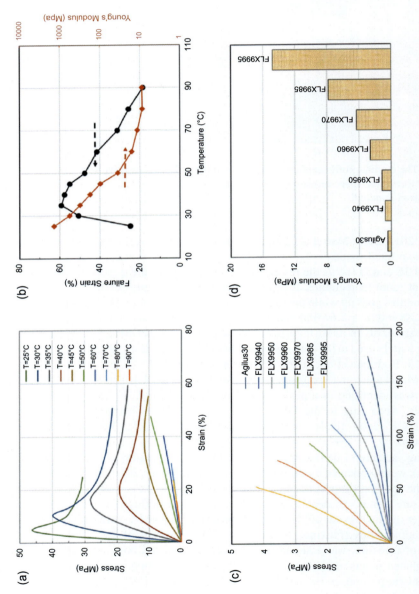

Fig. 14.3 Characterization of the mechanical properties of the printed materials. (A) VeroClear stress–strain curves versus temperatures. (B) VeroClear failure strain and Young's modulus versus temperature. (C) Stress–strain response of the elastomeric digital materials at room temperature that was used to print the soft (low-stiffness) segments of the actuators. (D) Young's modulus of the elastomeric digital materials at room temperature.

No permission required.

Simulation of actuation cycle

Finite element modeling (FEM) is a great tool to design and predict the deformation behavior of 4D-printed structures. We used the software package ABAQUS to perform 3D FEM simulations aiming at investigating the thermomechanical behavior of the fabricated soft actuators. The simulation objective was to present and verify a computational framework that can be employed for the further development of actuators exhibiting complex large deformations. Fig. 14.5 shows the 3D FE model of Actuator-1 with the details of soft and rigid segments as well as the displacement boundary conditions applied to one end.

The soft elastomeric layer, the SMP shape memory layer, and the SMA wires were simulated using three different material models. Based on the material characterization tests (Fig. 14.3), the soft layer was assumed to show a linear elastic behavior. In addition, a coupled thermomechanical, isotropic-based material model was used for the SMA wire. The model was created as an ABAQUS user-defined material (UMAT). Moreover, a second UMAT was used to simulate the SMP. It was developed based on a multibranch viscoelastic constitutive model. The model details are extensively discussed in Westbrook, Kao, Castro, Ding, and Jerry Qi (2011) and will be elaborated on in the following paragraphs.

As mentioned in the previous section, the elastomeric part of the actuator could be fabricated from different combinations of materials using Stratasys 3D printer to develop appropriate stiffness in the composite structure. In the current work, the so-called FLX9960 with an elastic modulus of $E = 2.6\,\mathrm{MPa}$ is used and implemented as a linear elastic material in the finite element model.

To model the mechanical behavior of the SMA in the composite actuator, the constitutive model presented by Lagoudas and Tadjbakhsh (1992) and Sakhaei and Thamburaja (2017) is used which is basically a thermomechanical isotropic model for the SMAs. Sakhaei and Thamburaja (Lagoudas & Tadjbakhsh, 1992; Sakhaei & Thamburaja, 2017) presented the Helmholtz free energy per unit reference volume, ψ, as

$$\psi = \hat{\psi}\left(\mathbf{E}^e, \theta, \xi\right) = \psi^e + \psi^\xi + \psi^\theta \tag{14.1}$$

$$\psi^e = \hat{\psi}^e\left(\mathbf{E}^e, \theta\right) = \frac{1}{2}\,\mathbf{E}^e.C[\mathbf{E}^e] - \mathbf{A}(\theta - \theta_0).C[\mathbf{E}^e] \tag{14.2}$$

$$\psi^\xi = \hat{\psi}^\xi\left(\theta, \xi\right) = \frac{\lambda_T}{\theta_T}\,(\theta - \theta_T)\xi, \tag{14.3}$$

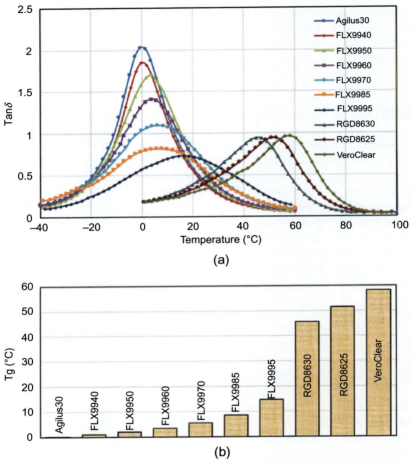

Fig. 14.4 The DMA characterization results for printed digital materials. (A) Tanδ as a function of temperature. (B) Glass transition temperature (Tg) of the printed materials. No permission required.

Table 14.1 Material properties of SMA wire.

Martensite Young's modulus (E_M)	Austenite Young's modulus (E_A)	Martensite start temperature (M_s)	Martensite finish temperature (M_f)	Austenite start temperature (A_s)	Austenite finish temperature (A_s)
28 GPa	75 GPa	52°C	42°C	68°C	78°C

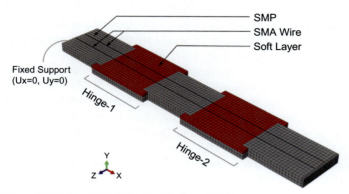

Fig. 14.5 FE mesh of Actuator-1 (described in Fig. 14.1A).
No permission required.

$$\psi^\theta = \hat{\psi}^\theta(\theta). \tag{14.4}$$

Above, ψ^e, ψ^ξ, and ψ^Θ are the thermoelastic, transformation, and thermal free energy densities, respectively, and each of them are functions of different material parameters. For example, ψ^e is a function of $\mathbf{E}^e = 1/2\,(\mathbf{C}^e - \mathbf{I})$ as the elastic Green–Lagrangian strain tensor, $\mathbf{C} = 2\mu I + (\kappa - 2\mu/3)\mathbf{I} \otimes \mathbf{I}$ as the fourth-order stiffness tensor and $\mathbf{A}$ as the second-order thermal expansion tensor. Moreover, ψ^ξ is a function of ξ as the martensitic volume fraction, which is always between 0 and 1, λ_T as the constant latent heat of phase transformation, and $\theta_T = 1/2(M_s + A_s)$ as the transformation temperature. In the latter equation, M_s and A_s are the martensite and austenite start temperatures, respectively. Then, the stress constitutive equation could be calculated from the Helmholtz free energy density function through

$$T = \frac{\partial \psi}{\partial \mathbf{E}^e} = C[\mathbf{E}^e - \mathbf{A}(\theta - \theta_0)] \tag{14.5}$$

As discussed earlier, $\mathbf{E}^e = 1/2\,(\mathbf{C}^e - \mathbf{I})$ is the elastic Green–Lagrangian strain tensor as a function of right Cauchy–Green elastic deformation tensor $\mathbf{C}^e = \mathbf{F}^{eT}\mathbf{F}^e$, where $\mathbf{F}^e$ is the elastic deformation gradient and is part of the total deformation gradient $\mathbf{F} = \mathbf{F}^e\,\mathbf{F}^{inel}$. Therefore, first, we should determine $\mathbf{F}^{inel}$ as the inelastic transformation gradient to calculate $\mathbf{F}^e$. This could be modeled as

$$\dot{\mathbf{F}}^{inel} = \left\{\sqrt{3/2}\bar{\varepsilon}_t\,\dot{\xi}\,\mathbf{S}\right\}\mathbf{F}^{inel} \tag{14.6}$$

where $\mathbf{S}$ is the inelastic flow direction, $\bar{\varepsilon}_t$ is a material parameter and describes the maximum transformation strain in the SMA martial. In the current work, the nickel–titanium SMA wires (commercially known as Flexinol) are used with $\bar{\varepsilon}_t = 5\%$. Furthermore, the elastic modulus of this alloy is 28 GPa and 75 GPa when the material is in the martensitic and austenitic phases, respectively. Finally, the martensite starts temperature and austenite start temperature of this SMA material are $M_s = 52°C$ and $A_s = 68°C$, respectively, which are then used to calculate transformation temperature in our equations ($\theta_T = 1/2(M_s + A_s)$).

The other material whose thermomechanical behavior should be modeled and implemented for the finite-element simulations is the SMP. This type of polymer material exhibits large recovery deformation after heating above the glass transition temperature, T_g. In this work, the multibranch viscoelastic model that is developed in Westbrook et al. (2011) and Yu, Ge, and Qi (2014) is used and implemented as a UMAT in ABAQUS FE package. This model is a spring-dashpot model that is based on equilibrium and multiple nonequilibrium branches; therefore, the Cauchy stress at any material point is developed as

$$\sigma = \sigma_{eq} + \sigma_g + \sum_{i=1}^{m} \sigma_r^i \tag{14.7}$$

where σ_r^i and σ_g are the Cauchy stress for nonequilibrium rubbery and glassy branches and σ_{eq} is the Cauchy stress in the equilibrium branch and represents the hyperelastic behavior of material above the glass transition temperature. Furthermore, based on the Arruda–Boyce model:

$$\sigma_{eq} = \frac{nk_BT}{3J_M} \frac{\sqrt{N}}{\lambda_{chain}} \mathcal{L}^{-1}\left(\frac{\lambda_{chain}}{\sqrt{N}}\right) \overline{\mathbf{B}}' + K(J_M - 1)\mathbf{I} \tag{14.8}$$

In Eq. (14.8), n, T, and N represent the crosslinking density, the temperature, and the number of Kuhn segments, respectively. k_B is the Boltzmann's constant and $J_M = \det \mathbf{F}_M$ where $\mathbf{F}_M$ is the deformation gradient, and K is the Bulk modulus. Noting that $\overline{\mathbf{B}} = \mathbf{F}_M^T \mathbf{F}_M$ is the isochoric left Cauchy–Green, the deviatoric part would be $\overline{\mathbf{B}}' = \overline{\mathbf{B}} - \frac{1}{3} tr(\overline{\mathbf{B}})\mathbf{I}$. Moreover, in Arruda–Boyce model, $\mathcal{L}(x) = \coth(x) - \frac{1}{x}$ is the Langevin function where $\lambda_{chain} = \sqrt{tr(\overline{\mathbf{B}})/3}$ describes the stretch of each chain. Please follow Westbrook et al. (2011) for further details of this model.

The actuator SMP substructures were created using 3568 of C3D8H (continuum linear hexahedral) elements, while the SMA wires were created using 109 continuum linear line elements of type B31H. In addition, a surface-to-surface frictionless contact was assumed at the interface between the SMA wire and the inner wall of the hole used for the placement of the SMA wires in the longitudinal direction.

Inspired by the experimental testing, the following loading steps were used sequentially in the simulation: first, the SMA wire was elongated 5% in the martensitic phase (room temperature). Then, the SMA wire loading was removed without changing the temperature. As a result of detwinning in the martensitic phase, the first two steps create 5% strain in the SMA wire.

Fig. 14.1 indicated that the temperature of each SMP substructure in the hinge area is separately controlled by the inserted resistive wires. The next part of the loading is to increase the temperature of the SMP segment in the desired hinge to over T_g in order to soften the SMP. This heating step is only limited to a specific hinge and will not influence the other SMP parts, and they will remain stiff (rigid) at lower temperatures. In the next loading step, the contact between the SMA wire and the surrounding matrix is activated, and the SMA temperature is heated up to a temperature higher than the austenitic finish temperature ($A_s = 78°C$, Table 14.1). This results in the recovery of 5% strain stored in the wire and bending deformation of the actuator due to the wire axial contraction. Subsequently, the temperature of the hinge SMP is lowered in order to increase its stiffness. In the next step, the SMA wire current is switched off. Then, the bent shape of the actuator is retained due to the high stiffness of the hinge SMP sections. Lastly, the hinge SMP is heated up again to recover the actuator as-fabricated configuration. These steps are discussed in more detail and compared with the experimental data in the following section.

Numerical and experimental results

For experimental characterization of Actuator-1 and Actuator-2 (Fig. 14.1), several samples of each design were 3D printed. Some preliminary testing was performed to measure the SMP parts heating time and bending stiffness as a function of the applied electrical current. To this end, an infrared camera was used to monitor the temperature change. The temperature was specifically monitored at the vicinity of the SMP substructures on Point C for Actuator-1 (Fig. 14.1A) and Point E for Actuator-2 (Fig. 14.1B). A range of currents was applied to both actuators. The main goal here was to measure the time at which each actuator reaches the target temperature of 60°C ($t60°C$) (see Fig. 14.6). As expected, the temperature rise was faster in Actuator-2 than Actuator-1. This is because the free surface from which heat dissipates was smaller for the SMP segments containing resistive wires in Actuator-2. As a result, the heat dissipation rate through convection (to the ambient air) was slow.

As observed in Fig. 14.6, in order to raise the SMP temperature to 60°C in less than 60 s, the electrical currents of roughly 1.1 A and 0.75 A must be applied to Actuator-1 and Actuator-2, respectively. We used these electrical currents in all the following experiments in order to create identical heating times for both actuators.

Moreover, the temperature change with time was monitored at the vicinity of the SMP and SMA (Points C and D in Fig. 14.1A and Points E and F in Fig. 14.1B). For this purpose, the electrical currents of 1.1 A and 0.75 A were applied for 60 s to Actuator-1 and Actuator-2, respectively.

It is important to ensure that the resistive heating in the SMP does not increase the SMA wire temperature to the actuation point. Fig. 14.7 shows that, as expected, the resistive heating in the SMP substructures also raises the temperature near the SMA wire. This temperature rise is more visible in Actuator-2 due to the small distance between the resistive and SMA wires. However, the maximum SMA temperature in both actuators never exceeds the austenite start temperature ($A_s = 68°C$; Table 14.1). Therefore, the SMA wire actuation due to resistive heating in the SMP segments is not a concern.

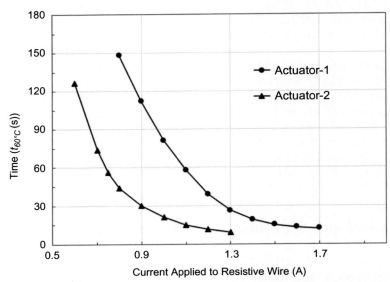

Fig. 14.6 The time required for the SMP substructures of (A) Actuator-1 and (B) Actuator-2 to reach the target temperature of 60°C as a function of the electrical current applied to the resistive wires. The temperature was monitored at Point C on Actuator-1 (Fig. 14.1A) and at Point E on Actuator-2 (Fig. 14.1B).
No permission required.

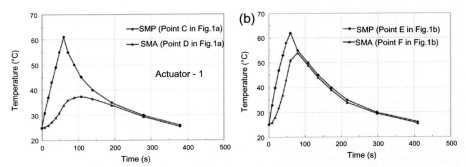

Fig. 14.7 The actuator's temperature changes at the vicinity of the SMA and SMP versus time. The electrical currents of 1.1 A and 0.75 A were applied for 60 s to the resistive wires inserted into the SMP segments of Actuator-1 and Actuator-2, respectively. (A) Actuator-1 (schematically described in Fig. 14.1A). (B) Actuator-2 (Fig. 14.1B).
No permission required.

Bending stiffness is the most important parameter affecting the actuator's strength. Compared with the SMP and elastomer segments, the resistive and SMA wires have a negligible volume; therefore, we assume they do not contribute to the actuator bending stiffness. So, the bending stiffness in the hinge area is only determined by the elastomer and SMP segments based on the following equation:

$$E_A I_A = E_{Soft} I_{Soft} + E_{SMP} I_{SMP} \qquad (14.9)$$

In Eq. (14.9), E_A, E_{Soft}, and E_{SMP} signify Young's moduli of the actuator, elastomer (soft), and SMP parts, respectively. I_A, I_{Soft}, and I_{SMP} show the second moment of area of the actuator, elastomer (soft), and SMP segments, respectively. All parameters shown in Eq. (14.9) are constant across the temperature ranges used in the actuation tests, except E_{SMP}, which has a higher value in the glassy state than the rubbery state. This is demonstrated clearly in Fig. 14.3A and B. Because the elastomeric (soft) material has a very low glass transition temperature (FLX9960, $T_g = 4°C$), it remains in a low-stiffness rubbery state throughout the operating temperature. Consequently, it is reasonable to assume that Young's modulus of the soft parts during the actuation cycle remains constant.

Fig. 14.8 indicates the bending stiffness change during the cooling process. The SMP thickness in the hinge area was 2 mm. The Young's moduli of the actuator's different soft and rigid substructures were obtained from Fig. 14.3A. Fig. 14.8 shows that the effect of temperature on bending stiffness is significant. The bending stiffness of Actuator-1 at low temperature (25°C) and high temperature (60°C) was around 3.32 and 0.13 GPa mm^4, respectively, demonstrating a 96.2% variation of bending stiffness at this temperature range. For Actuator-2, the bending stiffness increased from 0.33 GPa mm^4 at high temperature to 24.3 GPa mm^4 at room temperature, indicating a 98.6% increase in the bending stiffness. These findings show the significant stiffness variability of both actuators. This enables effective performance control for soft robotics applications.

In summary, Fig. 14.8 shows that Actuator-2 bending stiffness is one order of magnitude greater than Actuator-1 at lower temperatures and shows a more evident change in stiffness with temperature. This difference is simply because a larger volume of the SMP is included in the hinge areas of this actuator.

Each of the two hinges can be controlled independently for each actuator by applying a current to the resistive wires embedded into each hinge. This resistive heating softens the SMP. The SMA wire is subsequently heated up by applying an electrical

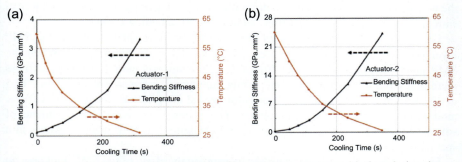

Fig. 14.8 Change of the actuators bending stiffness and temperature with time during the cooling process. (A) Actuator-1 (Fig. 14.1A). (B) Actuator-2 (Fig. 14.1B).
No permission required.

current to it to deform the actuator. Fig. 14.9 demonstrates the numerical and experimental results of the separate and simultaneous actuation of the hinges of Actuator-1 having the SMP thickness of 2 mm (Fig. 14.1A).

Several steps were followed sequentially in each actuation cycle. First, an electrical current of 1.1 A was applied to the resistive wire of the desired hinges for 60 s to raise the SMP temperature to 60°C (Fig. 14.9A, E, I). The resistive wire was then inactivated, and subsequently, an electrical current of 0.41 A was applied to the SMA to make it contract and deform the hinge, while the hinge was kept at a high-temperature rubbery state, as demonstrated in Fig. 14.9B, F and J. The SMA wire was activated for 120 s so that the SMP had enough time to cool down and enter into the glassy state. Fig. 14.9C, G and K indicate that the actuator managed to maintain the deformed shape after the SMA wire current was disconnected. But then, it slightly bounced back after the SMA wire disconnection. Finally, to recover the actuator's initial configuration, the electrical current of 1.1 A was applied again to the related resistive wire (Fig. 14.9D, H and L).

To quantify the shape fixity and recovery of the bending actuators, the shape fixing ratio (R_f) was used to determine their ability to retain a temporary shape, and the shape recovery ratio (R_r) to characterize its recovery performance, based on the following expressions:

$$R_f = \theta_2/\theta_1 \text{ and } R_r = 1 - \theta_3/\theta_2 \tag{14.10}$$

In the above equation, θ_1, and θ_2 and θ_3 are shown in Fig. 14.9B–D and defined as follows: θ_1 denotes the maximum bending angle of the actuator before the SMA wire is disconnected, θ_2 shows the bending angle after the SMA wire is disconnected, and θ_3 is the bending angle after recovery.

The SMP layer layout across the hinge area of Actuator-1 and Actuator-2 is depicted in Cross-Section A–A in Fig. 14.1A and Cross-Section B–B in Fig. 14.1B, respectively. As shown in Fig. 14.1, Actuator-2 hinges contain a larger volume of the SMP compared with Actuator-1. We repeated the actuator design to modify the SMP thickness across the hinge area in order to investigate its effect on the actuation performance. The experimental data were compared with the numerical simulations described in the section "Simulation of actuation cycle".

Fig. 14.10A shows the change of the actuators bending stiffness versus the SMP layer thickness across the hinge area. These are for room temperature. When the SMP layer thickness was raised from 0.5 to 3.0 mm, Actuator-1 bending stiffness increased from 2.3 to 6.3 GPa mm^4, while Actuator-2 stiffness changed from 16.5 to 46.8 GPa mm^4. This confirms that multimaterial 3D printing can achieve stiffness modulation in two ways: by printing various designs with complex arrangements of soft elastomer and rigid SMP segments and changing the SMP layer thickness in a special design.

Fig. 14.10B–D shows that the FEM can estimate the effect of the SMP thickness on the major actuation parameters with reasonable accuracy. Fig. 14.10B also indicates that the maximum bending angle for Actuator-1 is always larger than Actuator-2

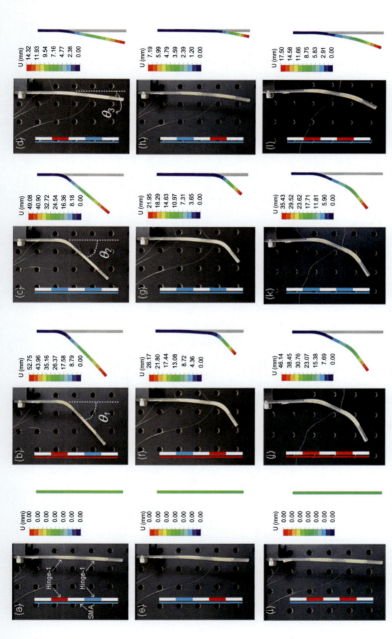

Fig. 14.9 Finite element and experimental results of the independent and simultaneous actuation of Hinge-1 and Hinge-2 of Actuator-1 (Figs. 14.1A and 14.5). Blue and red colors of the hinges in the insets show the glassy state at room temperature and rubbery state at high temperature, respectively, achieved by resistive heating. The blue and red colors of the SMA inset show the actuated and unactuated state of the SMA wire. The color contours show the actuator displacement in different configurations. (A–D) As-fabricated and deformed configurations of Hinge-1 (the top hinge) of Actuator-1. (E–H) As-fabricated and deformed states of Hinge-2 (the bottom hinge) of Actuator-1. (I–L) Simultaneous actuation of Hing-1 and Hinge-2 of Actuator-1.
No permission required.

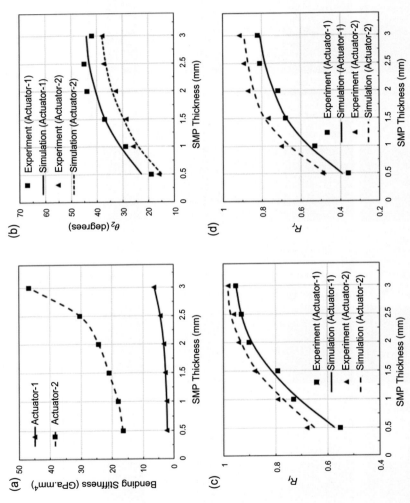

Fig. 14.10 Influence of the variation of the SMP layer thickness (from 0.5 mm to 3 mm) on the (A) bending stiffness, (B) rotation angle, (C) recovery ratio, and (D) fixity of Hinge-1 of Actuator-1 and Actuator-2. No permission required.

because Actuator-2 contains a larger volume of the SMP in the hinge area. This raises the stiffness of the hinge in the rubbery state and reduces its maximum rotation. At the same time, this causes that for a specific SMP thickness, the shape fixity and recovery of Actuator-2 be always greater than those of Actuator-1. In summary, more SMP means better maintenance of the actuated configuration and restoration of the as-fabricated straight shape. These findings confirm that the actuator layout can be redesigned to achieve the desired specific performance for each desired application.

This chapter presented a preliminary demonstration of 4D-printed structures combining SMAs and SMPs. We believe the following obstacles must be overcome for future development of this concept:

(a) Upon the SMA wire actuation, the polymer temperature at the SMA-SMP interface and at the SMA–soft layer interface can exceed the polymer degradation temperature for very short times. This can locally damage the polymer and form small cracks within it. Consequently, this may prevent a smooth force transition between the SMA wire and the polymer matrix at elevated temperatures, especially after several actuation cycles. Covering the SMA wire with a temperature-resistant polymer coating, such as polyimide, may resolve this issue.
(b) Prediction of the bending angle using simple analytical solutions can significantly facilitate the actuator design. This requires accurate calculation and integration of the contact forces at the interface of the SMA wire and the polymer matrix at the hinge area. The works of Lagoudas may be useful in this regard (Lagoudas & Tadjbakhsh, 1992, 1993).
(c) One drawback of the SMA-SMP 4D-printed structures presented in this chapter was that the SMA wires had to be inserted into the actuator body after the body is fabricated using inkjet printing. Therefore, it is a two-step fabrication process because the SMA could not be printed. Recently, some efforts have been made to 3D print SMAs. Especially, powder bed fusion technologies have received significant attention. They fuse powder particles of a shape memory alloy, such as nitinol, using a laser or electron beam energy source to form desired shape-memory structures (Sabahi, Chen, Wang, Kruzic, & Li, 2020; Van Humbeeck, 2018). Combining these powder bed fusion techniques with inkjet print-heads in a single printing machine can open new avenues to create complex shape-reversible 4D-printed structures in a single-step fabrication process.

Conclusions

This chapter discussed the remarkable capability of multimaterial inkjet 3D printing to fabricate variable stiffness composite actuators that can demonstrate shape retention and shape recovery. The stiffness adjustment of the SMA actuators was achieved by controlling the temperature of the printed SMP segments through Joule heating. The shape retention feature lowers the energy required to maintain the actuator in the deformed position. Besides, the shape memory behavior of the embedded SMP sub-structures enables recovery of the initial undeformed shape without applying any external mechanical load. The hinge original configuration was optimized to control maximum deformation, bending stiffness, and fixity and recovery ratios. Printing larger SMP segments in the hinge area improved the fixity and recovery but lowered the maximum deformation. The latter resulted from the increase of the bending stiffness in the hinge area.

It was confirmed that the fabrication of shape-reversible soft actuators by multimaterial 3D printing is significantly advantageous over conventional molding and casting methods used in the past. Multimaterial inkjet 3D printing allows for the placement of distinctive soft and stiff SMP parts at a voxel level with high precision and controllability. In addition, the geometric complexity of the 3D CAD designs has no effect on the time or cost of fabrication. It allows designers to create complicated substructures required to address various performance requirements. Although the printing strategy demonstrated in this chapter is limited to a few commercially available inks, further advancement of 3D printable materials in the coming years will increase the possibility of printing even more complex active structures.

References

Akbari, S., Sakhaei, A. H., Panjwani, S., Kowsari, K., & Ge, Q. (2021). Shape memory alloy based 3D printed composite actuators with variable stiffness and large reversible deformation. *Sensors and Actuators, A: Physical, 321.* https://doi.org/10.1016/j.sna.2021.112598.

Akbari, S., Sakhaei, A. H., Panjwani, S., Kowsari, K., Serjouri, A., & Ge, Q. (2019). Multimaterial 3D printed soft actuators powered by shape memory alloy wires. *Sensors and Actuators, A: Physical, 290,* 177–189. https://doi.org/10.1016/j.sna.2019.03.015.

Ashir, M., Nocke, A., Theiss, C., & Cherif, C. (2017). Development of adaptive hinged fiber reinforced plastics based on shape memory alloys. *Composite Structures, 170,* 243–249. https://doi.org/10.1016/j.compstruct.2017.03.031.

Barbarino, S., Flores, E., Ajaj, R., Dayyani, & Friswell, M. I. (2014). A review on shape memory alloys with applications to morphing aircraft. *Smart Materials and Structures, 23*(6).

Bruck, H. A., Moore, C. L., & Valentine, T. L. (2002). Repeatable bending actuation in polyurethanes using opposing embedded one-way shape memory alloy wires exhibiting large deformation recovery. *Smart Materials and Structures, 11*(4), 509–518. https://doi.org/10.1088/0964-1726/11/4/305.

Ge, Q., Dunn, C. K., Qi, H. J., & Dunn, M. L. (2014). Active origami by 4D printing. *Smart Materials and Structures, 23*(9). https://doi.org/10.1088/0964-1726/23/9/094007.

Ge, Q., Qi, H. J., & Dunn, M. L. (2013). Active materials by four-dimension printing. *Applied Physics Letters, 103*(13). https://doi.org/10.1063/1.4819837.

Ghosh, P., Rao, A., & Srinivasa, A. R. (2013). Design of multi-state and smart-bias components using shape memory alloy and shape memory polymer composites. *Materials and Design, 44,* 164–171. https://doi.org/10.1016/j.matdes.2012.05.063.

Holschuh, B., Obropta, E., & Newman, D. (2015). Low spring index NiTi coil actuators for use in active compression garments. *IEEE/ASME Transactions on Mechatronics, 20*(3), 1264–1277. https://doi.org/10.1109/TMECH.2014.2328519.

Hugo, R., Wei, W., Dong-Ryul, K., & Sung-Hoon, A. (2017). Curved shape memory alloy-based soft actuators and application to soft gripper. *Composite Structures,* 398–406. https://doi.org/10.1016/j.compstruct.2017.05.056.

Icardi, U., & Ferrero, L. (2009). Preliminary study of an adaptive wing with shape memory alloy torsion actuators. *Materials and Design, 30*(10), 4200–4210. https://doi.org/10.1016/j.matdes.2009.04.045.

Lagoudas, D. C., & Tadjbakhsh, I. G. (1992). Active flexible rods with embedded SMA fibers. *Smart Materials and Structures, 1*(2), 162–167. https://doi.org/10.1088/0964-1726/1/2/010.

Lagoudas, D. C., & Tadjbakhsh, I. G. (1993). Deformations of active flexible rods with embedded line actuators. *Smart Materials and Structures*, *167*, 71–81. https://doi.org/10.1088/0964-1726/2/2/003.

Lelieveld, C., Jansen, K., & Teuffel, P. (2016). Mechanical characterization of a shape morphing smart composite with embedded shape memory alloys in a shape memory polymer matrix. *Journal of Intelligent Material Systems and Structures*, *27*(15), 2038–2048. https://doi.org/10.1177/1045389X15620035.

Mao, Y., Ding, Z., Yuan, C., Ai, S., Isakov, M., Wu, J., et al. (2016). 3D printed reversible shape changing components with stimuli responsive materials. *Scientific Reports*, *6*. https://doi.org/10.1038/srep24761.

Mao, Y., Yu, K., Isakov, M. S., Wu, J., Dunn, M. L., & Jerry Qi, H. (2015). Sequential self-folding structures by 3D printed digital shape memory polymers. *Scientific Reports*, *5*. https://doi.org/10.1038/srep13616.

McEvoy, M. A., & Correll, N. (2015). Thermoplastic variable stiffness composites with embedded, networked sensing, actuation, and control. *Journal of Composite Materials*, *49*(15), 1799–1808. SAGE Publications Ltd https://doi.org/10.1177/0021998314525982.

Rodrigue, H., Wang, W., Han, M. W., Kim, T. J. Y., & Ahn, S. H. (2017). An overview of shape memory alloy-coupled actuators and robots. *Soft Robotics*, *4*(1), 3–15. https://doi.org/10.1089/soro.2016.0008.

Rodrigue, H., Wang, W., Han, M. W., Quan, Y. J., & Ahn, S. H. (2016). Comparison of mold designs for SMA-based twisting soft actuator. *Sensors and Actuators, A: Physical*, *237*, 96–106. https://doi.org/10.1016/j.sna.2015.11.026.

Sabahi, N., Chen, W., Wang, C. H., Kruzic, J. J., & Li, X. (2020). A review on additive manufacturing of shape-memory materials for biomedical applications. *JOM*, *72*(3), 1229–1253. https://doi.org/10.1007/s11837-020-04013-x.

Sakhaei, A. H., & Thamburaja, P. (2017). A finite-deformation-based constitutive model for high-temperature shape-memory alloys. *Mechanics of Materials*, *109*, 114–134. https://doi.org/10.1016/j.mechmat.2017.03.004.

Schubert, B. E., & Floreano, D. (2013). Variable stiffness material based on rigid low-melting-point-alloy microstructures embedded in soft poly(dimethylsiloxane) (PDMS). *RSC Advances*, *3*(46), 24671–24679. https://doi.org/10.1039/c3ra44412k.

Song, J. J., Chen, Q., & Naguib, H. E. (2016). Constitutive modeling and experimental validation of the thermo-mechanical response of a shape memory composite containing shape memory alloy fibers and shape memory polymer matrix. *Journal of Intelligent Material Systems and Structures*, *27*(5), 625–641. https://doi.org/10.1177/1045389X15575086.

Song, S. H., Lee, J. Y., Rodrigue, H., Choi, I. S., Kang, Y. J., & Ahn, S. H. (2016). 35 Hz shape memory alloy actuator with bending-twisting mode. *Scientific Reports*, *6*. https://doi.org/10.1038/srep21118.

Taya, M., Liang, Y., Namli, O. C., Tamagawa, H., & Howie, T. (2013). Design of two-way reversible bending actuator based on a shape memory alloy/shape memory polymer composite. *Smart Materials and Structures*, *22*(10). https://doi.org/10.1088/0964-1726/22/10/105003.

Tobushi, H., Hayashi, S., Sugimoto, Y., & Date, K. (2009). Two-way bending properties of shape memory composite with SMA and SMP. *Materials*, *2*(3), 1180–1192. https://doi.org/10.3390/ma2031180.

Van Humbeeck, J. (2018). Additive manufacturing of shape memory alloys. *Shape Memory and Superelasticity*, *4*(2), 309–312. https://doi.org/10.1007/s40830-018-0174-z.

Wang, W., & Ahn, S. H. (2017). Shape memory alloy-based soft gripper with variable stiffness for compliant and effective grasping. *Soft Robotics*, *4*(4), 379–389. https://doi.org/10.1089/soro.2016.0081.

Wang, Z., Hang, G., Wang, Y., Li, J., & Du, W. (2008). Embedded SMA wire actuated biomimetic fin: A module for biomimetic underwater propulsion. *Smart Materials and Structures*, *17*(2). https://doi.org/10.1088/0964-1726/17/2/025039.

Wang, W., Rodrigue, H., & Ahn, S. H. (2015). Smart soft composite actuator with shape retention capability using embedded fusible alloy structures. *Composites Part B: Engineering*, *78*, 507–514. https://doi.org/10.1016/j.compositesb.2015.04.007.

Wang, W., Rodrigue, H., Kim, H. I., Han, M. W., & Ahn, S. H. (2016). Soft composite hinge actuator and application to compliant robotic gripper. *Composites Part B: Engineering*, *98*, 397–405. https://doi.org/10.1016/j.compositesb.2016.05.030.

Westbrook, K. K., Kao, P. H., Castro, F., Ding, Y., & Jerry Qi, H. (2011). A 3D finite deformation constitutive model for amorphous shape memory polymers: A multi-branch modeling approach for nonequilibrium relaxation processes. *Mechanics of Materials*, *43*(12), 853–869. https://doi.org/10.1016/j.mechmat.2011.09.004.

Yu, K., Ge, Q., & Qi, H. J. (2014). Reduced time as a unified parameter determining fixity and free recovery of shape memory polymers. *Nature Communications*, *5*. https://doi.org/10.1038/ncomms4066.

Zolfagharian, A., Kaynak, A., Bodaghi, M., Kouzani, A. Z., Gharaie, S., & Nahavandi, S. (2020). Control-based 4D printing: Adaptive 4D-printed systems. *Applied Sciences*, *10*(9), 3020. https://doi.org/10.3390/app10093020.

Zolfagharian, A., Kaynak, A., & Kouzani, A. (2020). Closed-loop 4D-printed soft robots. *Materials & Design*, *188*, 108411. https://doi.org/10.1016/j.matdes.2019.108411.

Variable stiffness 4D printing

15

Yousif Saad Alshebly[a], Marwan Nafea[a], Khameel Bayo Mustapha[b], Mohamed Sultan Mohamed Ali[c], Ahmad Athif Mohd Faudzi[c], Michelle Tan Tien Tien[a], and Haider Abbas Almurib[a]

[a]Department of Electrical and Electronic Engineering, Faculty of Science and Engineering, University of Nottingham Malaysia, Semenyih, Selangor, Malaysia
[b]Department of Mechanical, Materials and Manufacturing Engineering, Faculty of Science and Engineering, University of Nottingham Malaysia, Semenyih, Selangor, Malaysia
[c]School of Electrical Engineering, Faculty of Engineering, Universiti Teknologi Malaysia, Johor Bahru, Johor, Malaysia

Introduction

There is a surge of interest in variable stiffness actuation (VSA) as a desired feature in many high-performance systems across different applications in recent years. The concept of variability in stiffness centers on the ability of a structure to exhibit a change in compliance within certain spatial-temporal states. Historical accounts of the origin of this concept point to the investigations of Japanese engineers around the issue of vibration control of large space structures (Böse, Rabindranath, & Ehrlich, 2012) and seismic response control of buildings (Takahashi, Kobori, Nasu, Niwa, & Kurata, 1998), where the idea is deployed for active attenuation of some undesired frequency-varying disturbances. Widely found in biological systems, variable stiffness has gained recognition in many latest technological advancements. Indeed, it has been reported to provide pliable functionality to applications that range from flexible bio-inspired robots to adaptable manipulators. For these applications and many others, VSA (i) provides nimble motion, (ii) facilitates safer robot-human interactions by minimizing the effect of impact with human operators, (iii) enhances energy-efficient actuation, (iv) enables embodied intelligence, and (v) allows flexible robots to exhibit different behavior than traditional structures that are designed exclusively with rigid members (Wolf et al., 2016).

On a broader scale, the work in Torrealba, Udelman, and Fonseca-Rojas (2017) positions VSA under the variable impedance actuators, an emerging class of actuators with close interaction between their morphology, sensory/motor control, and their environment. Along this direction, the most prominent approach pursued by various research groups to establish adjustable impedance is via a method called active impedance control (Kronander & Billard, 2016). In this approach, software-in-the-loop is used to manipulate the impedance of a stiff actuator. While this approach is considered mature and remains dominant (Al-Shuka et al., 2018), it has some disadvantages. For

Smart Materials in Additive Manufacturing, Volume 2: 4D Printing Mechanics, Modeling, and Advanced Engineering Applications
https://doi.org/10.1016/B978-0-323-95430-3.00015-4
Copyright © 2022 Elsevier Inc. All rights reserved.

one, the approach demands the use of actuator/sensor/software combination. Further, it is characterized by a complex, and at times bulky, impedance controller. On this premise, VSA has emerged to address the need to overcome the aforementioned weaknesses and the continuous demand for adjustable compliance for shock absorption, delicate material handling (Xie, Zhu, Yang, Okada, & Kawamura, 2021), fast and safe motion control (Hoang et al., 2021), bipedal robot (Luo, Wang, Zhao, & Fu, 2018), artificial muscles (Liu, Tang, Zhou, & Liu, 2020), adaptable grippers (Sun et al., 2020), and underwater robots (Park, Huh, Park, & Cho, 2014), among others.

Different implementations of variable stiffness with working principles have been proposed in the literature. It is suggested in the literature that the methods of implementation of the variability of stiffness can be categorized into four, namely, implementations based on (a) adjustable-stiffness elastic elements such as series of connected springs or nonlinear pretension springs (Guo & Tian, 2015), (b) jamming of granular materials (Wall, Deimel, & Brock, 2015); (c) use of magnetorheological or electrorheological fluids, and (d) stimuli-responsive shape memory alloys and polymers (Elgeneidy, Fansa, Hussain, & Goher, 2020).

The focus of the present chapter aligns closely with the last approach. Specifically, we combine the stimuli-responsive behavior of shape memory polymers (SMP) with the opportunity and recent progress achieved in the field of four-dimensional (4D) printing technology, a fast-growing category under the additive manufacturing (AM) process. Recognized for accelerating the advancement of soft robotics (Carrell, Gruss, & Gomez, 2020), AM offers more control over processing parameters, facilitates the design of complex shapes, and enables the tailoring of inertial and mechanical properties (Tong, Alshebly, & Nafea, 2021). This approach allows developing complex structures and actuation mechanisms when compared to types of fabrication methods (Abdul Kadir, Dewi, Jamaludin, Nafea, & Mohamed Ali, 2019; Nafea, Nawabjan, & Mohamed Ali, 2018). On the other hand, SMPs are a unique class of smart materials that are endowed with the ability to transition from soft to hard states using inherent shape memory effects (Lendlein & Kelch, 2002). SMPs enjoy admirable elastic deformation and can generate large reversible strain capacity (Barrie, Goher, & Elgeneidy, 2020).

4D printing allows the use of smart materials that can transform their shapes in preprogrammed ways as a response to external stimuli (Alshebly, Nafea, Mohamed Ali, & Almurib, 2021). Thus, the fourth dimension refers to the time that these structures take to change their shapes (Ryan, Down, & Banks, 2021). These smart materials have the ability to undergo changes in shape and functionality, using external stimuli, such as temperature (Yamamura & Iwase, 2021), solvent (Mulakkal, Trask, Ting, & Seddon, 2018), pH (Scarpa et al., 2019), magnetic (Bonifacich, Lambri, Recarte, Sánchez-Alarcos, & Pérez-Landazábal, 2021), and light (Kuksenok & Balazs, 2016). Such materials include hydrogels, liquid crystal elastomers, cellulose composites, and shape memory materials (SMM). SMMs are among the most commonly used materials in 4D printing applications, due to their shape memory effect property, which is the ability to recover their shape after deformation as a response to applied stimuli (Subash & Kandasubramanian, 2020). For instance, SMPs have the ability to transform to their original shape from single or multiple distorted shapes when stimulated by temperature, moisture, or radiation (Mohd Ghazali et al., 2020; Zolfagharian et al., 2016). In addition, SMPs can be magnetically induced by including nanoparticles within their structures (Falahati et al., 2020). Other attractive features of SMPs include their biodegradability, biocompatibility, and low cost while being derived from renewable resources (Gauss, Pickering, & Muthe, 2021). However, SMPs suffer from low tensile strength and stiffness, which have been addressed by developing various shape memory nanocomposites to improve their mechanical properties as well as their conductivity and electrical response (Zhang, Tan, & Li, 2018). One of the most commonly used SMPs is polylactic acid (PLA), which is a biodegradable thermoplastic based on lactic acid that can be produced from renewable resources, such as corn and sugar beets (Gauss et al., 2021).

Some of the most commonly used 4D printing methods for such materials include fused deposition modeling (FDM), direct ink writing (DIW), inkjet, and projection stereolithography (SLA). The FDM method relies on extruding a filament from a nozzle at a controlled rate from a heated nozzle to melt or soften the filament and deposit it on a tray to fabricate 3D structures in a layer-by-layer manner (Barletta, Gisario, & Mehrpouya, 2021). The DIW printing method uses a similar concept to the FDM method while depositing a viscoelastic ink or pastes instead of a solid filament (Peng et al., 2021). On the other hand, the inkjet method deposits tiny droplets of a low-viscosity ink on a build tray using thermal or piezoelectric technology. An ultraviolet (UV) light source is often combined with inkjet printers in one platform (Han, Kundu, Nag, & Xu, 2019). An alternative method is SLA, which polymerizes an entire layer using micromirror array devices. Cross-linking is achieved by projecting a patterned UV light on the surface of a photosensitive polymer resin that is synergized with an upward movement of a linear stage (González-Henríquez, Sarabia-Vallejos, & Rodriguez-Hernandez, 2019). Among the aforementioned methods, FDM is considered the most common method for 4D printing. This method allows printing relatively complex structures while being the cheapest printing method and offering the ease of use and the flexibility of choice of filament materials that can be easily manufactured.

According to their design, 4D-printed structures can be programmed to change their shapes using shapes with rolled, helix, twisted, nonzero Gaussian curvatures, or local double curves features (Bodaghi, Noroozi, Zolfagharian, Fotouhi, & Norouzi, 2019; Janbaz, Hedayati, & Zadpoor, 2016). These concepts have been used in various biomedical applications such as a bio-origami structure, artificial protein structure, smart valve, stents, smart gripper, adaptive scaffold, adaptive joints, drug delivery, smart textiles, soft robots, and tissues engineering (Hann, Cui, Nowicki, & Zhang, 2020). In a more recent work, variable stiffness structures have been used in energy absorption since the stiffness between an SMP and an elastomer provides superior and reversible energy absorption characteristics (Bodaghi et al., 2020). The performance presented in the work shows the capacity of SMEs to function as energy retainers when used in specific patterns of prints. Due to the aforementioned attractive features of 4D printing, this technology has emerged as a promising solution for various research problems while showing a growing fundamental influence on the economy and society (Jiang, Kleer, & Piller, 2017). Advances in this field have been demonstrated in robotics applications (Must, Sinibaldi, & Mazzolai, 2019; Yang, Chen, Li, & Zhiqiang Chen, 2016), due to the ability of VSAs to replicate mechanical movement. However, despite the significant work that has been carried out in this area, there is a lack of development when it comes to the variable stiffness 4D printing approach. In addition, the designs, printing methods, and materials used in this method are not well-explored, where the materials compatibility and the printing precision, time, and cost are among the major issues that need to be addressed. The results of this chapter are promising and show that this variable stiffness 4D printing can be realized using different approaches that are presented in this chapter and other approaches that will be explored in the near future.

Design and working principle

In this work, variable stiffness 4D printing is achieved by adjusting the printing method of an FDM 3D printer that prints a PLA filament. This concept has been explored with similar approaches in the literature, ranging from the use of stiffness difference for mechanical movement (Yang et al., 2016), grippers (Must et al., 2019), and energy absorption (Rahman, Yarali, Zolfagharian, Serjouei, & Bodaghi, 2021). The nozzle of the printer extrudes the materials at a much higher temperature than the melting temperature of the materials to ease the extrusion of the plastic. The extruded filament is then rapidly cooled by fans upon extrusion. The way that the printed material is extruded onto the plate adds internal strain to the material, due to the stretching and heating of the material, which are highly controllable (Bodaghi, Damanpack, & Liao, 2017). This has been observed in (Alshebly et al., 2021; Bodaghi, Damanpack, & Liao, 2017), where the programming of the SME in the structures allows for a proper prediction of the folding or shrinking of the actuators. By keeping the exact same settings of the print, the internal strain in the materials will remain the same for any print. This character can be used to assess the stiffness of the material by testing the SME caused by the internal strain of the material. As the stiffness of the material increases, the SME decreases. This can be used to confirm the difference in stiffness of the printed structures. The stiffness of the structures printed can be controlled in multiple ways. Printing of different patterns results in different stiffness for the material, which can be controlled to a great extent. The use of materials of different stiffness values is also used to allow the bending of materials in many favorable ways. The control available for the settings of the 3D printer is variant and precise, allowing a very accurate and precise SME of the prints.

Single-material actuators

This section presents the design and working principle of 15 actuators that are printed using a single material, where the infill percentage, infill patterns, and the use of hinges were implemented to create these different designs.

Variable infill percentages

The setting of the infill patterns allows for two main things, namely, the infill percentage and the infill patterns. For instance, a structure with a 20% infill means that 20% of the interior of the solid structure is printed with the material, while 80% is empty. The difference between the volume of materials and changes in structure because of the size of the pattern produces a difference in stiffness between the patterns, thus allowing control of the SME of the structure by printing at different infill percentages. A higher percentage used in the infill will produce a higher stiffness structure because more material has been used. The infill percentage can be increased by printing the infill patterns in smaller patterns, allowing more material to be used in the same space; thus, the higher the infill patterns, the more the material is used. The infill patterns can also be used to control the stiffness of the material in certain directions. Moreover, different settings of the print at different heights of the structure can be implemented. This capability can be used to print different infill patterns at different heights of the structure, allowing parts of the structure to have a different stiffness profile at different parts. This sort of print can be used to control the difference in the stiffness between the parts, thus using variable stiffness to benefit the SME of the structure.

The infill patterns used in this study are rectilinear, grid, triangular, and fast honeycomb, as shown in Fig. 15.1, where l_r, l_g, l_h, and l_t are the lengths of the sides of the rectilinear, grid, honeycomb, and triangle patterns, respectively. Each of the four patterns has two designs, one is with smaller patterns at the bottom and larger ones on top, meaning more material at the bottom as compared to the top, while the other one is the opposite, creating a total of eight designs. More specifically, each structure is printed with a bottom section of an infill percentage of 30% and a top section of an infill percentage of 20%, or vice versa. This was implemented due to the effect of the FDM printing process on the properties of the printed material. In fact, changing the position of the patterns from top to bottom is not the same as flipping the actuator. This is due to the strain loading of the printing, which generally causes bending of the materials upward. Thus, changing the positions will bring very different results for each actuator.

The first pattern that is utilized in this work is the rectilinear pattern, which is the most common in 3D printing because it is fast and provides good support for the structures. The pattern is made by printing lines at a 45-degree angle for the whole layer, then the next layer has the patterns printed at a -45 degrees, creating a full square at every two layers. The second pattern that is applied is the grid pattern, which is very similar to that of the rectilinear pattern, but it differs in that the full grid is printed on each layer, giving it a stronger structure. This is also a common infill pattern, and it would seem that less material is used for the same infill percentage, but it is not the case since a full grid is needed for each layer. This means that bigger patterns need to be printed to have the same infill percentage. The third pattern that is used is the honeycomb pattern, which is very common in the literature because of the unique support it provides, along with other characteristics associated with honeycomb structures. The pattern is created by printing sections of the honeycomb that are not fully

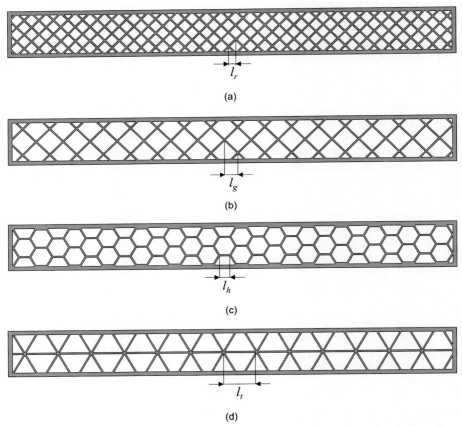

Fig. 15.1 Top views of the main patterns used to print the actuators. (A) Rectilinear. (B) Grid. (C) Honeycomb. (D) Triangle.
No permission required.

horizontal for one layer, then followed by the sections of the honeycomb that is not fully vertical, creating a full honeycomb of each two layers. The last pattern that is implemented is the triangular pattern, which is not commonly used in support structures because it has relatively large gaps. The full pattern is completed in each layer, which is done by printing triangles with two different directions to build a full layer of triangles mirrored about the centerline. All actuators in the single material section are of the same dimensions, that is, the length, width, and height are 90 mm, 9.6 mm, and 3 mm, respectively. Each actuator is made of two sections of different infill percentages at the bottom and top, each section having a height of 1.5 mm, for an equal 50% of the actuator height. Table 15.1 presents the pattern variation used in this section, along with the length of sides of the patterns for each of the bottom and top sections of the actuators, as demonstrated in Fig. 15.1.

Variable stiffness 4D printing | 413

Table 15.1 Summary of the parameters of the structures of Designs 1–8.

Design	Pattern	Bottom/top infill (%)	Pattern's side length of the bottom layer (mm)	Pattern's side length of the top layer (mm)
1	Rectilinear	30/20	1.20	2.20
2	Rectilinear	20/30	2.20	1.20
3	Grid	30/20	2.50	4.25
4	Grid	20/30	4.25	2.50
5	Honeycomb	30/20	1.50	2.50
6	Honeycomb	20/30	2.50	1.50
7	Triangular	30/20	3.25	6.50
8	Triangular	20/30	6.50	3.25

Variable infill patterns

The use of infill patterns allows the printed structures to have strong interior support, so it does not break on itself, without using too much material. By using the characteristics of the FDM printers, changes of these patterns would cause the structure to have a variable stiffness in certain directions, and thus using different infill settings will produce variable stiffness 4D-printed structures. This is also used in 4D printing to control the SME. At its simplest level, the SME characteristic of variable stiffness is made by the difference of strain between two layers and thus bending in certain directions. Using different stiffness profiles, although using the same printer settings and thus the same internal strain, the different stiffness in certain directions will produce a shape change in the structure. In this section, different patterns are printed in two sections, top and bottom. The same patterns from the previous section are used while being mixed in different orders. The same infill percentage of 20% is used for both patterns in each actuator. A total of six actuators are made, each with different patterns and arrangements. The dimensions of the actuators are the same as in the previous section. Table 15.2 presents the pattern variation for the two sections of the actuator with the length of the sides of the patterns for each section.

Patterns as hinges

An actuator can be used as a building block that is combined with other passive blocks to create a structure of shape-changing capabilities. One of the most common proofs of concept used in 4D printing is the design of hinges since it shows the most basic and yet most useful mechanical movement, which is folding. The patterns tested are used as hinges that are attached to passive blocks of PLA, where the hinges bend allowing the passive parts to have a folding movement. The pattern mix that is used in this work is Design 9, which had the highest bending angle (as demonstrated in "Results and discussion" section), providing the ability to demonstrate the effect clearly in the hinges. The hinging actuator that is made has dimensions of 100 mm × 10 mm × 2 mm for the length, width, and height, respectively. The actuator is made of five sections

Table 15.2 Summary of the parameters of the structures of Designs 9–14.

Design	Pattern (bottom-top)	Pattern's side length of the bottom layer (mm)	Pattern's side length of the top layer (mm)
9	Triangular—honeycomb	6.50	2.50
10	Honeycomb—grid	2.50	4.25
11	Honeycomb—triangular	2.50	6.50
12	Rectilinear—triangular	2.20	6.50
13	Grid—honeycomb	4.25	2.50
14	Triangular—rectilinear	6.50	2.20

along the length of the structure, and each section has a length of 20 mm. The design is made with interchanging sections of passive and active hinge sections, allowing a hinge section between every two passive sections.

Multimaterial actuators

A single structure can be printed in parts by printing with one material, and then changing the filament and printing over it with the new materials, thus achieving multimaterial prints with a single extruder. This simple method has its limitations though, which a double head extruder would not have. This includes the quick change of material and ease of printing complex mixes of the materials. FDM has the capability to print multiple materials into one structure, and this allows for structures with different characteristics in each part. A popular method of 4D printing of SME structures is using an SMP with an elastomer, which tends to enhance the SME of the structure greatly. The large difference in stiffness between the two materials at certain temperatures allows for a big shape change. This is also because the elastomer changes the force of the bending of the actuator despite being slightly rubbery. Since SMPs release their strain and have some change in their size at actuation, and that elastomers do not, this concept can be used to create various shape changes in simple structures. The printing of the structures in this work is made with a single extruder head FDM printer, which was done by changing the filament for each section of the print to change the material. All the actuators printed in this section are of the same settings for an infill percentage of 100% and a printing speed of 50 mm/s.

The first actuator in this section (Design 16) was printed entirely out of PLA (1.75 mm PLA Pro Filament 1.0 Kg Gray, Flashforge, Flashforge 3D Technology, Co., Ltd., Zhejiang, China), while the second (Design 17) of biodegradable polyethylene terephthalate glycol (Bio-PT) (1.75 mm Bio-PT Filament 1.0 Kg, Fabbxible

Technology, Penang, Malaysia), which were used as comparison actuators with the multimaterial structures. Design 18 was printed out of a bottom section of PLA and a top section of BIO-PT. The combination of the two materials causes the strain between the two layers to be different due to variation in the stiffness. The glass temperature of the BIO-PT is higher than that of PLA but may still be affected by the heat. However, since BIO-PT is not an SMP, then it does not have significant SME characteristics. The BIO-PT acts as an elastomer, which can be used to enhance the SME behavior of the printed actuators. All actuators in this section are of the same dimensions of 60 mm × 10 mm × 2 mm for the length, width, and height, respectively. The third actuator is made of a 1-mm-height section of PLA and a 1-mm-height section of BIO-PT printed on top of it. The fourth actuator is made of a 1-mm-height section of PLA, and a top section of a BIO-PT part printed between two PLA parts, where all three sections are 20 mm in length.

Fabrication process and experimental setup

3D models were created using SolidWorks software, which was then used to generate STL files for the Slicer software. The Slicer software used is Simplify3D, which allows full control of the whole printing operation. The slicer divides the actuator into layers of specified height and generates a G-code file that the 3D printer can read and follow throughout the printing process. The fabrication of the actuators used for the study is carried out using FDM printer Ender 3 V2 (Creality, Shenzhen Creality 3D Technology Co., Ltd.). The printer is a single extruder head FDM printer, which was used for both single-material and multimaterial prints. In this study, two types of thermoplastics were used, which are PLA and Bio-PT, as mentioned previously. The printing parameters used to print the PLA and BIO-PT are presented in Table 15.3. The slicer allows changing the infill patterns, of which some are available on the software itself. In addition, the slicer increases the infill percentage by printing the infill patterns in smaller patterns. The aforementioned features can also be implemented using different settings at different heights of the structure. Although the printer used is a single-head extruder printer, the printer, however, can be used to print structures of multimaterials that usually require using a double extruder head printer. A single structure can be printed in parts by printing with one material, and then changing the filament and printing over it with the new materials, thus achieving multimaterial

Table 15.3 Printing parameters of the PLA and BIO-PT filaments

Material	PLA	BIO-PT
Nozzle temperature (°C)	200	230
Nozzle speed (mm/s)	50	50
Bed temperature (°C)	45	45
Environment temperature (°C)	24	24
Layer height (mm)	0.1	0.1

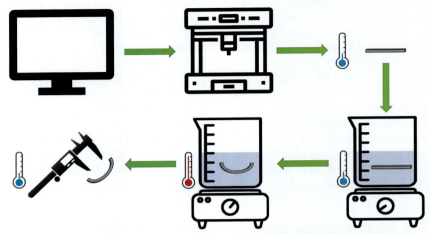

Fig. 15.2 The fabrication process and experimental setup for the 4D-printed actuators. No permission required.

prints with a single extruder. This simple method has its limitations though, which a double head extruder would not have. This includes the quick change of material and ease of printing complex mixes of the materials.

The fabrication process and experimental setup are presented in Fig. 15.2. The designs are developed on the computer and transferred to the printer as G-code files on an SD card. The printer is fed the material from a filament roll to print. After the completion of the print, the actuator is removed and kept at room temperature. The actuator is then activated by submerging it in hot water at a temperature of 80°C, which is 20°C above the glass temperature of PLA. This is to ensure that the full structure is heated, and thus, all the stored internal strain is released in the activation process. The temperature of the water is constantly monitored so that it does not divert from 80°C. Then, the structure is removed carefully so that the shape is not disturbed while it is in a rubbery state. The structure is rested on its side on a towel so that the SME is not affected by gravity while still in a rubbery state. When the structure is cool and dry, measurements for the assessment of the bending angle are taken using a digital caliper.

Results and discussion

This section characterizes the properties of the materials used to fabricate the actuators. Then, the section presents 4D printing results of the actuators using a printing pattern for each actuator with different infill percentages (Designs 1–8). This is followed by 4D printing results of the actuators with two different printing patterns (Designs 9–14). After that, 4D printing results of a proof-of-concept design that utilizes a hinge for a structure are presented (Design 15). Lastly, the section presents 4D printing results of actuators that use two different materials, PLA and BIO-PT, that are embedded into the other (Designs 16–19).

Material properties

The material properties that are vital for the characterization of the actuators are found using a dynamic mechanical analysis (DMA) machine. The machine generates the storage modulus and tan delta (tan δ), which are commonly used to assess the glass temperature of the materials. The DMA is done for both PLA and BIO-PT, and to an actuator that is a mix of the two materials to find their activation temperatures as one structure. Thermal characteristics of both PLA and BIO-PT are needed to find out their response to strain and thermal stimulus to change their state. A dynamic mechanical analyzer (DMA 8000, Perkin Elmer, Massachusetts, United States) is used to test the thermal response of the materials. The testing is done with three different actuators that are printed out of PLA, Bio-PT, and an actuator of two layers one with each material. The DMA test is done using axial tensile clamping, and the frequency used is 1 Hz, at a 5°C/min from 30 to 130°C. The DMA results for the PLA actuator are shown in Fig. 15.3A. It can be seen that at the peak of the tan δ, the Tg is at a temperature of 70.9°C. In addition, the figure also illustrates the storage modulus, which is at about 1730 MPa and lowers down to 4 MPa beyond transition temperature. The storage modulus shows the elastic response of the material, which is used to determine the elastic property of the material.

The DMA results of the BIO-PT are presented in Fig. 15.3B. As is presented by the peak of the tan δ in the graph, the Tg of the material is at 87.7°C, which is considerably higher than that of the PLA actuator. This characteristic of the materials can be used to control the shape change in the materials by activating the PLA only (which possesses a lower Tg) using an activation temperature that is higher than the PLA and lower than that of the BIO-PT. The temperature that can be used is between 70°C and 87°C, but for a safer option, a temperature of about 70°C, or slightly lower, should be used to accommodate any variations in the temperature during the actuation. This will activate the PLA releasing the stored internal strain causing a shape change of the whole structure. A second shape can be achieved by increasing the temperature of the actuator higher than the Tg of the BIO-PT, which is at 87.7°C, where a temperature of 107°C can be used since it is 20°C above BIO-PT Tg. At that temperature, both the PLA and BIO-PT are activated with both releasing their internal strain, but since PLA is an SMP, it is expected to have more strain induced in it than the BIO-PT, causing more deformation at the PLA part than that of the BIO-PT.

The actuator that is printed with both materials was also analyzed by the DMA machine, as shown in Fig. 15.3C. It can be seen from the tan δ of the graph that there are two peaks, which are associated with two separate Tg values, one for each material. The two peaks are located at temperatures of 66.9°C and 83.73°C, which are of course, for the PLA and BIO-PT, respectively. As seen in this graph, if the activation of the PLA is needed alone, then the temperature used needs to be before the dip between the two peaks, which is located at a temperature of around 74.63°C. However, since the midpoint is too close to the Tg of the BIO-PT, a lower temperature

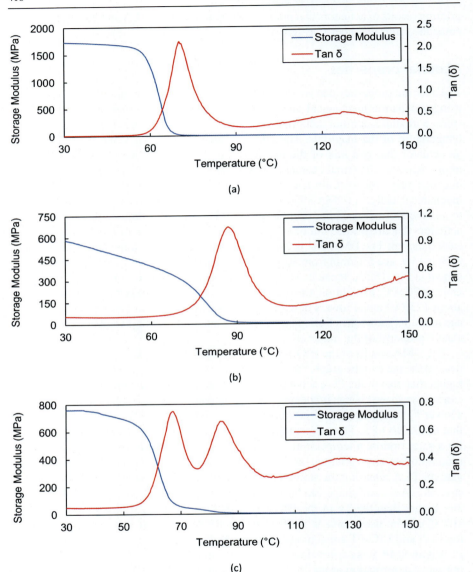

Fig. 15.3 DMA results of the actuators printed using (A) PLA, (B) BIO-PT, and (C) PLA and BIO-PT (multimaterial).
No permission required.

should be used. Keeping the temperature at the Tg of PLA or a little lower would ensure that the BIO-PT is not activated. At that range, the PLA can be activated without interfering too much with the BIO-PT, and thus keeping the PLA at a rubbery state while the BIO-PT is in a glassy form. This provides a solid elastomer with a greatly higher stiffness than the PLA at the same temperature.

Single-material actuators

The fabrication and deformation results of Designs 1–15, which were printed using a single material using different infill percentage values, infill patterns, and hinges, are presented in this section.

Variable infill percentages

In this work, Designs 1 and 2 utilized the rectilinear pattern. The rectilinear pattern designs top and side views can be seen in Fig. 15.4A and B, respectively. The first two structures are of the pattern printed at 20% infill before and after actuation, respectively. The top actuator is of the structure at 20% before deformation, which is used as a reference to compare the deformation of the other designs to the original printed shape (this is used throughout this section). Although only two sides of the squares are printed at a time, after the combination, there does seem to be any gaps between the layer. This is because the PLA melts and expands to the layer below creating virtually solid walls. On the other hand, Designs 3 and 4 were printed using the grid pattern. Despite the similarity between the rectilinear and grid patterns, this does not mean that the grid pattern will have the same stiffness and strain profile as that of rectilinear since the material does not need to expand to accommodate the gaps. The grid pattern designs top and side views can be seen in Fig. 15.4C and D, respectively.

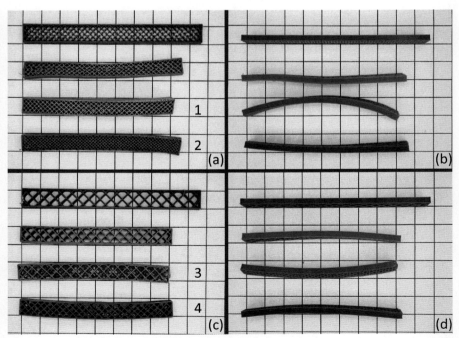

Fig. 15.4 (A) and (B) Top and side views of Designs 1 and 2. (C) and (D) Top and side views of Designs 3 and 4.
No permission required.

As can be seen in the figure, Design 1 has a strong bending angle of about 12 degrees downward, while Design 2 has a small bending angle of about 2-degree angle upward. Moreover, Design 3 has a moderate bending angle of about 5 degrees upward, while Design 4 has a small bending angle of about 4-degree angle downward.

Designs 5 and 6 were printed using the honeycomb pattern. Much like the rectilinear pattern, the PLA has to expand to compensate for some gaps in the full structures. The honeycomb patterns top view designs can be seen in Fig. 15.5A and B, respectively. Designs 7 and 8 utilized the triangular pattern. The triangular patterns top view designs can be seen in Fig. 15.5C and D, respectively. As can be seen in the figure, Design 5 has a moderate bending angle of about 8 degrees downward, while Design 6 has a similar bending angle of about 7-degree angle downward. Furthermore, Design 7 has a moderate bending angle of about 6 degrees downward, while Design 8 has a stronger bending angle of about 11-degree angle downward.

The measurements of the printed Designs 1–8 are summarized in Table 15.4. Lengths A and B are units measured from the actuators after actuation, where length A is the distance from one end to the other of the actuator, while length B is the height of the structure when placing the two ends on a flat surface at the highest point in the actuator. Lengths A and B are also used to determine the bending angle of

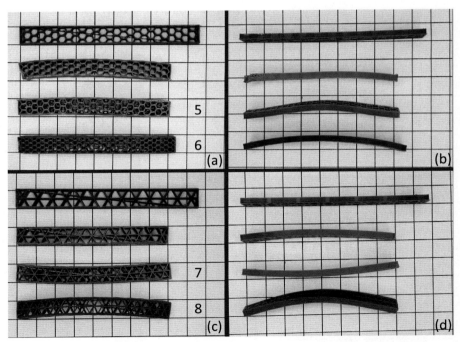

Fig. 15.5 (A) and (B) Top and side views of Designs 5 and 6. (C) and (D) Top and side views of Designs 7 and 8.
No permission required.

Table 15.4 Summary of the bending results of Designs 1–8.

Design	Length A (mm)	Length B (mm)	Bending angle (degrees)	Bending direction
1	75.95	11.29	12.49	Downward
2	79.98	4.31	2.05	Upward
3	76.02	6.51	5.46	Upward
4	77.09	5.39	3.73	Downward
5	76.46	8.19	7.91	Downward
6	79.05	7.69	6.94	Downward
7	76.13	5.72	4.27	Upward
8	75.45	9.95	10.61	Downward

each actuator. The bending direction is based on which way the two ends of the actuator move.

Variable infill patterns

The first actuator of mixed patterns (Design 9) was printed out of a mixture of a triangular pattern on the bottom and a honeycomb pattern on top. The actuator is shown in Fig. 15.6A and B, where the top structure is before actuation while the bottom one is after actuation. As can be seen in the figure, the bending angle of the actuator is a

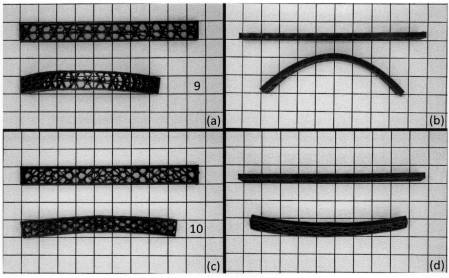

Fig. 15.6 (A) and (B) Top and side views of Design 9. (C) and (D) Top and side views of Design 10.
No permission required.

strong one with an angle of about 27 degrees, with a downward bending direction. Design 10 was printed out of a honeycomb pattern on the bottom and a grid pattern on top, as demonstrated in Fig. 15.6C and D. As can be seen in the figure, the bending angle of the actuator is a small one with an angle of about 5 degrees, with an upward bending direction.

Design 11 was printed out of a honeycomb pattern on the bottom and a triangular pattern on top. The actuator is shown in Fig. 15.7A and B. As can be seen in the figure, the bending angle of the actuator is a moderate one with an angle of about 10 degrees and an upward bending direction. Design 12 was printed out of a rectilinear pattern on the bottom and a triangular pattern on top, which is presented in (C) and (D). As can be seen in the figure, the bending angle of the actuator is a moderate one with an angle of about 10 degrees and an upward bending direction.

Design 13 was printed out of a grid pattern on the bottom and a honeycomb pattern on top. As can be seen in Fig. 15.8A and B, the bending angle of the actuator is a moderate one with an angle of about 17 degrees, with a downward bending direction. Design 14 was printed out of a triangular pattern on the bottom and a rectilinear pattern on top. As can be seen in Fig. 15.8C and D, the bending angle of the actuator is a strong one with an angle of about 25 degrees, with an upward bending direction. The measurements of the printed Designs 9–14 are summarized in Table 15.5.

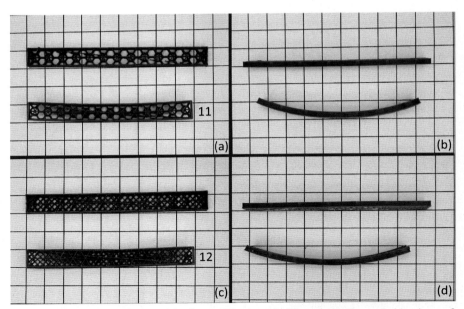

Fig. 15.7 (A) and (B) Top and side views of Design 11. (C) and (D) Top and side views of Design 12.
No permission required.

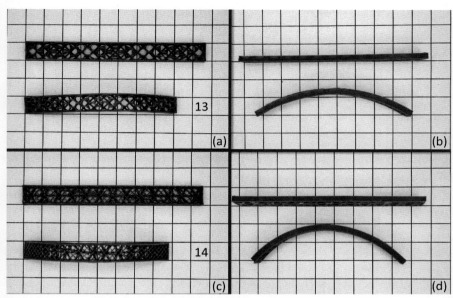

Fig. 15.8 (A) and (B) Top and side views of Design 13. (C) and (D) Top and side views of Design 14.
No permission required.

Table 15.5 Summary of the bending results of Designs 9–14.

Design	Length A (mm)	Length B (mm)	Bending angle (degrees)	Bending direction
9	70.5	20.66	26.77	Downward
10	78.23	6.39	5.13	Upward
11	79.93	10.01	10.12	Upward
12	80.42	9.86	9.84	Upward
13	75.93	14.27	16.70	Downward
14	72.52	19.68	24.86	Downward

Patterns as hinges

From the tested patterns and pattern variation, many applications can be made. Any significant actuation that is observed in a design can be implemented by using the actuator as a building block and then combining it with other passive blocks to create a structure of shape-changing capabilities. The patterns tested are used as hinges that are attached to passive blocks of PLA, where the hinges bend allowing the passive parts to have a folding movement. The pattern mix that is used in this work is Design 9, which had the highest bending angle, providing the ability to demonstrate the effect clearly in the hinges. The top and side views of the hinge design can be seen in

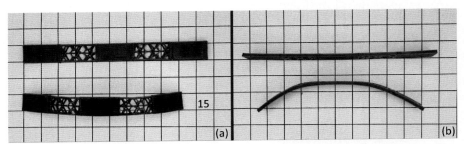

Fig. 15.9 (A) and (B) Top and side views of Design 15 using Design 9 as hinges for an actuator. Two hinges are used along with three passive sections to create a folding SME.
No permission required.

Fig. 15.9A and B, respectively, where the top and bottom structures represent the actuator before and after actuation, respectively. From the figure, it can be seen that a great deformation has been achieved at the hinges and not at the passive parts, which comes to show the difference between the effect of the printing patterns on the SME of the materials. The bending angle of the structure on each side is very similar to that of Design 9 because the same patterns are used, where the two angles are 26 degrees and 24 degrees.

Multimaterial actuators

The first actuator in this section (Design 16) was printed entirely out of PLA while the second (Design 17) of BIO-PT, which were used as comparison actuators with the multimaterial structures. The PLA actuator is shown in Fig. 15.10A and B while the BIO-PT is shown in Fig. 15.10C and D. As can be seen in the figure, the bending of the PLA is moderate, which is expected of an actuator of this size and thickness, and the bending angle is about 20 degrees. While the BIO-PT is small, which is expected too of an actuator of this size and thickness of an elastomer, which is not an SMP, thus not holding a lot of strain upon printing, with a bending angle of about 4 degrees.

Design 18 was printed out of a bottom section of PLA and a top section of BIO-PT. The combination of the two materials causes the strain between the two layers to be different due to variation in the stiffness. The two-section actuator is shown in Fig. 15.11A and B. As can be seen in the figure, the bending of the material is very high (the bending angle is about 88 degrees), which is nowhere compared to the other actuators. The degree of bending of the actuator proves that the difference in stiffness between two materials can create a huge SME. As can be seen in Fig. 15.3 from the DMA results, the difference in the storage modulus of the two materials at 80°C is huge, where PLA has a modulus of about 4.6 MPa while BIO-PT has a modulus of about 109 MPa. Design 19 was printed out of a bottom section of PLA and a top section of BIO-PT embedded between two sections of PLA and is presented in Fig. 15.11C and D. As can be seen in the figure, the bending at the part where the BIO-PT is the layer above the PLA is very high, causing a folding effect on the parts of the PLA without BIO-PT. The actuator had a bending angle of about 82 degrees. This actuator is a good indicator of the different SME shapes that can be achieved

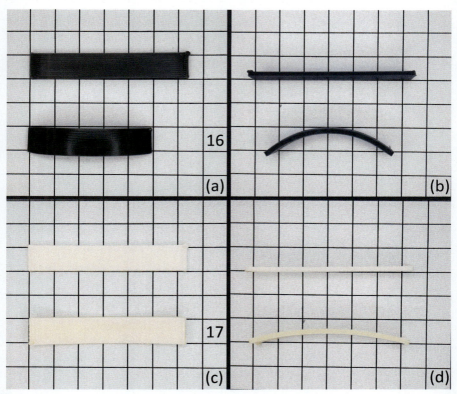

Fig. 15.10 (A) and (B) Top and side views of the PLA actuator (Design 16). (C) and (D) Top and side views of the BIO-PT actuator (Design 17).
No permission required.

by multimaterial variable stiffness prints when designed properly. A folding effect can be made as the one made by the hinges of variable stiffness patterns. The measurements of the printed and tested actuators above are summarized in Table 15.6.

From Table 15.6, it can be seen that all the actuators bend downward, which is expected in the printing of nonpatterned structures. Since the structure is smooth and straight, the printing is done in straight lines across the whole structure for all the layers. This causes a strain that bends in the only direction which is vertical along the direction of the print. This effect is explained in the infill percentage section.

Discussion

This section offers a more in-depth discussion of the mechanism that is involved in the patterns, based on the bending angles of the actuators and the patterns used in these designs. The results of the actuators with variable stiffness and variable patterns are summarized in Table 15.7.

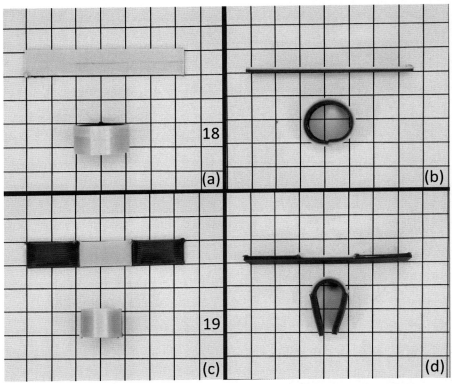

Fig. 15.11 (A) and (B) Top and side views of a structure of a bottom section of PLA and a top section of BIO-PT (Design 18). (C) and (D) Top and side views of a structure of a bottom section of PLA and a top section of BIO-PT embedded between PLA (Design 19).
No permission required.

Table 15.6 Summary of the bending results of Designs 16–19.

Design	Material	Length A (mm)	Length B (m)	Bending angle (degrees)	Bending direction
16	PLA strip	46.89	10.42	20.04	Downward
17	BIO-PT strip	58.24	6.29	4.31	Downward
18	PLA and BIO-PT strip	0.50	18.26	88.25	Downward
19	Embedded BIO-PT in PLA	2.88	22.36	81.99	Downward

Variable stiffness 4D printing 427

Table 15.7 Summary of all the actuators' patterns and bending angles.

Design	Bottom pattern	Top pattern	Bending angle (degrees)	Bending direction
1	Small rectilinear	Rectilinear	12.49	Downward
2	Rectilinear	Small rectilinear	2.05	Upward
3	Small grid	Grid	5.46	Upward
4	Grid	Small grid	3.73	Downward
5	Small honeycomb	Honeycomb	7.91	Downward
6	Honeycomb	Small honeycomb	6.94	Downward
7	Small triangles	Triangles	4.27	Upward
8	Triangles	Small triangles	10.61	Downward
9	Triangles	Honeycomb	26.77	Downward
10	Honeycomb	Grid	5.13	Upward
11	Honeycomb	Triangles	10.12	Upward
12	Rectilinear	Triangles	9.84	Upward
13	Grid	Honeycomb	16.7	Downward
14	Triangles	Rectilinear	24.86	Downward
15	Triangles	Honeycomb	25.77	Downward
16	PLA strip	PLA strip	20.04	Downward
17	BIO-PT strip	BIO-PT strip	4.31	Downward
18	PLA strip	BIO-PT strip	88.25	Downward
19	PLA strip	Embedded BIO-PT in PLA	81.99	Downward

By assessing the results in the table, there does not seem to be a direct relationship between the infill percentage and the direction of bending. Since as is seen in the rectilinear and grid patterns, the effect of the infill percentage has caused an opposite reaction in the shape change for each pattern. In the rectilinear pattern, the bending was downward with a higher infill at the bottom, while the same setting caused the grid pattern to bend upward. The same effect is observed in the honeycomb and triangular patterns. The changes in the infill percentage of the section in the actuator cause an obvious shape change, but the direction of the bending will need further analysis on the pattern profile and how it causes the material to react. The rectilinear pattern of Design 1 has the highest bending angle, which can be explained by the way the material is printed. Although the same amount of material is used for all patterns, the rectilinear has the most density of patterns, which means the most stretch in the material. Stretching of the material upon printing causes the internal strain in the actuator to increase. There are some minor deformations sideways in some actuators, such as Designs 7 and 8. This is attributed to the pattern and the method of printing since both of these actuators are of triangular patterns. The printing of the triangles is made in three angles, 45 degrees, −45 degrees, and 0 degrees, where the first two are for the sides of the triangle. Each printed line of the FDM material has a strain that causes bending in the material in the vertical direction on the line along the printing direction. Due to the reheating of the layers as every new layer is printed over it, the strain reduces in the lower layers. If a line is printed at 45 degrees, then there would be a pulling force, very similar to that of contraction of a spring, in the 45-degree direction. Minor differences in the print, such as air bubbles and traces of other materials, coupled with the environmental condition cause small changes in the profile of the lines of the print, leading to slightly different characteristics in the print. The small irregularities in the print along with the weaker strain in certain areas of a layer caused by the thermal effect of the print will cause some lines to have a stronger shape change than others, causing the shape to bend in certain, sometimes unwanted, directions.

From the summary of the results of Designs 9 through 14, it is very clear that the highest bending angles are achieved when the triangular patterns are on the bottom of the actuator, which is confirmed in two designs (Designs 9 and 14). Both the designs bend downward, while their top patterns are different, which stands to reason that the bending is caused by the presence of the triangular patterns in the bottom section. The difference in stiffness profiles and strain between the top and bottom sections is causing a large deformation, both in the downward direction. The shape change in the bottom section is much higher than that in the top one, caused by the shrinking of the patterns at a higher rate at the bottom. However, the triangular patterns did not have the highest shape change in the previous section, which is due to the difference in strain and not the total strain. If a similar amount of strain is present in the top and bottom sections, even if high, the shape change would be high because the whole structure is changing as a similar amount. In another case, such as the one with Designs 9 and 14, the large shape change can be attributed to the difference in strain, even though the total strain is not very large. This means that the difference in strain between the triangular patterns and the other patterns printed on top of it is the largest, thus causing the largest bending in the downward direction. With Designs 11 and 12, the triangular patterns are printed on the top, causing the bending to happen in the upward direction, which confirms the reasoning in the strain difference. Since the triangular patterns are on top, the strain difference is reversed. Design 11 is the reverse of Design 9, while Design 12 is the reverse of Design 14, meaning that the patterns are the same printed in the opposite order. Although the case is so, the bending angles are far from each other, meaning the strain is not the same. This effect is expected because of the changes in the strain that happens in the prints at each layer, and the profile of the patterns, as is explained in the previous sections.

For the multimaterial Designs (18 and 19), the actuators have very large deformation angle values, which are caused slightly by the internal strain of the printed materials, and largely by the strain difference between the PLA and BIO-PT materials. In Design 18, the bending is regular along the length of the actuator since the strain difference is the same along the boundary of the PLA and BIO-PT, causing the actuator to bend into a full circle. With Design 19, the bending at the part of both materials (the hinge) has the same characteristic of a smooth regular bend, but in the sections of only PLA, the bending is much smaller. The bending in the PLA section is caused only by the strain difference between each PLA layer, which is little since the settings are the same. This effect is explained in previous sections. Since the bending in the hinge is much higher than that of the PLA, the structures start to fold, allowing for certain sections in the actuators to remain straight while changing their position.

VSA that is demonstrated in this study shows a wide range of control of the actuation, and various patterns for similar actuation, each with its own stiffness profile. The work presented brings favorable characteristics to 4D printing since the use of structurally stable actuators without the use of excessive materials. This paper tested a variety of patterns that have not been used previously as well as tested the difference in the SME that each design can have. Future work could investigate the use of different patterns and their effects on the actuation. The patterns used in this study, although cover all the main patterns used in 3D printing, still have a plethora of patterns to be tested. Beyond that, the use of multiple patterns on a single actuator can be expanded into different infill percentages and printing directions. Along with the use of passive blocks and direction of printing, these VSAs can be used to perform complex shape changes that would allow plenty of alternative designs for 4D printing fields, these include bio-mimetic (Jeong, Woo, Kim, & Jun, 2020; Zafar & Zhao, 2020) and biomedical applications (Kashyap, Kishore Kumar, & Kanagaraj, 2018; Miao, Zhu, Castro, Leng, & Zhang, 2016).

Conclusion

Variable stiffness in 4D printing can be used to program the SME of the structures in many ways. There is a plethora of control that can be achieved by adjusting the stiffness of the prints. There are multiple ways to influence the variation in stiffens of a single printed structure. In this study, three types of variable stiffness 4D-printed structures were achieved whether using single or multiple materials. The first type was realized by changing the amount of infill percentage inside the materials to change the stiffness of the actuator. This method can be used to produce moderate bending in the actuators, with the highest tested being that of a rectilinear pattern achieving a 12.49-degree bend. On the other hand, the second type was attained by using different printing patterns of different profiles to change the directional stiffness of the materials. This approach brought about much stronger shape changes due to the strain difference of the patterns, with the highest bending angle of about 26.77 degrees achieved by using triangular and honeycomb patterns on the bottom and top sections, respectively. These single-material designs can be used for many applications, one of which was designed and tested as a proof of concept, the double-hinge structure that was made using hinges made of different printed patterns. Then, the final method of variable stiffness 4D printing was accomplished by using multiple materials with different stiffness values on the same structures, which generated a very strong SME effect in the structures. The bending of the multimaterial has reached a maximum bending angle of about 88.25 degrees, where the actuator has bent all the way to form a circle. The use of multimaterials in a single structure can be used when large deformations are needed. Future work may involve implementing other infill patterns and percentages, using different printing materials, and investigating the effect of the thickness of layers as well as the printing speed and temperature on the bending performance.

Acknowledgment

This work was supported by the Ministry of Higher Education Malaysia under the Fundamental Research Grant Scheme (FRGS/1/2019/TK05/UNIM/02/2), the University of Nottingham Malaysia, and the International and Industry Incentive Grant (IIIG 02M33 & 02M36) from Universiti Teknologi Malaysia.

References

Abdul Kadir, M. R., Dewi, D. E. O., Jamaludin, M. N., Nafea, M., & Mohamed Ali, M. S. (2019). A multi-segmented shape memory alloy-based actuator system for endoscopic applications. *Sensors and Actuators A: Physical, 296*, 92–100. https://doi.org/10.1016/j.sna.2019.06.056.

Alshebly, Y. S., Nafea, M., Almurib, H. A. F., Mohamed Ali, M. S., Mohd Faudzi, A. A., & Tan, M. T. T. (2021). Development of 4D printed PLA actuators with an induced internal strain upon printing. In: *2021 IEEE international conference on automatic control & intelligent systems (I2CACIS)* (pp. 41–45). https://doi.org/10.1109/I2CACIS52118.2021.9495898.

Alshebly, Y. S., Nafea, M., Mohamed Ali, M. S., & Almurib, H. A. F. (2021). Review on recent advances in 4D printing of shape memory polymers. *European Polymer Journal, 159*, 110708. https://doi.org/10.1016/j.eurpolymj.2021.110708.

Al-Shuka, H. F. N., Leonhardt, S., Zhu, W. H., Song, R., Ding, C., & Li, Y. (2018). Active impedance control of bioinspired motion robotic manipulators: An overview. *Applied Bionics and Biomechanics, 2018*. https://doi.org/10.1155/2018/8203054.

Barletta, M., Gisario, A., & Mehrpouya, M. (2021). 4D printing of shape memory polylactic acid (PLA) components: Investigating the role of the operational parameters in fused deposition modelling (FDM). *Journal of Manufacturing Processes, 61*, 473–480. https://doi.org/10.1016/j.jmapro.2020.11.036.

Barrie, D. D., Goher, K., & Elgeneidy, K. (2020). *3D printed variable infill soft fingers for the SIMPA prosthetic arm* (pp. 83–85). https://doi.org/10.31256/wj4jc8q.

Bodaghi, M., Damanpack, A. R., & Liao, W. H. (2017). Adaptive metamaterials by functionally graded 4D printing. *Materials and Design, 135*, 26–36. https://doi.org/10.1016/j.matdes.2017.08.069.

Bodaghi, M., Noroozi, R., Zolfagharian, A., Fotouhi, M., & Norouzi, S. (2019). 4D printing self-morphing structures. *Materials, 12*(8). https://doi.org/10.3390/ma12081353.

Bodaghi, M., Serjouei, A., Zolfagharian, A., Fotouhi, M., Rahman, H., & Durand, D. (2020). Reversible energy absorbing meta-sandwiches by FDM 4D printing. *International Journal of Mechanical Sciences, 173*. https://doi.org/10.1016/j.ijmecsci.2020.105451, 105451.

Bonifacich, F. G., Lambri, O. A., Recarte, V., Sánchez-Alarcos, V., & Pérez-Landazábal, J. I. (2021). Magnetically tunable damping in composites for 4D printing. *Composites Science and Technology, 201*. https://doi.org/10.1016/j.compscitech.2020.108538.

Böse, H., Rabindranath, R., & Ehrlich, J. (2012). Soft magnetorheological elastomers as new actuators for valves. *Journal of Intelligent Material Systems and Structures, 23*(9), 989–994. https://doi.org/10.1177/1045389X11433498.

Carrell, J., Gruss, G., & Gomez, E. (2020). Four-dimensional printing using fused-deposition modeling: A review. *Rapid Prototyping Journal*, *26*(5), 855–869. https://doi.org/10.1108/RPJ-12-2018-0305.

Elgeneidy, K., Fansa, A., Hussain, I., & Goher, K. (2020). Structural optimization of adaptive soft fin ray fingers with variable stiffening capability. In *2020 3rd IEEE international conference on soft robotics, RoboSoft 2020* (pp. 779–784). Institute of Electrical and Electronics Engineers Inc. https://doi.org/10.1109/RoboSoft48309.2020.9115969.

Falahati, M., Ahmadvand, P., Safaee, S., Chang, Y. C., Lyu, Z., Chen, R., et al (2020). Smart polymers and nanocomposites for 3D and 4D printing. *Materials Today*, *40*, 215–245. https://doi.org/10.1016/j.mattod.2020.06.001.

Gauss, C., Pickering, K. L., & Muthe, L. P. (2021). The use of cellulose in bio-derived formulations for 3D/4D printing: A review. *Composites Part C: Open Access*, *4*. https://doi.org/10.1016/j.jcomc.2021.100113, 100113.

González-Henríquez, C. M., Sarabia-Vallejos, M. A., & Rodriguez-Hernandez, J. (2019). Polymers for additive manufacturing and 4D-printing: Materials, methodologies, and biomedical applications. *Progress in Polymer Science*, *94*, 57–116. https://doi.org/10.1016/j.progpolymsci.2019.03.001.

Guo, J., & Tian, G. (2015). Conceptual design and analysis of four types of variable stiffness actuators based on spring pretension. *International Journal of Advanced Robotic Systems*, *12*. https://doi.org/10.5772/60580.

Han, T., Kundu, S., Nag, A., & Xu, Y. (2019). 3D printed sensors for biomedical applications: A review. *Sensors (Switzerland)*, *19*(7). https://doi.org/10.3390/s19071706.

Hann, S. Y., Cui, H., Nowicki, M., & Zhang, L. G. (2020). 4D printing soft robotics for biomedical applications. *Additive Manufacturing*, *36*. https://doi.org/10.1016/j.addma.2020.101567.

Hoang, T. T., Quek, J. J. S., Thai, M. T., Phan, P. T., Lovell, N. H., & Do, T. N. (2021). Soft robotic fabric gripper with gecko adhesion and variable stiffness. *Sensors and Actuators A: Physical*, *323*. https://doi.org/10.1016/j.sna.2021.112673.

Janbaz, S., Hedayati, R., & Zadpoor, A. A. (2016). Programming the shape-shifting of flat soft matter: From self-rolling/self-twisting materials to self-folding origami. *Materials Horizons*, *3*(6), 536–547. https://doi.org/10.1039/c6mh00195e.

Jeong, H. Y., Woo, B. H., Kim, N., & Jun, Y. C. (2020). Multicolor 4D printing of shape-memory polymers for light-induced selective heating and remote actuation. *Scientific Reports*, *10*(1). https://doi.org/10.1038/s41598-020-63020-9.

Jiang, R., Kleer, R., & Piller, F. T. (2017). Predicting the future of additive manufacturing: A Delphi study on economic and societal implications of 3D printing for 2030. *Technological Forecasting and Social Change*, *117*, 84–97. https://doi.org/10.1016/j.techfore.2017.01.006.

Kashyap, D., Kishore Kumar, P., & Kanagaraj, S. (2018). 4D printed porous radiopaque shape memory polyurethane for endovascular embolization. *Additive Manufacturing*, *24*, 687–695. https://doi.org/10.1016/j.addma.2018.04.009.

Kronander, K., & Billard, A. (2016). Stability considerations for variable impedance control. *IEEE Transactions on Robotics*, *32*(5), 1298–1305. https://doi.org/10.1109/TRO.2016.2593492.

Kuksenok, O., & Balazs, A. C. (2016). Stimuli-responsive behavior of composites integrating thermo-responsive gels with photo-responsive fibers. *Materials Horizons*, *3*(1), 53–62. https://doi.org/10.1039/c5mh00212e.

Lendlein, A., & Kelch, S. (2002). Shape-memory polymers. *Angewandte Chemie International Edition*, *41*(12), 2034–2057. https://doi.org/10.1002/1521-3773(20020617)41:12<2034::aid-anie2034>3.0.co;2-m.

Liu, S., Tang, X., Zhou, D., & Liu, Y. (2020). Fascicular module of nylon twisted actuators with large force and variable stiffness. *Sensors and Actuators A: Physical*, *315*. https://doi.org/10.1016/j.sna.2020.112292.

Luo, J., Wang, S., Zhao, Y., & Fu, Y. (2018). Variable stiffness control of series elastic actuated biped locomotion. *Intelligent Service Robotics*, *11*(3), 225–235. https://doi.org/10.1007/s11370-018-0248-y.

Miao, S., Zhu, W., Castro, N. J., Leng, J., & Zhang, L. G. (2016). Four-dimensional printing hierarchy scaffolds with highly biocompatible smart polymers for tissue engineering applications. *Tissue Engineering Part C: Methods*, *22*(10), 952–963. https://doi.org/10.1089/ten.tec.2015.0542.

Mohd Ghazali, F. A., Hasan, M. N., Rehman, T., Nafea, M., Mohamed Ali, M. S., & Takahata, K. (2020). MEMS actuators for biomedical applications: A review. *Journal of Micromechanics and Microengineering*, *30*(7). https://doi.org/10.1088/1361-6439/ab8832.

Mulakkal, M. C., Trask, R. S., Ting, V. P., & Seddon, A. M. (2018). Responsive cellulose-hydrogel composite ink for 4D printing. *Materials and Design*, *160*, 108–118. https://doi.org/10.1016/j.matdes.2018.09.009.

Must, I., Sinibaldi, E., & Mazzolai, B. (2019). A variable-stiffness tendril-like soft robot based on reversible osmotic actuation. *Nature Communications*, *10*(1). https://doi.org/10.1038/s41467-018-08173-y.

Nafea, M., Nawabjan, A., & Mohamed Ali, M. S. (2018). A wirelessly-controlled piezoelectric microvalve for regulated drug delivery. *Sensors and Actuators A: Physical*, *279*, 191–203. https://doi.org/10.1016/j.sna.2018.06.020.

Park, Y. J., Huh, T. M., Park, D., & Cho, K. J. (2014). Design of a variable-stiffness flapping mechanism for maximizing the thrust of a bio-inspired underwater robot. *Bioinspiration & Biomimetics*, *9*(3). https://doi.org/10.1088/1748-3182/9/3/036002.

Peng, X., Kuang, X., Roach, D. J., Wang, Y., Hamel, C. M., Lu, C., et al. (2021). Integrating digital light processing with direct ink writing for hybrid 3D printing of functional structures and devices. *Additive Manufacturing*, *40*. https://doi.org/10.1016/j.addma.2021.101911.

Rahman, H., Yarali, E., Zolfagharian, A., Serjouei, A., & Bodaghi, M. (2021). Energy absorption and mechanical performance of functionally graded soft–hard lattice structures. *Materials*, *14*(6). https://doi.org/10.3390/ma14061366.

Ryan, K. R., Down, M. P., & Banks, C. E. (2021). Future of additive manufacturing: Overview of 4D and 3D printed smart and advanced materials and their applications. *Chemical Engineering Journal*, *403*. https://doi.org/10.1016/j.cej.2020.126162.

Scarpa, E., Lemma, E. D., Fiammengo, R., Cipolla, M. P., Pisanello, F., Rizzi, F., et al. (2019). Microfabrication of pH-responsive 3D hydrogel structures via two-photon polymerization of high-molecular-weight poly(ethylene glycol) diacrylates. *Sensors and Actuators B: Chemical*, *279*, 418–426. https://doi.org/10.1016/j.snb.2018.09.079.

Subash, A., & Kandasubramanian, B. (2020). 4D printing of shape memory polymers. *European Polymer Journal*, *134*. https://doi.org/10.1016/j.eurpolymj.2020.109771.

Sun, T., Chen, Y., Han, T., Jiao, C., Lian, B., & Song, Y. (2020). A soft gripper with variable stiffness inspired by pangolin scales, toothed pneumatic actuator and autonomous controller. *Robotics and Computer-Integrated Manufacturing*, *61*. https://doi.org/10.1016/j.rcim.2019.101848.

Takahashi, M., Kobori, T., Nasu, T., Niwa, N., & Kurata, N. (1998). Active response control of buildings for large earthquakes - seismic response control system with variable structural characteristics. *Smart Materials and Structures*, *7*(4), 522–529. https://doi.org/10.1088/0964-1726/7/4/012.

Tong, K. J., Alshebly, Y. S., & Nafea, M. (2021). Development of a 4D-Printed PLA Microgripper. In: *2021 IEEE 19th Student Conference on Research and Development (SCOReD)*, 207–2011. https://doi.org/10.1109/SCOReD53546.2021.9652719.

Torrealba, R. R., Udelman, S. B., & Fonseca-Rojas, E. D. (2017). Design of variable impedance actuator for knee joint of a portable human gait rehabilitation exoskeleton. *Mechanism and Machine Theory*, *116*, 248–261. https://doi.org/10.1016/j.mechmachtheory.2017.05.024.

Wall, V., Deimel, R., & Brock, O. (2015). Selective stiffening of soft actuators based on jamming. In *Proceedings—IEEE international conference on robotics and automation (Vols. 2015–, issue June)* (pp. 252–257). Institute of Electrical and Electronics Engineers Inc. https://doi.org/10.1109/ICRA.2015.7139008.

Wolf, S., Grioli, G., Eiberger, O., Friedl, W., Grebenstein, M., Hoppner, H., et al. (2016). Variable stiffness actuators: Review on design and components. *IEEE/ASME Transactions on Mechatronics*, *21*(5), 2418–2430. https://doi.org/10.1109/TMECH.2015.2501019.

Xie, M., Zhu, M., Yang, Z., Okada, S., & Kawamura, S. (2021). Flexible self-powered multifunctional sensor for stiffness-tunable soft robotic gripper by multimaterial 3D printing. *Nano Energy*, *79*. https://doi.org/10.1016/j.nanoen.2020.105438.

Yamamura, S., & Iwase, E. (2021). Hybrid hinge structure with elastic hinge on self-folding of 4D printing using a fused deposition modeling 3D printer. *Materials and Design*, *203*. https://doi.org/10.1016/j.matdes.2021.109605.

Yang, Y., Chen, Y., Li, Y., & Zhiqiang Chen, M. (2016). 3D printing of variable stiffness hyperredundant robotic arm. In *Proceedings—IEEE international conference on robotics and automation (Vols. 2016–)* (pp. 3871–3877). Institute of Electrical and Electronics Engineers Inc. https://doi.org/10.1109/ICRA.2016.7487575.

Zafar, M. Q., & Zhao, H. (2020). 4D printing: Future insight in additive manufacturing. *Metals and Materials International*, *26*(5), 564–585. https://doi.org/10.1007/s12540-019-00441-w.

Zhang, X., Tan, B. H., & Li, Z. (2018). Biodegradable polyester shape memory polymers: Recent advances in design, material properties and applications. *Materials Science and Engineering C*, *92*, 1061–1074. https://doi.org/10.1016/j.msec.2017.11.008.

Zolfagharian, A., Kouzani, A. Z., Khoo, S. Y., Moghadam, A. A. A., Gibson, I., & Kaynak, A. (2016). Evolution of 3D printed soft actuators. *Sensors and Actuators A: Physical*, *250*, 258–272. https://doi.org/10.1016/j.sna.2016.09.028.

Index

Note: Page numbers followed by *f* indicate figures and *t* indicate tables.

A

ABAQUS, 7, 55–56, 56*f*, 58–64, 59*t*, 62–66*f*, 62*t*
  4D printing modeling using, 3–4
ABAQUS user-defined material (UMAT), 393
Acoustic metamaterials, 213–214
Active materials manufacturing, 311–312
Actuators, 417, 418*f*, 425*f*
Adaptive dynamic structures, 213–224
  adaptive diagonal structure, 217–220, 218–222*f*
  adaptive parallel structure, 220–224, 222–224*f*
  design adaptive periodic structures, 216–217, 217*f*
  wave propagation formulation, 215–216, 215–216*f*
Additive manufacturing (AM), 195–196, 311–313, 329, 347, 408
Additive manufacturing technologies (AMTs), 347
Adhesion, mechanical, 239–240
Adhesion theories, 239, 240*f*
Analysis of variance (ANOVA), 83–85, 85–87*f*
Analytical model, 75–78
ANSYS, 4, 7
Artificial neural network (ANN), 74
  machine learning (ML) modeling, 89–92, 91–92*t*
Auxetic cellular materials, 144–145
Auxetic materials, 144
  concaved cavities, 144–145

B

Bimaterial beam, 335, 336*f*
Biodegradable natural hydrogels, 8–9, 251–252

Biodegradable polyethylene terephthalate glycol (Bio-PT), 414–417, 424–425
Biodegradable soft robots, 251

C

Cartesian FDM printers, 234–235
Cellular materials, auxetic, 144–145
Chipless radio frequency identification (RFID), 368
Chitosan, 252–253
  electroactive polymer actuator, 251–252
  hydrogel, 251–252
Closed-loop control of 4D-printed hydrogel soft robots, 8–9
Compliance, 407–408
Computer-aided design (CAD), 195–196
  4D-printed hydrogel soft robots, 252, 254
  model, 9
COMSOL Multiphysics, 7, 207, 370, 371*f*, 372–373
Conductive filaments, 2–3, 20–21
Conductive-layer placement influence on performance, 42–43, 43*f*
Conductive PLA (CPLA), 3, 19
  electrical contacting of, 28–30
  layer in shape memory polymers (SMP) structure, 42
  printing parameters, 34
    filament, 31*t*
Connector, electrical resistance of, 30, 31*f*
Copolymers, 231

D

Degrees of freedom (DOFs), 331–335, 337–339
Deposition, 197–198
Desktop 3D printers, 319–320
Diamond-shaped reinforcement, 246
Dielectric reflectarrays, 374
Differential scanning calorimetry (DSC), 290–293, 292*f*

436 Index

Digital light processing (SLA/DLP) process, 22–23
Direct ink writing (DIW) method, 409
Distribution computation component, 342–343, 343–344t
DLP 3D-printers, 114
Dynamic mechanical analysis (DMA), 389–391, 394f, 417
Dynamic mechanical thermal analysis (DMTA), 201, 210–211, 290

E

Edible actuators, 251
Elastane fibers, 231
Elasticity, 231
Elastic skin, 320f
   integrated with twisted coiled polymer muscles, 10–11
Electrical connector types, 30f
Electrical resistance, 30
   of connector, 30, 31f
   printing parameter influence on, 31–33, 32f
Electroactive polymer actuator, 251–252
Electro-chemo-mechanical model, of 3D-printed polyelectrolyte actuator, 270f, 271–274
Electro-induced shape memory polymers, 4D printing, 2–3

F

Fabricated 6-ply actuator, 314–316
Fabricated prototypes, 108–110
Fabrication, 74–75, 111–114, 111f, 112t
Fabrics, 232, 232f
FDM. *See* Fused deposition modeling (FDM)
FEM. *See* Finite element method (FEM)
Fibers, 231
Filaments, 415t
   conductive, 20–21
Finite element method (FEM), 3–4, 7
   model of thermal-mechanical coupling in ABAQUS, 58–64, 59t, 62–66f, 62t
Finite element modeling (FEM), 107, 393
   of 4D-printed pneumatic soft actuators, 107, 107–109f
   modeling using hyperplastic material constitutive laws, 78–83, 79–82f, 80t

training data acquisition from, 81–83, 83t, 83–84f
Finite-element modeling (FEM), 1–2
Formlabs
   Flexible 80A material, 374
   FLGR02 flexible material, 362f, 363–365
4D additive manufacturing process, 319–320, 320f, 321t
4D-printable soft actuators, 5
4D-printed auxetic structures, with tunable mechanical properties, 5–6
4D-printed chipless RFID pressure sensors, for wireless sensor network (WSN) applications, 368–373
4D-printed composite actuators, 388
4D-printed elements
   fabrication and modeling, 204–208
      finite element modeling, 206–208, 207–208t, 207–208f
      4D printing elements, 205–206, 205–206f
   materials, 204
4D-printed honeycomb structures, 142–143
4D-printed hydrogel soft robots
   actuation performance, 266–268, 266–269f
   characterizations, 259
   closed-loop control of, 8–9
   computer-aided design (CAD) model, 252, 254
   controller design, 270–271
   experimental setup and image processing, 262–263, 262–263f
   fabrication of actuator, 254
   geometrical effects, 263–266, 264–265f
   ionic strength effect, 263, 264f
   mechanical tests results, 260, 260f
   optimizing the printing parameters, 255–257, 256f
      3D printing, 257–259, 257t, 258f, 259t
   swelling measurements, 261, 261f
   T-S fuzzy system formulations, 271–274, 271–273f, 275f
4D-printed linear vacuum soft actuators, 107
4D-printed origami-inspired frequency selective surfaces, 363–367, 365–369f
4D-printed one-shot deployable dielectric reflectarray antenna
   for mm-wave applications, 374–376, 375–378f

Index

4D-printed origami-inspired deployable and reconfigurable antennas, 373–381

4D-printed origami-inspired radio frequency structures
fabrication process of, 359–362
inkjet-print SU-8 buffer layer, 361–362, 362–363f
inkjet-print conductive layer, 362, 363–364f
3D-printing of substrate, 360–361, 361f

4D-printed periodic cellular solid structures, 144–145

4D-printed planar pressure sensor
using metamaterial absorber, 369–372, 370–371f
using substrate integrated waveguide (SIW) technology, 372–373, 372–373f

4D-printed pneumatic soft actuators, 105–125
applications, 119–125, 119f
soft artificial muscles, 123, 124f
soft assistive wearable and medical devices, 123–125
soft grippers and parallel manipulators, 121–123, 122f
soft locomotion robots, 119–121, 120–121f
challenges of, 125–127, 125f
mass production and lifetime, 126–127
noise and vibration, 126
portability, 125–126
3D-printing materials and printing time, 126
modeling, fabrications, and control, 4–5
requirements for, 127
scalability and customizability, 118
self-healing properties, 117
modularity, 118
multimodal and programmable actuation, 118–119
scalability and customizability, 118
for soft robotic systems, 111–113, 111f
3D printing technologies, 111–113, 112t

4D-printed reconfigurable origami antennas, liquid-metal-alloy microfluidic-based, 377–381, 379–383f

4D-printed shape memory polymers, modeling and fabrication, 6–7

4D-printed soft actuator, 4–5, 105

shape classification using ML, 94–95, 97–98f

4D-printed soft pneumatic actuator (SPA), 73, 90–91, 111–113

4D-printed soft robots, 251

4D-printed specimens
stability and functional properties of, 146–187
geometric stability following heat exposure, 146–159, 147–148f, 150f, 151t, 152–158f, 160f
stress-free shape recovery, 159–167, 161–164f, 163t, 166f, 167t, 168f
tunable mechanical properties, 167–187, 169f
heterogeneities development, 182–187, 182–186f, 186–187t
tunability in complex periodic structures, 171–181, 172f, 174–181f
tunability in simple structures, 167–171, 170t, 170–171f

4D-printed structures, 1, 2f, 19
shape-morphing of, 21

4D printing, 53–54, 73–74, 141, 143, 196, 313, 329–330, 330f
computational design for, 330–335
rationales and theoretical background, 330–334, 331–332f
electro-induced shape memory polymers, 2–3
mechanism and design, 54–55, 55f
modeling, 54
using ABAQUS, 3–4
using machine learning, 4, 74, 75f
multimaterial, 341–342, 388
parameters, 64
polyelectrolyte soft robot actuator, 254, 255f
potentials and challenges of, 230t
programming, 197–198
simulation procedures for, 62–64
variable stiffness, 13–14

4D textiles, 7–8, 229, 242, 243f, 247–248
applications
finger, 243–245, 244–245f
orthosis, 245–247, 246–247f
hybrid structure, 229, 230f
meshed structures, 232
stiffness, 230

Frequency selective surfaces (FSSs), 352–353, 354*t*
  origami-inspired inkjet-printed FSS structures, 353–359, 355*f*
    fabrication process, 353–356, 356–357*f*
    single-layer Miura-FSS structure, 357–358, 357–359*f*
Full-field strain assessment, 151–155
Fused deposition modeling (FDM), 108–110, 195–196, 232–233, 312, 409, 414
  polymers used in, 22–23
  printer, 22–23, 232–233
    conductive filaments, 27
    3D printer, 410, 413, 415–416
  printing, 411
  technology, 197–198
Fused filament fabrication (FFF), 117

G

Gaussian support vector machines (SVM), 94–95
Geometric code, 1
Gripper actuator, self-folding structures, 209–210, 209–210*f*

H

Heating–cooling process, 198
Hierarchical motion of 4D-printed structures using temperature memory effect, 9–10
Hook's law, 59
Hydrogels, 21–22, 53, 251–252
  actuator, 335–336, 336*f*
  biodegradable natural, 8–9, 251–252
  chitosan, 251–252
  polyelectrolyte, 8, 260
Hydrostatic pressures, porous materials to, 144–145
Hyperelastic materials, behavior of, 78–79
Hyperplastic material constitutive laws, finite element model (FEM) modeling using, 78–83, 79–82*f*, 80*t*

I

Inkjet-printing technology, 348–350, 349*f*, 351*t*
  conductive layer, 362, 363–364*f*
  SU-8 buffer layer, 361–362, 362–363*f*

In-plane stresses, 24
Interfaces of surface and reinforcement, 229–230, 239–240
Ionic strength effect, 4D-printed hydrogel soft robots, 263, 264*f*

J

Joule heating, 13, 27

L

Layer configurations analysis, 92–94, 92*t*
  activation functions analysis, 93–94, 93–94*t*, 94–96*f*
Light stimulation, 53
Linear regression, 85–89, 88*f*, 88*t*
  model, 98–99
Liquid crystal gel-phase transition, 53
Liquid crystalline elastomers (LCE) material, 21–22
Liquid metal-alloy (LMA), 377–380
  microfluidic-based 4D-printed reconfigurable origami antennas, 377–381, 379–383*f*

M

Machine learning (ML)
  4D-printed soft actuator shape classification using, 94–95, 97–98*f*
  4D printing modeling via, 4
  model, 4
    using artificial neural network, 89–92, 91–92*t*
Magnetostrictive actuator, 337, 337–338*f*
Material distribution
  generation, 337–341, 339*f*
    attempt to retrieve a known distribution, 337–339, 339*f*
    distribution computation, 340–341, 341*f*
  modeling and simulation
    bimaterial beam, 335, 336*f*
    hydrogel actuator, 335–336, 336*f*
    magnetostrictive actuator, 337, 337–338*f*
Material heterogeneity, 329, 333*f*
MATLAB
  image processing, 262–263
  Neural Network Toolbox on, 91
Maxwell model, 199
Mechanical adhesion, 239–240

Index

Meshed structures, 232
Metamaterial absorber, 4D-printed planar
    pressure sensor using, 369–372,
    370–371f
Miura-FSS
  EM behavior of, 353
  unit cell of, 355f
Miura-Ori tessellation, 348–352, 352f
MJ 3D-printers, 114
mm-wave applications, 4D-printed one-shot
    deployable dielectric reflectarray
    antenna for, 374–376, 375–378f
Mooney Rivlin model, 79
Motion mechanism of soft actuator, 252–254,
    253f
Multilayer perceptron (MLP), 90–91, 92t
Multimaterial actuators, variable stiffness 4D
    printing, 414–415
Multimaterial printing, 229
  4D printing, 341–342, 389–391
    inkjet, 388
    simulation using grasshopper plugin, 11

N

Natural hydrogels, biodegradable, 8–9,
    251–252
Natural polymers, 251–252
Near infrared (NIR) light, heat generation and
    temperature rise due to, 56–58
Negative Poisson's ratio (NPR), 143–144,
    171–172, 182–183
Negative pressure soft actuators, 105–106
Neural Network Toolbox on MATLAB, 91
Nicolson-Ross-Weir (NRW) methodology,
    360–361
Nonlinear material model, 78–79
Nozzle, 195–198
NPR. See Negative Poisson's ratio (NPR)

O

Ogden 3-parameter model, 79
  material coefficients for, 80t
Optimization theory, 1–2
Optimization topology, 1–2
Origami-inspired 4D radio frequency (RF),
    and wireless structures and modules,
    12
Origami-inspired frequency selective surface
    (FSS) substrate, 363–365

Origami-inspired inkjet-printed
    frequency selective surfaces (FSSs)
    structures, 353–359, 355f
    fabrication process, 353–356, 356–357f
    single-layer Miura-FSS structure,
    357–358, 357–359f
Origami-inspired radio frequency (RF)
    structures, 348–349

P

Paraboloidal shape of surface, 229
Passive electromagnetic components,
    368–369
Periodic cellular solids, 143, 145
Photo-thermal stimulus, 71
Pneumatic soft actuators, 4–5, 104
  4D-printed, 105–125
  with programmable motion, 113
  proportional-integral-derivative (PID)
    controllers for controlling, 116–117
  in soft robotic systems, 105–106
Poisson's ratios, 143–144
Polyelectrolyte actuators
  bending motion in, 253, 253f
  3D-printed, electro-chemo-mechanical
    model of, 270f, 271–274
Polyelectrolyte hydrogels, 8, 260
Polyelectrolyte soft robot actuator, 4D-
    printing, 254, 255f
PolyJet, 22–23
Polylactic acid (PLA), 3, 19, 22, 204, 205f,
    323
  behavior of, 23
  conductive (see Conductive PLA (CPLA))
  influence on deformation of, 23–26
  integration of conductive, 26–33
    materials and equipment, 27–28
Polymers, 6–7
  natural, 251–252
  used in FDM, 22–23
  viscoelastic nature of, 6–7
Polystyrene (PS) film, 54
  material properties of, 58–59, 59t
  specimen's physical and material
    properties of, 58, 58t
  surface temperature characteristics during
    IR irradiation, 57–58, 57f
  Young's modulus of, 62, 62f
Porous materials, 1–2
  to hydrostatic pressures, 144–145

Positive pressure soft pneumatic actuators, 105

Pressure parameters, for rotational symmetrical pressure, 235, 238t

Pressure sensor, 368–369

Prestressing technologies, 233–234, 234–236f

Printing, 232–242

Printing parameters, 23–26, 23f, 237–239, 239t
  influence on resistance, 31–33, 32f

Printing speed, 23–24, 23f, 25f

Proportional-integral-derivative (PID) controllers, 116
  for controlling pneumatic soft actuators, 116–117

Prototypes, fabricated, 108–110

Pulse width modulation (PWM), 262

Push-button principle, 239–241, 240f, 242f

R

Radial basis function (RBF) neural networks, 90–91

Radio frequency identification (RFID) technologies, 368

Radio frequency (RF) structures
  origami-inspired, 348
  3D, 348
  tunable, 348

Reflectarray, 374
  antennas, traditional, 374
  dielectric, 374

Reinforcements, 230
  diamond-shaped, 246
  interface of surface and, 229–230, 239
  printed, 229, 241–242, 241f

Resins, synthesized, 113–114

Robotics, 73

Room-temperature-vulcanizing (RTV) silicone elastomer, 316, 317f, 320f
  mechanical properties of, 316–319, 318f

Rotational symmetrical pressure, pressure parameters for, 235, 238t

Rotational symmetric substrate, 234–236, 237–238f

S

Selective laser melting (SLM), 195–196

Self-folding mechanism, 196

Self-folding smart composites, 210–213, 211–214f

Self-folding structures, 209–213
  Gripper actuator, 209–210, 209–210f
  self-folding smart composites, 210–213, 211–214f

Self-locking mechanisms, 300–302

Sensing and control, 114–117, 115f

Sensors, 145
  pressure, 368–369

Shape memory alloys (SMAs), 12–13, 387
  actuators design, 389, 390–391f
  disadvantages of, 387
  experimental procedure, 389–393
  numerical and experimental results, 397–403, 398–399f, 401–402f
  printed actuators, 388
  and shape memory polymers (SMPs), 403
  shape-reversible 4D printing aided by, 12–13
  simulation of actuation cycle, 393–397, 395f
  soft actuators, 387–388
  wire, 388–389, 393
    material properties, 391, 394t

Shape memory effect (SME), 22, 279, 329

Shape memory materials (SMM), 408–409

Shape memory polymers (SMPs), 6–7, 12–13, 53, 142, 196, 279, 281, 408–409
  advantage of, 21–22
  constitutive equations, 199–204
    phase transformation approach, 200–204
    thermoviscoelastic approach, 199–200
  cycle of, 197f
  functional properties of, 142–143
  material, 146
  and shape memory alloys (SMAs), 403
  thermoplastic, 142

Shape memory polymers (SMPs) structures, 3, 19, 20f
  behavior at different activation voltages, 39–42, 40–42f
  blocking force of, 36–38, 46–47, 46f
  change in electrical resistance, 44–45, 45f
  conductive PLA (CPLA) layer in, 42
  design of, 34, 35f, 37f, 43, 43f
  different segments of, 36t
  disadvantage of, 20
  free bending of, 43–46, 44–45f

Index

investigation of, 33–47
manufacturing of, 34–36, 38*f*
measurement setup for, 36–39, 38–39*f*
with single-stimulus responsive material, 33
3D-printed, 20*f*
Shape-reversible 4D printing, aided by shape memory alloys, 12–13
Shrink Film, 54
Silicone, 316–319, 323
Silicone elastomer, 319–320
room-temperature-vulcanizing (RTV), 316, 317*f*, 320*f*
mechanical properties of, 316–319, 318*f*
Silver-bonded cables, 29
Silver conductive lacquer, 29
Single-material actuators
variable stiffness 4D printing, 410–414
patterns as hinges, 413–414
variable infill patterns, 413, 414*t*
variable infill percentages, 411–412, 412*f*, 413*t*
Sliding mode control system (SMC), 270–271
SLM. *See* Selective laser melting (SLM)
SLS 3D-printers, 114
Smart constituent material (SMP), 144–145
Soft actuator, 4–5, 103
fabrication of, 9, 252
4D-printable, 4–5, 105
geometry of, 76*f*
motion mechanism of, 252–254, 253*f*
pneumatic, 4–5
Soft modular actuators, 118
Soft pneumatic actuator (SPA), 74–75
development of novel 4D-printable soft materials for, 110, 110*f*
fabrication, printing parameters used for, 77*t*
4D-printed, 73, 90–91
positive pressure, 105
Soft robots, 4–5, 73, 103–104, 104*f*, 323–325
actuator, 73
development, 311–312
4D printing of, 74
4D-printed pneumatic soft actuators for, 111–113, 111*f*
pneumatic soft actuators in, 105–106
Soft sensing technologies, 115–116
Soft sensors, 115–116

Standard linear viscoelastic (SLV) model, 199–200, 199*f*
Stereolithography (SLA), 22–23
3D printers, 113–114
3D-printing technology, 360–362
Stiffness-tunable structures, 13–14
Strain energy density function, 78–79
Stratasys 3D printer, 393
Subtractive manufacturing technologies (SMTs), 347
Sunflowers, 53
Sunlight, 117
Surface tessellation, 242, 243*f*
Synthesized resins, 113–114

T

TA instruments, 291–292, 297–298
Tangoplus FLX930 (Stratasys), 114
Temperature memory effect (TME), 280–281
constitutive modeling
model formulation, 298–299
model parameters identification, 299–300, 300*t*, 301–302*f*
evaluation of
material parameters, 291
mechanical behavior, 290
sequential shape memory response, 290–291
shape memory behavior, 290
temperature memory behavior, 290
thermal and thermomechanical properties, 290
experimental activity for generation of input data for numerical simulation, 297–298, 298*f*
experimental testing, 282–283
exploitation of, 284–290, 286–289*f*
4D-printed self-locking clamp, case study, 300–305, 304*f*
hierarchical motion of 4D-printed structures using, 9–10
modeling and simulation, 283–284
preliminary experimental activity to assess the possibility, 289–290
evaluate and model shape memory response, 290–291
screening protocol to assess the possibility to exploit, 291–296

442 Index

Temperature memory effect (TME)
(Continued)
    mechanical testing, 293–296, 293f
    thermomechanical testing, 291–293,
        292f
    testing protocol, 289–291
        based on possibility to exploit
            temperature memory effect (TME),
            296, 297f
Tensile testing, 260, 260f
Tension energy, 7–8, 229
Tessellation, 247f
    form giving through surface, 242, 243f
Textiles, 231
    adhesion design of 3D printing on, 240
    fabrics, 232, 232f
    high bending stiffness of, 231
    3D printing on, 232
Textile surface, porosity of, 231
Thermal activation, 34
Thermally stimulated recovery (TSR), 282,
    295f
    test, 294–296
Thermal-structural finite element method
    (FEM) model, 71
Thermo-mechanical 4D-printed structure
    modeling, 55–56
Thermoplastic continuous filament, 195–196
Thermoplastic shape memory polymers
    (SMPs), 142
3D-bioplotter, 257t, 259, 259t
3D computer-aided design (CAD) models,
    108–110
3D-printed polyelectrolyte actuator, electro-
    chemo-mechanical model of, 270f,
    271–274
3D-printed robotic biomimetic structure,
    321–323, 322f, 324f
3D-printed shape-memory polymer (SMP)
    structures, 20f
3D printers, 233f
    desktop, 319–320
    digital light processing (DLP), 114
    stereolithography (SLA), 113–114
3D printing, 19, 73–74, 195–196, 252
    benefits of, 313
    technologies, 74–75, 113, 196, 198f
        with 4D-printed pneumatic soft
            actuators, 111–113, 112t

    stereolithography (SLA), 360–361
    on textiles, 232
3D radio frequency (RF) structures, 348
Topology optimization, 1–2
Traditional reflectarray antennas, 374
T-S fuzzy system formulations, 4D-printed
    hydrogel soft robots, 271–274,
    271–273f, 275f
Tunable radio frequency (RF) structures, 348
Twisted and coiled polymer (TCP), 314–316,
    317f
    actuator, 314
    fabrication setup and process, 314–316,
        315f
    fibers, 313
    muscles, 10, 312–314, 322f, 324f
    in room-temperature-vulcanizing (RTV)
        silicone, 321

V

Vacuum soft actuators, 4D-printed linear
    vacuum, 107
Variable stiffness actuation (VSA), 407–408,
    429
Variable stiffness 4D printing
    actuators' patterns and bending angles, 425,
        427t
    design and working principle, 410–415
    fabrication process and experimental setup,
        415–416, 415t, 416f
    material properties, 417–418, 418f
    multimaterial actuators, 414–415,
        424–425, 425–426f, 426t
    single-material actuators, 410–414,
        419–424
        patterns as hinges, 413–414, 423–424,
            424f
        variable infill patterns, 413, 414t,
            421–422, 421–423f, 423t
        variable infill percentages, 411–412,
            412f, 413t, 419–421, 419–420f, 421t
Variable stiffness structures, 13–14
Vector network analyzer (VNA), 363–365
VeroClear, 389–391, 392f
Versatility, 141
Viscoelastic models, 283–284
Viscoelastic nature of polymers, 6–7
Viscosity, 6–7

Index

Viscous material, 237–239
Voxel object (VO), 331–332
VoxSmart, 332–335, 334*f*

W

Wave propagation formulation, 215–216,
  215–216*f*
Williams-Landel-Ferry (WLF) equation, 284
Wireless sensor network (WSN), 368

applications, 4D-printed chipless RFID
  pressure sensors, 368–373
Wireless structures and modules, origami-
  inspired 4D RF and, 12

Y

Yarn formation, 231
Young's modulus, for 3D-printed polylactic
  acid (PLA), 206–207, 207*t*

Printed in the United States
by Baker & Taylor Publisher Services